INTERPRETING THE *NATIONAL* ELECTRICAL CODE®

Based on the 1999
NATIONAL ELECTRICAL CODE®

INTERPRETING THE *NATIONAL* ELECTRICAL CODE®

Fifth Edition

Truman C. Surbrook
Professor
Michigan State University

Jonathan R. Althouse
Instructor
Michigan State University

Delmar Publishers

an International Thomson Publishing company

Albany • Bonn • Boston • Cincinnati • Detroit • London • Madrid
Melbourne • Mexico City • New York • Pacific Grove • Paris • San Francisco
Singapore • Tokyo • Toronto • Washington

NOTICE TO THE READER

Cover Design: Brucie Rosch

Delmar Staff:
Publisher: Alar Elken
Acquisitions Editor: Mark Huth
Project Editor: Barbara Diaz
Production Manager: Mary Ellen Black
Art and Design Coordinator: Cheri Plasse
Editorial Assistant: Dawn Daugherty

COPYRIGHT © 1999
By Delmar Publishers

an International Thomson Publishing company

The ITP logo is a trademark under license
Printed in the United States of America

For more information contact:

Delmar Publishers
3 Columbia Circle, Box 15015
Albany, New York 12212-5015

International Thomson Publishing Europe
Berkshire House
168-173 High Holborn
London, WC1V 7AA
United Kingdom

Nelson ITP, Australia
102 Dodds Street
South Melbourne,
Victoria, 3205 Australia

Nelson Canada
1120 Birchmont Road
Scarborough, Ontario
M1K 5G4, Canada

International Thomson Publishing France
Tour Maine-Montparnasse
33 Avenue du Maine
75755 Paris Cedex 15, France

International Thomson Editores
Seneca 53
Colonia Polanco
11560 Mexico D. F. Mexico

International Thomson Publishing GmbH
Königswinterer Strasße 418
53227 Bonn
Germany

International Thomson Publishing Asia
60 Albert Street
#15-01 Albert Complex
Singapore 189969

International Thomson Publishing Japan
Hirakawa-cho Kyowa Building, 3F
2-2-1 Hirakawa-cho, Chiyoda-ku,
Tokyo 102, Japan

ITE Spain/Paraninfo
Calle Magallanes, 25
28015-Madrid, Espana

Library of Congress Cataloging-in-Publication Data

Surbrook, Truman C.
    Interpreting the National Electrical Code / Truman C. Surbrook,
Jonathan R. Althouse. — 5th ed.
        p. cm.
    "Based on the 1999 National Electrical Code."
    Includes bibliographical references and index.
    ISBN 0-7668-0187-X (alk. paper)
    1. Electrical engineering—Insurance requirements—United States.
2. National Fire Protection Association. National Electrical Code
(1999) I. Althouse, Jonathan R.   II. Title.
TK260.S87 1999
621.319'24'021873—DC21

                                                        98-39311
                                                        CIP

BRIEF CONTENTS

CONTENTS

Unit 2   Wire, Raceway, and Box Sizing 45

Unit 9   Hazardous Location Wiring.................................291

Unit 13   Commercial Wiring Applications 397

Unit 14   Special Applications Wiring 415

PREFACE

INTERPRETING THE NATIONAL ELECTRICAL CODE® is a text and workbook designed as a self-contained study course for learning to read and interpret the meaning of the Code, and to find information about how to do wiring installations. This text is intended for use by personnel with some experience in electrical wiring and use of the *National Electrical Code*®.[1] The text is organized into 15 units covering the entire Code. The individual units cover a particular subject and show how to find information about that subject in the entire Code.

PREREQUISITES

Electrical wiring experience and a basic familiarity with the *National Electrical Code*® are desirable for the student to maximize learning from this text and the course in which it is used. Basic arithmetic, written communication, and reading skills are required for the course. Students with an interest in electrical wiring and the Code will, after completion of this text and course, be able to find information from the Code needed to do residential, commercial, farm, and industrial wiring.

EACH ARTICLE DISCUSSED

A brief description of the type of information found in every article of the Code is included in the text. Code article discussions are included in the unit of the text where the article contains important information for the subject of that unit. The Contents provides a list of Code articles included in each unit to provide a quick reference for the student to find information in this text.

CODE CHANGES DISCUSSED

This text contains a discussion of the *National Electrical Code*® changes of significance since the last edition of the Code. The articles that pertain to the subject of the unit as listed in the Contents are contained in each unit. This discussion of changes is intended to explain the meaning and application of the change. Every article of the Code is included in this discussion of Code changes.

---

[1]*National Electrical Code*® and *NEC*® are Registered Trademarks of the National Fire Protection Association, Inc., Quincy, MA. Excerpts are reprinted with permission from NFPA 70-1999, the *National Electrical Code*®, Copyright © 1998, National Fire Protection Association, Quincy, Massachusetts 02269. This reprinted material is not complete and official position of the National Fire Protection Association on the referenced subject, which is represented only by the standard in its entirety.

USEFUL WIRING INFORMATION INCLUDED

Each unit of this text contains useful information for understanding the meaning behind Code requirements, or information on the subject of the unit useful for applications in the field that is not included in the Code. This text can, therefore, be used as a handy reference for making wiring installations.

SAMPLE CALCULATIONS

Many types of calculations are needed for making electrical installations. This text includes a discussion of the Code requirements, and then provides sample calculations as would be typically encountered in everyday wiring. Methods have been developed in this text that help the electrician better understand the Code requirements for making electrical calculations.

PRACTICAL CODE QUESTIONS AND PROBLEMS

Learning to find information in the *National Electrical Code®* requires looking up answers to practical everyday wiring installation questions. Each unit of this text contains 25 practical questions or problems pertaining to the subject of the unit. The answers to the questions are found in the Code articles listed for that particular unit. Space is provided to write the answer and the Code sections where the answer was found. When calculations are required to obtain an answer, space is provided to write the solution in the text. The instructor will give the answer and work the calculation during the class session. An answer sheet is also provided for the student to write the answer and the Code reference to hand in for grading.

INSTRUCTOR'S GUIDE

An *INSTRUCTOR'S GUIDE* is available that gives a detailed lesson plan for each of the 15 sessions. It is also possible to offer the course in segments of less than 15 sessions, and suggestions for a shorter format are provided. The *INSTRUCTOR'S GUIDE* contains all answers to the student homework questions and problems as well as complete calculations. Code references also are given for each homework question or problem. Some questions or problems have multiple Code references. All possible Code references are provided with a background discussion to help the instructor explain the reasoning behind the answer and Code reference.

ILLUSTRATIONS

The illustrations in this text provide numerous references to Code sections to help the reader find important information in the Code. The purpose of the illustrations is to help the reader gain an understanding of a typical application of a particular Code section. It is not possible for an illustration to provide all information contained in a particular Code section. It is important that the reader study the particular Code section referenced to gain a complete understanding of the meaning of the Code section. The illustrations were provided by David T. Surbrook.

TO THE STUDENT

This is a text and workbook that can be used for self-study, but it was developed primarily for use with three to four hours of instruction for each unit. The greatest benefit will be derived from the text and course if the student will make a concerted effort to work all problems and answer all questions. The only way to learn to find information in the *National Electrical Code®* and to interpret the information is to use the Code to find answers to daily wiring questions. The class discussion of the answers to questions will be of greatest meaning when the student has made an effort to find the answer and Code reference.

The instructor will briefly review the main points of the Code material that will be used to answer the homework questions. These pointers may save time lost searching for answers in the Code. The instructor will review the changes in the Code of greatest significance since the last edition. This will help to avoid confusion with the previous Code.

The student should quickly read the Code discussion portion of the text that describes the type of information found in each article. This will help to make question look-up more efficient. Then the student should read carefully the portion of the text that discusses Code material and gives sample calculations. Most problems will be similar to these sample calculations. Then work the homework problems. It will take four to eight hours to answer the questions and work the homework problems. It usually works best to spend some time on several occasions rather than trying to work the homework all at one time.

Before returning to class, write the answer and the Code reference on the answer sheet to hand in at the beginning of class.

ABOUT THE AUTHORS

One of the authors of *INTERPRETING THE NATIONAL ELECTRICAL CODE®*, Truman C. Surbrook, has extensive practical experience in electrical wiring, as well as many years of experience as an instructor in the electrical trade.

Truman C. Surbrook, Ph.D., is a Professor of Agricultural Engineering at Michigan State University, a registered Professional Engineer, and Master Electrician. Dr. Surbrook developed an Electrical Apprenticeship Program at Michigan State University, and later served as chair for curriculum development for a statewide electrical inspector training short course. Dr. Surbrook developed a highly successful and comprehensive *National Electrical Code®* training course through Lifelong Education Programs of Michigan State University.

Dr. Surbrook has spent his professional career teaching, authoring textbooks, training bulletins, and papers, working with youths, electricians, contractors, and inspectors, and conducting research in the areas of electrical wiring and electronics. He has been an active member and officer of several professional organizations and has served on numerous technical committees. Organizations in which he has been most active are the Institute of Electrical and Electronic Engineers, American Society of Agricultural Engineers, International Association of Electrical Inspectors, and the National Fire Protection Association. Dr. Surbrook has served on several Ad Hoc Technical Committees of the National Electrical Code, and he has been a member of National Electrical Code-Making Panels 13 and 19. For 15 years, Dr. Surbrook was writer and producer of a weekly radio program dealing with electrification. He has written many bulletins, articles, and technical publications on various subjects related to electrical wiring.

Jonathan R. Althouse, M.S., who is Dr. Surbrook's coauthor, is an Instructor of Agricultural Engineering at Michigan State University, a licenced Master Electrician and Electrical Contractor. Mr. Althouse is the Coordinator of the Electrical Technology Apprenticeship Program at Michigan State University.

Mr. Althouse has developed an electrical awareness program for high school students entitled *Teaching Electrification in Agribusiness Classes in High Schools* (T.E.A.C.H.S.). He has also developed and provided instruction for *National Electrical Code®* Update Courses and Journey and Master Electrician examination prep classes. Mr. Althouse has been involved with organizations such as the International Association of Electrical Inspectors, the American Society of Agricultural Engineers, and the Michigan Agricultural Electric Council.

REVIEWERS

The following instructors provided recommendations during the writing and production of *INTERPRETING THE NATIONAL ELECTRICAL CODE®*:

Mr. Richard Ericson, JATC of Nassau and Suffolk Co., Melville, NY
Ms. Deanna Hanieski, Professor, Lansing Community College, Lansing, MI
Mr. Michael McBurney, New Castle School of Trades, Pulaski, PA
Mr. David Nelson, North Montco AVTS, Lansdale, PA
Mr. Ronald G. Oswald, Dean of the Institute of Technology, Pittsburgh, PA
Mr. Mel Sanders, Electrical Consultant, Ankeny, IA
Mr. Jimmie L. Sigmon, Gaston College, Dallas, NC
Mr. Wayne Tanner, Franklin City VTS, Chambersburg, PA
Mr. Gerald W. Williams, Oxnard College, Oxnard, CA

ACKNOWLEDGMENTS

It is with sincere gratitude that the authors express appreciation to the following individuals who have either helped with some aspect of the development of this text or have provided field testing and critical review of the material to ensure completeness and accuracy:

Lori A. Althouse, for encouragement
and editing
Dennis Cassady, Electrical Inspector
Sparta, MI
George Chorny, Electrical Inspector
and Contractor
Muskegon, MI
John P. Donovan, Standard Electric Co.
Lansing, MI
William Fetter, Electrical Consultant
Alpena, MI
George Little, Electrical Inspector
Farmington, MI
John Negri, Electrical Consultant
Bloomingdale, MI

Jim O'Donnell, Electrical Inspector
Marquette, MI
Ron Purvis, Georgia Power Company
Atlanta, GA
Mary Surbrook, for encouragement
and editing
Pete VanPutten, Electrical Consultant
Byron Center, MI
Francis Whiting, Electrical Consultant
Hillsdale, MI
James Worden, Electrical Inspector
Port Huron, MI

Delmar Publishers Is Your Electrical Book Source!

Whether you're a beginning student or a master electrician, Delmar Publishers has the right book for you. Our complete selection of proven best-sellers and all-new titles is designed to bring you the most up-to-date, technically-accurate information available.

NATIONAL ELECTRICAL CODE

National Electrical Code® 1999/NFPA
Revised every three years, the *National Electrical Code®* is the basis of all U.S. electrical codes.
Order # 0-8776-5432-8
Loose-leaf version in binder
Order # 0-8776-5433-6

National Electrical Code® Handbook 1999/NFPA
This essential resource pulls together all the extra facts, figures, and explanations you need to interpret the 1999 *NEC®*. It includes the entire text of the Code, plus expert commentary, real-world examples, diagrams, and illustrations that clarify requirements.
Order # 0-8776-5437-9

Illustrated Changes in the 1999 National Electrical Code®/O'Riley
This book provides an abundantly-illustrated and easy-to-understand analysis of the changes made to the 1999 *NEC®*.
Order # 0-7668-0763-0

Understanding the 1999 NEC®, 3E/Holt
This book gives users at every level the ability to understand what the *NEC®* requires, and simplifies this sometimes intimidating and confusing code.
Order # 0-7668-0350-3

Illustrated Guide to the National Electrical Code®/Miller
Highly-detailed illustrations offer insight into Code requirements, and are further enhanced through clearly-written, concise blocks of text which can be read very quickly and understood with ease. Organized by classes of occupancy.
Order # 0-7668-0529-8

Interpreting the National Electrical Code®, 5E/Surbrook and Althouse
This updated resource provides a process for understanding and applying the *National Electrical Code®* to electrical contracting, plan development, and review.
Order # 0-7668-0187-X

Electrical Grounding, 5E/O'Riley
Electrical Grounding is a highly illustrated, systematic approach for understanding grounding principles and their application to the 1999 *NEC®*.
Order # 0-7668-0486-0

ELECTRICAL WIRING

Electrical Raceways and Other Wiring Methods, 3E/Loyd
The most authoritive resource on metallic and non-metallic raceways, provides users with a concise, easy-to-understand guide to the specific design criteria and wiring methods and materials required by the 1999 *NEC®*.
Order # 0-7668-0266-3

Electrical Wiring Residential, 13E/Mullin
Now in full color! Users can learn all aspects of residential wiring and how to apply them to the wiring of a typical house from this, the most widely-used residential wiring book in the country.
Softcover Order # 0-8273-8607-9
Hardcover Order # 0-8273-8610-9

House Wiring with the NEC®/Mullin
The focus of this new book is the applications of the *NEC®* to house wiring.
Order # 0-8273-8350-9

Electrical Wiring Commercial, 10E/Mullin and Smith
Users can learn commercial wiring in accordance with the *NEC®* from this comprehensive guide to applying the newly revised 1999 *NEC®*.
Order # 0-7668-0179-9

Electrical Wiring Industrial, 10E/Smith and Herman
This practical resource has users work their way through an entire industrial building—wiring the branch-circuits, feeders, service entrances, and many of the electrical appliances and sub-systems found in commercial buildings.
Order # 0-7668-0193-4

Cables and Wiring, 2E/AVO
This concise, easy-to-use book is your single-source guide to electrical cables—it's a "must-have" reference for journeyman electricians, contractors, inspectors, and designers.
Order # 0-7668-0270-1

ELECTRICAL MACHINES AND CONTROLS

Industrial Motor Control, 4E/Herman and Alerich
This newly-revised and expanded book, now in full color, provides easy-to-follow instructions and essential information for controlling industrial motors. Also available are a new lab manual and an interactive CD-ROM.
Order # 0-8273-8640-0

Electric Motor Control, 6E/Alerich and Herman
Fully updated in this new sixth edition, this book has been a long-standing leader in the area of electric motor controls.
Order # 0-8273-8456-4

Introduction to Programmable Logic Controllers/Dunning
This book offers an introduction to Programmable Logic Controllers.
Order # 0-8273-7866-1

Technician's Guide to Programmable Controllers, 3E/Cox
Uses a plain, easy-to-understand approach and covers the basics of programmable controllers.
Order # 0-8273-6238-2

Programmable Controller Circuits/Bertrand
This book is a project manual designed to provide practical laboratory experience for one studying industrial controls.
Order # 0-8273-7066-0

*Electronic Variable Speed Drives/*Brumbach

Aimed squarely at maintenance and troubleshooting, *Electronic Variable Speed Drives* is the only book devoted exclusively to this topic.

Order # 0-8273-6937-9

*Electrical Controls for Machines, 5E/*Rexford

State-of-the-art process and machine control devices, circuits, and systems for all types of industries are explained in detail in this comprehensive resource.

Order # 0-8273-7644-8

*Electrical Transformers and Rotating Machines/*Herman

This new book is an excellent resource for electrical students and professionals in the electrical trade.

Order # 0-7668-0579-4

*Delmar's Standard Guide to Transformers/*Herman

Delmar's Standard Guide to Transformers was developed from the best-seller *Standard Textbook of Electricity* with expanded transformer coverage not found in any other book.

Order # 0-8273-7209-4

DATA AND VOICE COMMUNICATION CABLING AND FIBER OPTICS

*Complete Guide to Fiber Optic Cable System Installation/*Pearson

This book offers comprehensive, unbiased, state-of-the-art information and procedures for installing fiber optic cable systems.

Order # 0-8273-7318-X

*Fiber Optics Technician's Manual/*Hayes

Here's an indispensable tool for all technicians and electricians who need to learn about optimal fiber optic design and installation as well as the latest troubleshooting tips and techniques.

Order # 0-8273-7426-7

*A Guide for Telecommunications Cable Splicing/*Highhouse

A "how-to" guide for splicing all types of telecommunications cables.

Order # 0-8273-8066-6

*Premises Cabling/*Sterling

This reference is ideal for electricians, electrical contractors, and inspectors needing specific information on the principles of structured wiring systems.

Order # 0-8273-7244-2

ELECTRICAL THEORY

*Delmar's Standard Textbook of Electricity, 2E/*Herman

This exciting full-color book is the most comprehensive book on DC/AC circuits and machines for those learning the electrical trades.

Order # 0-8273-8550-1

*Industrial Electricity, 6E/*Nadon, Gelmine, and Brumbach

This revised, illustrated book offers broad coverage of the basics of electrical theory and industrial applications. It is perfect for those who wish to be industrial maintenance technicians.

Order # 0-7668-0101-2

EXAM PREPARATION

*Journeyman Electrician's Exam Preparation, 2E/*Holt

This comprehensive exam prep guide includes all of the topics on the journeyman electrician competency exams.

Order # 0-7668-0375-9

*Master Electrician's Exam Preparation, 2E/*Holt

This comprehensive exam prep guide includes all of the topics on master electrician's competency exams.

Order # 0-7668-0376-7

REFERENCE

ELECTRICAL REFERENCE SERIES

This series of technical reference books is written by experts and designed to provide the electrician, electrical contractor, industrial maintenance technician, and other electrical workers with a source of reference information about virtually all of the electrical topics that they encounter.

*Electrician's Technical Reference—Motor Controls/*Carpenter

Electrician's Technical Reference—Motor Controls is a source of comprehensive information on understanding the controls that start, stop, and regulate the speed of motors.

Order # 0-8273-8514-5

*Electrician's Technical Reference—Motors/*Carpenter

Electrician's Technical Reference—Motors builds an understanding of the operation, theory, and applications of motors.

Order # 0-8273-8513-7

*Electrician's Technical Reference—Theory and Calculations/*Herman

Electrician's Technical Reference—Theory and Calculations provides detailed examples of problem-solving for different kinds of DC and AC circuits.

Order # 0-8273-7885-8

*Electrician's Technical Reference—Transformers/*Herman

Electrician's Technical Reference—Transformers focuses on the theoretical and practical aspects of single-phase and 3-phase transformers and transformer connections.

Order # 0-8273-8496-3

*Electrician's Technical Reference—Hazardous Locations/*Loyd

Electrician's Technical Reference—Hazardous Locations covers electrical wiring methods and basic electrical design considerations for hazardous locations.

Order # 0-8273-8380-0

*Electrician's Technical Reference—Wiring Methods/*Loyd

Electrician's Technical Reference—Wiring Methods covers electrical wiring methods and basic electrical design considerations for all locations, and shows how to provide efficient, safe, and economical applications of various types of available wiring methods.

Order # 0-8273-8379-7

*Electrician's Technical Reference—Industrial Electronics/*Herman

Electrician's Technical Reference—Industrial Electronics covers components most used in heavy industry, such as silicon control rectifiers, triacs, and more. It also includes examples of common rectifiers and phase-shifting circuits.

Order # 0-7668-0347-3

RELATED TITLES

*Common Sense Conduit Bending and Cable Tray Techniques/*Simpson

Now geared especially for students, this manual remains the only complete treatment of the topic in the electrical field.

Order # 0-8273-7110-1

*Practical Problems in Mathematics for Electricians, 5E/*Herman

This book details the mathematics principles needed by electricians.

Order # 0-8273-6708-2

*Electrical Estimating/*Holt

This book provides a comprehensive look at how to estimate electrical wiring for residential and commercial buildings with extensive discussion of manual versus computer-assisted estimating.

Order # 0-8273-8100-X

*Electrical Studies for Trades/*Herman

Based on *Delmar's Standard Textbook of Electricity,* this new book provides non-electrical trades students with the basic information they need to understand electrical systems.

Order # 0-8273-7845-9

UNIT 1

General Wiring and Fundamentals

OBJECTIVES

Upon completion of this unit, the student will be able to:

- name the sponsoring organization of the *National Electrical Code®*.
- know how and when to submit a proposal to change the *National Electrical Code®*.
- explain what is a tentative interim amendment.
- describe direct current and alternating current.
- calculate the unknown quantity if only two of these quantities are known: volts, current, or resistance.
- determine the current in a circuit or equipment if power, voltage, and power factor are known.
- determine the Circular Mil area of a conductor if the conductor diameter is given.
- calculate the total resistance of a circuit if the values of resistance in either series or in parallel are given.
- choose the correct ampacity table from *Article 310* to determine the minimum size of conductor required for a wiring application.
- answer wiring installation questions relating to *Articles 90, 100, 110, 200, 300, 305, 310, 324, 326, 328, 330, 333, 334, 336, 338, 339, 340, 342,* or *Chapter 9 Tables 8* or *9*.
- state at least five significant changes that occurred from the 1996 to the 1999 Code for *Articles 90, 100, 110, 200, 210 Part A, 300, 305, 310, 324, 326, 328, 330, 333, 334, 336, 338, 339, 340, 342,* or *Chapter 9 Tables 8* or *9*.

ORIGIN OF THE *NATIONAL ELECTRICAL CODE®*

The first *National Electrical Code®* was developed in 1897. In 1911, the National Fire Protection Association (NFPA) became the sponsor, and the Code has been revised on numerous occasions since that date. Now it is revised every three years. The time schedule for revising the 1999 Code is listed in the front of the Code. The next revision will be the 2002 edition.

The *National Electrical Code®* is available for adoption as the electrical law in a governmental jurisdiction. That governmental jurisdiction may add one or more amendments to allow for local needs, preferences, or conditions. It is not the intent of this text to cover amendments to the Code made by local or state jurisdictions, but an instructor using this material for a course may cover these amendments as an addendum to this material.

Process of Revising the Code

The process of revising the Code begins with any interested person who submits a proposal for changing the Code not later than the closing date for accepting proposals. This closing date is given in each edition of the Code. Proposals are mailed to the Vice President of the Technical Council, National Fire Protection Association, Batterymarch Park, Quincy, MA 02269. A blank form for submitting proposals is included at the end of this unit, and there is an example of a proposal in the Code.

The proposals for changing a particular section or adding a new section or article are reviewed by the appropriate Code-Making Panel (listed in the front of the Code) at the date listed in the Code. The Code-Making Panel actions on proposals are published in a document titled the *Committee Report on Proposals,* which is usually available to the public in July of the year shown in the timetable of events in the Code. Any interested person can obtain a copy of this document and make public comment that must be received by the NFPA not later than the closing date listed for public comment. A comment form is in the front of the *Committee Report on Proposals.*

Code-Making Panels meet in December of every third year to act on public comments. The Panel is only permitted to take action on sections of the Code for which public comment was received. The Code-Making Panel's actions at these two meetings constitute the changes that occur in the Code. The *Committee Report on Comments* reports on the Panel action taken on public comment. The electrical section meeting of the National Fire Protection Association in May 2001 results in the adoption of the 2002 Code. Members present at that meeting are permitted to make motions to change Code-Making Panel action taken on a particular proposal published in the *Committee Report on Proposals* or change the action taken on a comment as published in the *Committee Report on Comments.* If the motion receives a majority vote of members present at the meeting, the change becomes a part of the new Code. A person who submitted a proposal that was rejected and made public comment that was rejected, can make a motion for adoption of the proposal at the annual meeting. Following this meeting of the electrical section of NFPA, the change process is completed, and NFPA publishes the Code. The new edition of the Code is available for public sale by September of the year of the NFPA Annual Meeting when the Code was adopted.

There is one additional means by which the Code may change in addition to the process previously described. This is by the issuing of a Tentative Interim Amendment or TIA. These are generally changes of such importance that they should not wait until the next Code. There is a process through which they are approved, and if approved, a TIA is added to the Code only for the duration of that edition. A TIA will automatically be submitted as a proposal for the next Code change, and it must go through the same process as all other proposals to remain as a part of the Code.

ELECTRICAL CALCULATIONS WITH A CALCULATOR

An electronic calculator is a valuable tool for making calculations required for electrical installations. It is important to become familiar with the proper use of your calculator. The procedures for making calculations may be different from one calculator to another. The following procedure is common for many calculators, but it may be different for your calculator. If this procedure does not work for your calculator, then check the calculator instructions for each example problem. Practice with your calculator until you become comfortable with its operation. This will help you with the problems in this text as well as eliminating errors in electrical calculations on the job. There are more efficient methods of making the following calculations on some calculators; however, the following method works on most calculators.

Example of String Multiplication

Calculate power (P) when the voltage (V), current (A), and power factor (pf) are known.

$$P = V \times A \times pf = 480 \text{ V} \times 22 \text{ A} \times 0.75 = 7,920 \text{ W}$$

Step	Press	Display
1	480	480
2	×	480
3	22	22
4	=	10,560
5	×	10,560
6	.75	.75
7	=	**7,920** ←

Example of Addition and Multiplication

Determine the cost of several runs of conduit. The runs are 22 ft., 74 ft., and 15 ft., and the cost is \$1.15 per foot.

$$\text{Conduit cost} = (22 \text{ ft.} + 74 \text{ ft.} + 15 \text{ ft.}) \times \$1.15 = \$127.65$$

Step	Press	Display
1	22	22
2	+	22
3	74	74
4	=	96
5	+	96
6	15	15
7	=	111
8	×	111
9	1.15	1.15
10	=	**127.65** ←

Example of String Multiplication and Division

Determine the full-load current of a 3-phase, 37.5-kVA transformer with a 208-volt secondary winding.

$$A = \frac{kVA \times 1,000}{1.73 \times V} = \frac{37.5 \times 1,000}{1.73 \times 208} = 104.21 \text{ A}$$

Step	Press	Display
1	37.5	37.5
2	×	37.5
3	1,000	1,000
4	=	37,500
5	÷	37,500
6	1.73	1.73
7	=	21,676.3
8	÷	21,676.3
9	208	208
10	=	**104.21** ←

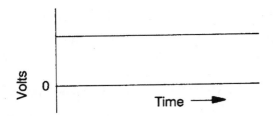

Figure 1.1   A plot of voltage in a direct current circuit over a period of time is a straight line.

FUNDAMENTALS OF ELECTRICITY

An electrician must have a basic understanding of the fundamentals of electricity to perform most effectively in the field and also to pass electrical examinations. This section of the text is a brief review of minimum basic electrical fundamentals.

AC and DC

The Code refers to alternating current (ac) and direct current (dc). Direct current travels only in one direction in the circuit. A typical source of direct current is a battery or a rectifier. Figure 1.1 shows how the voltage in a direct current circuit would look if plotted on a graph over a period of time. An oscilloscope is an electrical instrument that can show the voltage and time plot of a direct current circuit. The voltage will be either a positive or a negative constant value. The level of voltage remains at a constant or a nearly constant magnitude. The polarity will remain either positive or negative. As long as the polarity does not change from positive to negative or from negative to positive, the electrical current will flow in one direction in the circuit. The polarity must change if the current is to reverse direction of flow in the circuit.

In the case of an alternating current circuit, the voltage is constantly changing. The voltage increases to a maximum, then decreases back to zero. Then the voltage builds to a maximum with opposite polarity, then decreases in magnitude back to zero. This completes one cycle. Normal electrical power operates at 60 cycles per second or hertz (Hz). A graph of an ac voltage varying with time is shown in Figure 1.2. This is what an ac voltage would look like if viewed on the screen of an oscilloscope, and the waveform is known as a sine wave from the mathematics of trigonometry. If an analog voltmeter with a needle or a digital voltmeter is connected to measure voltage across a component of an ac circuit, the voltmeter will indicate the root-mean-square (rms) voltage as indicated in Figure 1.2. The actual voltage is changing as shown in Figure 1.2, but the analog or digital voltmeter will register only a single value of voltage. The rms voltage is compared with the alternating voltage in Figure 1.2. The rms voltage is 0.707 times the peak of the varying voltage (sine wave) of the circuit. Or, the

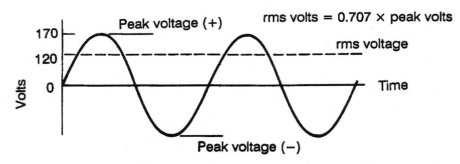

Figure 1.2.   The plot of voltage in an alternating current circuit over a period of time is a sine wave, and the voltage that would show on a meter would be the rms voltage.

peak voltage of the sine wave is 1.414 times the rms voltage. Root-mean-square is used in the definition of voltage in *Article 100* of the Code. Alternating current frequencies other than 60 Hz are sometimes used for special applications in electrical equipment.

Many voltmeters are not true rms (root-mean-square) meters. The common hand-held voltmeter actually measures the average value of the alternating sine wave at 60 Hz, which is 0.637 times the peak value. The voltmeter output is calibrated to increase the average voltage value by 1.11 so the output will be equal to the rms voltage ($0.637 \times 1.11 = 0.707$). If the voltage is something other than 60 Hz or if the power source is being altered by solid-state switching, the voltage indicated by the meter will be in error. It is also possible in some electrical systems for more than one frequency to be present at the same time. This is the case when harmonic frequencies are present on the same 3-phase, 4-wire circuits as mentioned in the fine print note to *Section 220-22*. Under these conditions, an averaging voltmeter will not indicate an accurate rms voltage. The amount of the error will depend on the amount of harmonic distortion of the 60-Hz sine wave. True rms reading voltmeters are available.

Wire Dimensions

Wire dimensions are given in American Wire Gauge (AWG) and in thousands of Circular Mils (kcmil). The **k** is the metric symbol for 1,000. The **c** represents **circular** for Circular Mils. The **mil** is 1/1,000. In the case of electrical wires, the mil represents 1/1,000 of an inch. The area of a wire is determined by taking the diameter of a wire in inches, multiplying by 1,000, and then squaring that number. Squaring a number means the number is multiplied by itself. The old designation for thousands of Circular Mils is (MCM). The **M** is the Roman numeral for 1,000. Some wires encountered in electrical wiring will be marked **kcmil** and some will be marked **MCM.** Both of these designations represent thousands of Circular Mils. Area of a wire is expressed as Circular Mils, not as a true area. Common sizes of wire are listed in *Table 8, Chapter 9* of the Code. The area of the wire in Circular Mils is given in the table.

The diameter of a wire is given in inches as shown in *Table 8, Chapter 9* of the Code. The diameter can be converted to mils by multiplying the diameter in inches by 1,000. A mil is 0.001 inch. An AWG number 14 solid wire has a diameter of 0.064 inch, and therefore the diameter would be 0.064 times 1,000, or 64 mils. This is illustrated in Figure 1.3.

The Circular Mil area of a wire is obtained by squaring the diameter of the wire in mils, or multiplying the diameter by itself. The area of an AWG number 14 solid wire would be 64 mils times 64 mils, or 4096 Circular Mils (4.096 kcmil) as shown in Figure 1.3. The values given in *Table 8, Chapter 9* of the Code are approximate, so they may not match exactly with a calculation.

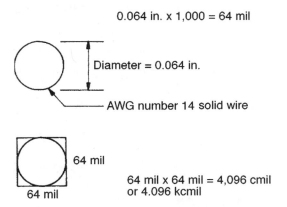

0.064 in. x 1,000 = 64 mil

Diameter = 0.064 in.

AWG number 14 solid wire

64 mil

64 mil

64 mil x 64 mil = 4,096 cmil
or 4.096 kcmil

Figure 1.3   The diameter of a wire in mils is squared to obtain the Circular Mil area.

Table 1.1   Metric equivalent conductor sizes for common AWG and kcmil wire sizes.

AWG	mm²	kcmil	mm²
18	0.8	250	127
16	1.3	300	152
14	2.1	350	177
12	3.3	400	203
10	5.3	500	253
8	8.4	600	304
6	13.3	700	355
4	21.2	750	380
3	26.7	800	405
2	33.6	900	456
1	42.4	1000	507
1/0	53.5	1250	633
2/0	67.4	1500	760
3/0	85	1750	887
4/0	107	2000	1013

In other parts of the world, wire sizes are in square millimeters (mm²), but the Code does not give standard metric wire sizes or the metric equivalent sizes for standard AWG and kcmil sizes. Table 1.1 gives the metric equivalent sizes in mm² for standard AWG and kcmil sizes.

Wire Resistance

Conductor resistance is measured in ohms. The resistance of a wire is different for different types of materials such as copper and aluminum. A wire with a larger cross-sectional area than another wire has a lower resistance for a given length. For example, the resistance listed for 1,000 feet of AWG number 10 solid copper wire is 1.21 ohms. AWG number 14 copper wire is smaller than AWG number 10, and a 1,000 foot length of AWG number 14 solid copper wire has a resistance of 3.07 ohms. The resistance of a wire decreases as the cross-sectional area of the wire increases. This is illustrated in Figure 1.4.

The resistance of a wire increases as the length of the wire increases. If one wire of the same size is twice as long as another wire, the longer wire will have twice the resistance as the shorter wire. This is illustrated in Figure 1.5. The resistance of 1,000 feet of wire at a temperature of 75°C is given in *Table 8, Chapter 9* of the Code. If the resistance of the length of wire other than 1,000 feet is desired, Equation 1.1 will be useful in determining the actual resistance of the wire.

$$\text{Resistance}_{\text{Desired length}} = \frac{\text{Desired length (ft.)}}{1,000 \text{ ft.}} \times \text{Resistance}_{1,000 \text{ ft.}} \qquad \textbf{Eq. 1.1}$$

Temperature has an effect on the resistance of an electrical conductor. As the temperature of a copper or an aluminum wire increases, the resistance will also increase. The change in

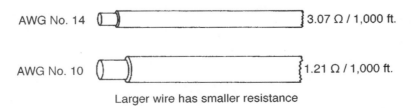

Figure 1.4   A wire with a larger cross-sectional area has a lower resistance for a given length.

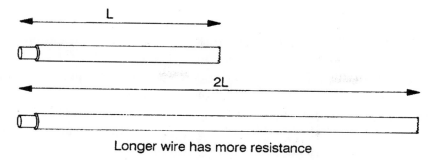

Figure 1.5   A long wire has a higher resistance than a shorter wire of the same size.

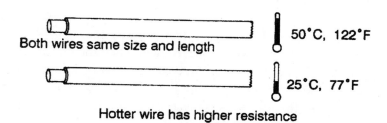

Figure 1.6   The resistance of a wire increases as the temperature of the wire increases.

resistance of a wire as temperature increases is illustrated in Figure 1.6. A wire carrying electrical current for several hours will have a higher temperature and a higher resistance than a wire not carrying electrical current. The temperature of the area where a wire is to be installed is frequently different from one area to another. For copper and aluminum wires, and most other wires, the resistance increases as the temperature increases. The resistivity (unit value of resistance of a material) of copper and aluminum wire is different, and it increases as the temperature increases. Table 1.2 gives values of resistivity for copper and aluminum at three different temperatures. Note that the resistivity increases as the temperature increases.

The resistance of a given length of a particular size of copper or aluminum wire also can be determined at a particular temperature using Equation 1.2. The equation gives the relationship between wire resistance, the resistivity of the wire, the cross-sectional area of the wire, and the length of the wire. The Circular Mil area of the wire is found in *Table 8, Chapter 9.*

$$R = K \times \frac{L}{A}$$

Eq. 1.2

R = resistance in ohms
K = resistivity of material in ohms Circular Mils per foot
A = Circular Mil area of wire
L = length of wire in feet

Table 1.2   Resistivity of copper and aluminum conductors at different temperatures.

Wire type	Resistivity, K (ohm Cir. Mils/ft.)		
	25°C	50°C	75°C
Copper	10.8	11.8	12.7
Aluminum	17.0	19.0	21.0

Ohm's Law

It is important to understand the relationship between electrical current, voltage, and resistance of an electrical circuit. This relationship is known as Ohm's Law. Equation 1.3 is Ohm's Law arranged to solve for current. The current in amperes is equal to the voltage divided by resistance in ohms. Assume the resistance of a circuit is held constant but the voltage is changed. According to Equation 1.3, if the voltage is increased, the current will increase. If the voltage is decreased, the current will decrease. This is illustrated in Figure 1.7. Note when the voltage is doubled from 120 volts to 240 volts, the current in the circuit of Figure 1.7 doubles from 12 amperes to 24 amperes.

$$\text{Current} = \frac{\text{Voltage}}{\text{Resistance}} \qquad I = \frac{E}{R} \qquad \text{Eq. 1.3}$$

Ohm's Law can be written to solve for the voltage when the current flow and resistance are known. Equation 1.4 is Ohm's Law arranged to solve for voltage. If the current flow through a wire increases, the voltage drop along the wire will increase. Figure 1.8 illustrates Ohm's Law for a 10-ampere flow through wires with two different resistances. The voltage drop will be doubled for the wire with twice as much resistance when the current flow remains constant.

$$\text{Voltage} = \text{Current} \times \text{Resistance} \qquad E = I \times R \qquad \text{Eq. 1.4}$$

The resistance of a circuit or a device can be determined using Equation 1.5. If the current flow and the voltage are known, the resistance can be calculated.

$$\text{Resistance} = \frac{\text{Voltage}}{\text{Current}} \qquad R = \frac{E}{I} \qquad \text{Eq. 1.5}$$

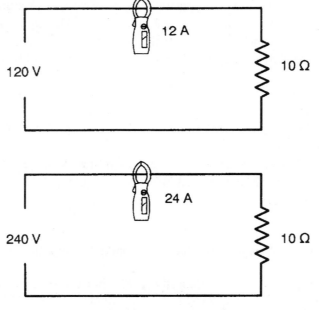

Figure 1.7   According to Ohm's Law, doubling the voltage will result in a doubling of the current in the circuit if the resistance remains constant.

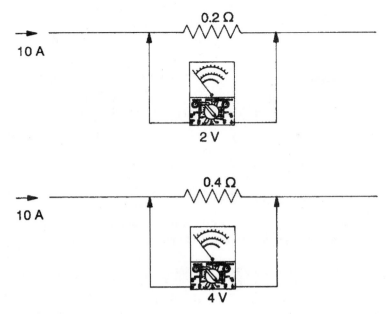

Figure 1.8 Increasing the resistance for a constant current flow will result in an increase in voltage along the circuit.

Example 1.1 The voltage drop across an incandescent lightbulb is 120 volts, and the current flow through the lightbulb is 0.8 amperes. Determine the resistance of the lightbulb filament when it is lighted.

Answer: The resistance is desired, and the voltage and current are known; therefore, use Equation 1.5.

$$\text{Resistance} = \frac{120\ V}{0.8\ A} = 150\ \Omega$$

Some equations for Ohm's Law may read impedance (Z) instead of resistance (R). Impedance is in ohms, and it is the combined effect of resistance, inductive reactance, and capacitive reactance in a circuit.

Circuits

Electricity must always travel in a circuit from a source of voltage back to the source of voltage. A simple circuit contains a voltage source, resistance of the wires and the load, and electrical current flow. The relationship that governs these quantities is Ohm's Law, which was just discussed. Most electrical circuits actually consist of several resistances in series or several resistances in parallel. Series and parallel circuits with three resistances are shown in Figures 1.9 and 1.10. Troubleshooting electrical problems is much easier if the electrician has a basic understanding of series and parallel circuits.

A series circuit has all of the loads or resistances connected so there is only one path through the circuit. Therefore, if there is only one path, the electrical current is the same everywhere in the circuit, as illustrated with Equation 1.6. The combined resistance of the series circuit is the sum of the individual resistances, as represented by Equation 1.7. The voltage at the source of the circuit will drop across the individual series resistances. The sum of the voltage drops across all of the resistances in series will add up to the circuit source voltage, as indicated by Equation 1.8. Refer to the circuit of Figure 1.9 of a series circuit.

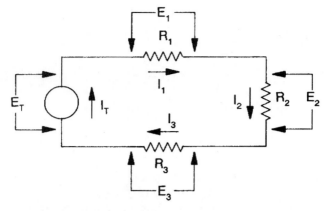

Figure 1.9   There is only one path through a circuit where the loads or resistances are arranged in series.

$$I_T = I_1 = I_2 = I_3 \qquad \text{Eq. 1.6}$$

$$R_T = R_1 + R_2 + R_3 \qquad \text{Eq. 1.7}$$

$$E_T = E_1 + E_2 + E_3 \qquad \text{Eq. 1.8}$$

Example 1.2   Assume the resistances in the series circuit of Figure 1.9 are 4 ohm, 8 ohm, and 12 ohm, and the circuit is powered at 120 volts. Determine: (1) the total circuit resistance, (2) the current flow in the circuit, and (3) the voltage drop across the 8-ohm resistor.

Answer: (1) The resistors are in series; therefore, use Equation 1.7 to find the total resistance.

$$R_T = 4\ \Omega + 8\ \Omega + 12\ \Omega = 24\ \Omega$$

(2) Total voltage and total resistance are known; therefore, use Ohm's Law (Equation 1.3) to solve for current.

$$\text{Current} = \frac{120\ V}{24\ \Omega} = 5\ A$$

(3) Ohm's Law can be used to determine the voltage drop across the 8-ohm resistor (Equation 1.4). The current determined in (2) flows through each resistor because this is a series circuit.

$$\text{Voltage} = 5\ A \times 8\ \Omega = 40\ V$$

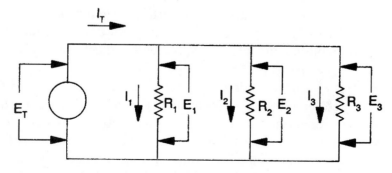

Figure 1.10   There are multiple paths through a circuit where the loads or resistances are arranged in parallel.

This is the voltage drop across the 8-ohm resistor. The voltage drop across the 4-ohm resistor is 20 volts, and across the 12-ohm resistor is 60 volts. The voltage drop across all series resistors adds to 120 volts.

A parallel circuit has all of the loads or resistances arranged so there are as many paths for the electrical current to follow as there are resistances in parallel, as shown in Figure 1.10. The total electrical current flowing in the circuit is the sum of the electrical current in each path of the circuit, as indicated by Equation 1.9. The total circuit resistance is the most difficult quantity to determine. The reciprocal of the total resistance is the sum of the reciprocals of each resistance in parallel, as indicated in Equation 1.10. The total resistance of the circuit will always be less than the value of the smallest resistance in parallel. The voltage across each parallel resistance will be equal and the same as the source voltage, as illustrated in Equation 1.11.

$$I_T = I_1 + I_2 + I_3 \qquad\qquad \text{Eq. 1.9}$$

$$\frac{1}{R_T} = \frac{1}{R_1} + \frac{1}{R_2} + \frac{1}{R_3} \qquad\qquad \text{Eq. 1.10}$$

$$E_T = E_1 = E_2 = E_3 \qquad\qquad \text{Eq. 1.11}$$

Example 1.3   Assume the parallel resistances in Figure 1.10 have the values of 4 ohm, 6 ohm, and 12 ohm, and the circuit is operated at 120 volts. Determine the following for the circuit: (1) total circuit resistance, (2) total circuit current flow, and (3) the current flow through the 6-ohm resistor.

Answer: (1) This is a circuit with all resistances in parallel. The total resistance of the circuit can be determined using Equation 1.10.

$$\frac{1}{R_T} = \frac{1}{4} + \frac{1}{6} + \frac{1}{12} = 0.250 + 0.167 + 0.083 = 0.500$$

$$R_T = \frac{1}{0.500} = 2\,\Omega$$

(2) Ohm's Law can be used to determine the total current flow in the circuit (Equation 1.3).

$$\text{Current} = \frac{120\ \text{V}}{2\ \Omega} = 60\ \text{A}$$

(3) All of the resistances are in parallel, and there are 120 volts across each resistor. The current flowing through the 6-ohm resistor is found by using Equation 1.3, and dividing 120 volts by 6 ohms to get 20 amperes. Using Equation 1.3, the current through the 4-ohm resistor is 30 amperes, and 10 amperes through the 12-ohm resistor. Note the current through each resistor adds to the total current, which was 60 amperes (20 A + 30 A + 10 A = 60 A).

Power and Work Formulas

Power is the rate of doing work, as illustrated by Equation 1.12. It will take twice as much power to lift a weight a given number of feet in one minute as it will to lift the same weight the distance in two minutes. Horsepower (hp) is a common unit of measure of power, but watts (W) is also a unit of measure of power. The conversion from horsepower to several other units of power is shown below:

$$\text{Power} = \frac{\text{Work}}{\text{Time}} \qquad \text{Eq. 1.12}$$

$$1 \text{ hp} = 33{,}000 \text{ ft-lb/min}$$
$$= 44{,}760 \text{ J/min}$$
$$= 746 \text{ W}$$

Electrical power in watts can be determined if the volts, amperes, and power factor of the circuit are known. For a direct current circuit or an alternating current circuit where the loads are resistance type, such as incandescent lights or electric resistance heaters, the power factor is 1.0. In the case of electric discharge lights, motors, and many other types of equipment, the power factor will be less than 1.0. A power factor meter can be used to determine the power factor of a particular circuit. Equation 1.13 is used to determine the single-phase power of a circuit, and Equation 1.14 is used to determine the power of a 3-phase circuit.

$$\text{Power}_{\text{Single-phase}} = \textbf{Volts} \times \textbf{Amperes} \times \textbf{Power Factor} \qquad \text{Eq. 1.13}$$

$$\text{Power}_{\text{3-phase}} = \textbf{1.73} \times \textbf{Volts} \times \textbf{Amperes} \times \textbf{Power Factor} \qquad \text{Eq. 1.14}$$

Example 1.4   A circuit of fluorescent lighting fixtures draws 15.4 amperes at 120 volts with a circuit power factor of 0.72. Determine the single-phase power in watts drawn by the circuit.

Answer: This is a single-phase power problem. Therefore, Equation 1.13 is used.

$$\text{Power} = 120 \text{ V} \times 15.4 \text{ A} \times 0.72 = 1{,}331 \text{ W}$$

It is often useful to determine the current drawn by a particular load. This can be done by rearranging the power (Equation 1.13) to solve for current. The new form of the equation to determine current when the watts and voltage are known is given by Equation 1.15.

$$\textbf{Amperes} = \frac{\textbf{Power}}{\textbf{Volts} \times \textbf{Power Factor}} \qquad \text{Eq. 1.15}$$

The power factor is 1.0 for dc loads and resistance ac loads such as incandescent lamps and resistance heating elements. The use of Equation 1.15 is illustrated by an example.

Example 1.5   Determine the current drawn by a 100-watt incandescent lightbulb operating at its rated 120 volts.

Answer: Use Equation 1.15 and note that the power factor is 1.0 for an incandescent lightbulb.

$$\text{Amperes} = \frac{100 \text{ W}}{120 \text{ V} \times 1.0} = 0.83 \text{ A}$$

Heat is a form of work, and work is power multiplied by time. Heat is produced as electrical current flows through an electrical conductor. Equation 1.16 shows the relationship between heat (in watt-seconds), electrical current (in amperes), resistance (in ohms), and time (in seconds).

$$\text{Heat} = I^2 \times R \times \text{Time} \qquad \text{Eq. 1.16}$$

The kilowatt-hour (kWh) is a unit of measure of electrical work. A smaller unit of work is the watt-second. The watt-second is equal to one joule (J), which is the metric unit of measure of heat. The British thermal unit (Btu) is also a unit of measure of heat. The following are conversion factors for different units of measure of work or heat:

$$
\begin{aligned}
1 \text{ kWh} &= 3{,}413 \text{ Btu} = 3{,}600{,}000 \text{ joules} \\
1 \text{ watt-sec.} &= 1 \text{ joule} \\
1 \text{ watt-hr.} &= 3{,}600 \text{ joules} \\
1 \text{ kilojoule} &= 0.948 \text{ Btu}
\end{aligned}
$$

Example 1.6. A wire supplying power to an electric motor has a resistance of 0.2 ohm. Assume that a short-circuit occurs at the electric motor controller, and the short-circuit current of 3,500 amperes flows for one second. Determine the heat produced in the wire by the short-circuit.

Answer: The heat produced is determined by using Equation 1.16. For simplicity, assume the resistance of the wire remains constant during the short-circuit.

$$\text{Heat} = 3{,}500 \text{ A} \times 3{,}500 \text{ A} \times 0.2\ \Omega \times 1 \text{ s} = 2{,}450{,}000 \text{ J}$$

This is 2,450 kilojoules or 2,323 Btu.

Efficiency

The efficiency of a system or equipment is the output divided by the input. Efficiency of an electric motor can be calculated using Equation 1.17 provided it is known exactly how much power is being developed by the motor. The watts draw by the motor can be determined by using Equation 1.13 for a single-phase circuit, or Equation 1.14 for a 3-phase circuit.

$$\text{Eff} = \frac{\text{Power Out}}{\text{Power In}} = \frac{\text{Horsepower Developed} \times 746}{\text{Watts}} \qquad \text{Eq. 1.17}$$

Example 1.7 A single-phase electric motor draws 9.8 amperes at 115 volts with a power factor of 0.82. The motor is developing 1/2 horsepower. Determine the efficiency of the motor.

Answer: Efficiency is determined by dividing the power developed by the input power to operate the motor using Equation 1.17. The power out of the motor is in horsepower; therefore, it must first be converted to electrical power.

$$\text{Efficiency} = \frac{0.5 \text{ hp} \times 746 \text{ W/hp}}{9.8 \text{ A} \times 115 \text{ V} \times 0.82} = 0.40 \text{ or } 40 \ \%$$

Student Practice Problems

These practice problems on electrical fundamentals will help improve skills at working with the basic concepts of electricity previously discussed. Look for the equation from the previous discussion that relates to the problem. It is helpful to list the known information and the quantity desired. Then try to find an equation using that information. Refer to the example problems for help.

1. The resistance of 320 feet of AWG number 4 uncoated copper wire with resistance as given in Code *Table 8, Chapter 9* is:
   A. 0.050 ohms.   B. 0.099 ohms.   C. 0.257 ohms.   D. 1.250 ohms.

2. You observe a technician using an oscilloscope to measure the voltage at an outlet. The value of the peak voltage of the sine wave on the screen is 167 volts. If you were to measure this same voltage with your rms meter, the rms voltage would be:
   A. 95 volts.   B. 110 volts.   C. 118 volts.   D. 130 volts.

3. A baseboard electric heater operates at 240 volts and draws 3.33 amperes. The resistance of the heating element is:
   A. 72 ohms.   B. 95 ohms.   C. 110 ohms.   D. 240 ohms.

4. The current drawn by a 200-watt incandescent lamp operating at 120 volts is:
   A. 0.50 ampere.   B. 1.00 ampere.   C. 1.36 amperes.   D. 1.67 amperes.

5. An 18-ohm load is connected to a 240-volt power source using wires that each have a resistance of 0.8 ohm. The circuit is shown in Figure 1.11. The voltage across the terminals of the load is:
   A. 120 volts.   B. 130 volts.   C. 220 volts.   D. 240 volts.

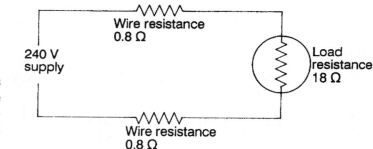

Figure 1.11   A typical series circuit is one in which the resistance of the wire supplying the load is in series with the resistance of the load.

6. A 120-volt circuit supplies two loads connected in parallel. One load has a resistance of 20 ohms, and the other load has a current flow of 4 amperes. The circuit is shown in Figure 1.12. The total current flowing in the circuit is:
   A. 4 amperes.   B. 5 amperes.   C. 7.5 amperes.   D. 10 amperes.

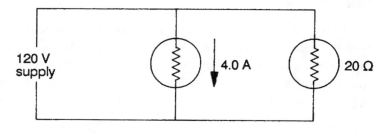

Figure 1.12   A typical parallel circuit is one in which the loads are all connected to the same voltage source, such as the light fixtures in a room.

7. The total resistance of the circuit with the parallel loads shown in Figure 1.12 is:
   A. 6 ohms.   B. 12 ohms.   C. 18 ohms.   D. 24 ohms.

8. A 3-phase electric motor draws 68 amperes and operates at 230 volts with a power factor of 0.78. The power in watts drawn by the motor is:
   A. 11,200 watts.   B. 16,350 watts.   C. 21,105 watts.   D. 28,654 watts.

9. The 230-volt 3-phase electric motor of Problem 8 is developing 25 horsepower. The efficiency of the motor is:
   A. 56 percent.   B. 78 percent.   C. 81.6 percent.   D. 88.4 percent.

10. A house is operating six 100-watt incandescent lamps, four 60-watt lamps, and four 40-watt lamps for six hours. The amount of electrical energy used during the six hours is:
    A. 6 kWh.   B. 8.4 kWh.   C. 10 kWh.   D. 12.5 kWh.

CODE DISCUSSION

A person in the electrical trade who makes proper use of the Code must understand the background of the Code. Read all of the introductory pages. Several important questions about the Code in this first set of questions are answered in the introductory pages.

Article 90, the introduction, contains information necessary for the use of the Code. It is very important for the electrical tradesperson to read this article, especially *Sections 90-1, 90-2, 90-4, 90-5, 90-6, 90-7,* and *90-8.* Installations covered and not covered by the Code are discussed in *Section 90-2.* The authority having jurisdiction is given the responsibility of making interpretations when there is a difference of opinion about the meaning of a particular Code section. This is stated in *Section 90-4.*

Article 100 provides definitions of terms that are generally used in more than one article of the Code. If the term is used only in a specific article, the definition is usually provided in that article. These definitions are frequently necessary for a proper understanding of a Code section. This point will be illustrated in this first set of questions. Several helpful definitions for this first question set are approved: branch circuit, multiwire, continuous load, labeled, listed, receptacle, and receptacle outlet.

Article 110 contains general information about wiring installations. *Sections 110-2, 110-7, 110-9, 110-12, 110-14,* and *110-16* are of particular importance. *Section 110-26* deals with the working clearances around electrical equipment.

Article 200 primarily covers the means of identification of the grounded circuit conductor (usually the neutral) and grounded circuit conductor terminations on devices. The use of white or gray conductor insulation, *Section 200-7,* is illustrated in Figure 1.13 for a switch loop using cable.

Article 300 is another section that covers general wiring installation methods. *Section 300-4* covers the protection of conductors passing through structural members. There are also minimum clearance requirements for the installation of cables and some raceways run on the surface of studs and joists. The issue is the protection of the conductors from damage from

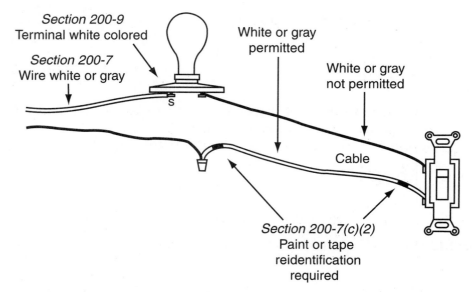

Section 200-9
Terminal white colored

Section 200-7
Wire white or gray

White or gray
permitted

White or gray
not permitted

Cable

Section 200-7(c)(2)
Paint or tape
reidentification
required

Figure 1.13   Grounded circuit conductors shall be white or gray, but a white or gray wire is permitted to be hot when cable is used for a switch loop.

nails and other fasteners. Underground installations are covered in *Section 300-5. Table 300-5* gives the minimum depth of burial permitted for different types of occupancies and different conditions. For farm and some commercial installations, *Section 300-7* is particularly important for conditions in which a raceway is exposed to different temperatures. There can be problems of condensation in the raceway system, or there can be problems with damage to the conduit or separation of the conduit due to expansion and contraction.

A multiwire branch circuit is defined in *Article 100* as a circuit that has two or more ungrounded conductors with a potential between them and sharing a common grounded conductor. Typical single-phase and 3-phase multiwire branch circuits are illustrated in Figure 1.14. A requirement that deals with multiwire circuits is *Section 300-13(b)*. Removing a device in a multiwire circuit is not permitted to interrupt the neutral wire. This is illustrated in Figure 1.15. Numerous other miscellaneous, but important, requirements are covered such as supporting conduit, preventing induced currents, traversing ducts, and preventing fire spread.

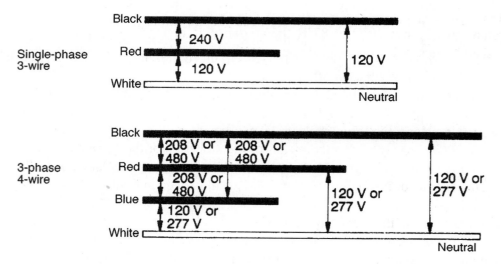

Figure 1.14   A common neutral wire is run with two or more ungrounded wires in multiwire circuits.

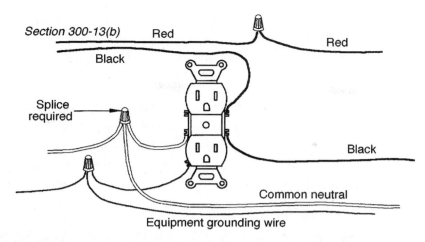

Figure 1.15   Device removal in multiwire circuits is not permitted to open the neutral.

Article 305 covers temporary wiring for such purposes as decorative lighting and construction site power. Ground-fault circuit-interrupter (GFCI) requirements are contained in this short article.

Article 310 is one of the most frequently used articles of the Code. This article contains the ampacity tables used to determine the minimum size of wire permitted for a circuit or load. Several methods can be used to determine the ampacity of electrical conductors, as stated in *Section 310-15*. Generally, *Table 310-16, Table 310-17, Table 310-18, Table 310-19,* and *Table 310-20* will be used to determine the minimum size of wire for a particular circuit. *Table 310-16* can be used for all situations of not more than three wires in cable or raceway or in the earth. *Table 310-17* applies to single insulated wires in free air. This would be the case for single overhead conductors mounted on insulators. *Table 310-20* is for insulated conductors supported on a messenger such as triplex cable. *Section 310-15(c)* provides a method permitted to be used to determine the ampacity of a conductor when the calculations are performed by or checked by an engineer. A number of ampacity tables are contained in *Appendix B* of the Code. These tables are not generally permitted to be used to determine the ampacity of a conductor unless the governmental jurisdiction specifically permits the *Appendix B* ampacity tables to be used. Several examples will help in understanding which table is to be used for which situation. Keep in mind that the ampacity tables only apply directly to raceways and cables that contain not more than three current-carrying conductors. If there are more than three wires, or if the wires are exposed to a temperature greater than 30°C or 87°F, the ampacity as found in the *Table 310-16* or *Table 310-17* shall be reduced. Conductor derating for ambient temperature and number of wires in raceway or cable is covered in Unit 2. There is a minimum size of wire permitted regardless of the ampere rating. This minimum size is given in *Section 310-5*.

The allowable ampacity of a conductor depends upon the size of the conductor, the type of insulation, and the temperature rating of the splices and terminations. The allowable current must be limited to a value that will not allow the temperature of the conductor to exceed the withstand temperature rating of the insulation. At points where conductors are spliced or terminated into a lug or under a screw terminal, the conductor operating temperature is not permitted to be higher than the maximum temperature limit of the termination. If a termination operates at more than the maximum temperature rating, it most likely will deteriorate and eventually fail. A fire may result. If a circuit-breaker, for example, has a maximum termination temperature of 75°C, then the 75°C column of the allowable ampacity table must be used even if the conductor insulation has a higher rating such as type THHN which is rated at 90°C.

There are sometimes exceptions to this rule when applying the derating factors, but those cases will be discussed in the next unit.

Another factor that determines the actual ampere rating of a conductor is the rating of the overcurrent device (fuse or circuit-breaker) used to protect the conductor from overheating. The actual load or calculated load for a circuit is not permitted to exceed the rating of the over-current device no matter what size of wire is used. If a 20-ampere circuit-breaker protects a circuit where the conductor is AWG number 10 copper with type THHN insulation, the calculated load is not permitted to exceed 20 amperes, and if the actual load is continuous, it is not permitted to exceed 16 amperes for this circuit-breaker. All of these factors must be considered when selecting the minimum size conductor for a particular circuit. This issue will be studied in the first units.

Copper conductor sizes AWG numbers 14, 12, and 10, and aluminum conductor sizes AWG numbers 12 and 10, when used for most circuits have an overcurrent device maximum limit. In previous editions of the Code, this limit was given in a footnote at the bottom of *Table 310-16* and *Table 310-17*. Those overcurrent device limits are now in *Section 240-3(d)*. The overcurrent device rating for AWG number 14 copper wire is 15 amperes, for AWG number 12 it is 20 amperes, and for AWG number 10 it is 30 amperes.

Example 1.8   Three AWG number 10, type THWN insulated copper conductors are contained in a conduit in a building. Determine the ampacity of the conductors if the ambient temperature does not exceed 30°C and all terminations are rated 75°C.

Answer: *Table 310-16* applies to this situation. Find the type of insulation in the copper conductor part of the table (75°C column), and find the ampacity that corresponds to AWG number 10 wire, which is 35 amperes.

Example 1.9   Type USE aluminum cable, size AWG number 2, consisting of three current-carrying conductors, is directly buried in the earth at a depth of 24 inches. Determine the ampacity of the cable if all terminations are rated 75°C.

Answer: *Table 310-16* applies in this situation. Find the ampacity of AWG number 2 aluminum in the column headed with conductor insulation type USE. The cable is rated at 90 amperes.

Example 1.10   Type XHHW aluminum wire, size AWG number 4, is installed overhead in free air as single conductors on insulators. Assume that the ambient temperature does not exceed 30°C and that at times this would be considered a wet location. Assume the terminations at the drip loop are rated at 90°C. Determine the ampacity of the aluminum wire in this situation.

Answer: *Table 310-17* is used in the case of single conductors in free air. Note that there are two columns of aluminum wire with type XHHW insulation listed, the 75°C column and the 90°C column. It will be necessary to refer to type XHHW insulation in *Table 310-13* for an explanation of when to use the two columns. Because this is considered at times to be a wet location, the 75°C column is used. The AWG number 4 aluminum wire is rated at 100 amperes.

Sections 310-7, 310-8, and *310-9* specify that wires and cables shall be rated for the conditions and environment. *Section 310-10* discusses the conditions that produce heat that can damage the insulation if the maximum operating temperature is exceeded. For example, a wire with type THW insulation has a maximum operating temperature of 75°C. Required markings on wires are given in *Section 310-11*. *Section 310-12* states the color code for conductor insulation. *Section 310-14* specifies the type of alloy required to be used for aluminum conductors. *Section 310-4* gives the rules for the installation of conductors where there are two or more conductors in parallel for each phase or neutral. This subject is discussed in detail in Unit 2.

Article 324 covers a type of wiring called concealed knob-and-tube wiring. This type of wiring was used in buildings in the early years of wiring. Single insulated conductors were held in place with insulators called knobs and cleats. The conductor would then pass through a structural member and would be run inside a porcelain tube. This type is encountered in existing older buildings.

Article 326 provides basic specifications and uses permitted for medium-voltage cable that is intended for use where energized at from 2,001 to 35,000 volts. Type MV cable is available as a single conductor cable or as a multiconductor cable. The cable is permitted to be installed in raceway, cable trays, direct burial according to *Section 310-7*, and supported by a messenger in air as specified in *Article 321*. When installed in tunnels, the rules for installation are found in *Article 110, Part C*. General installation requirements and clearances from live parts and equipment are given in *Part B* of *Article 110*. The method of determining the ampacity of the conductors rated 2,001 to 35,000 volts is specified in *Section 310-60(b)*. It is permitted to use *Table 310-67* through *Table 310-86*, or the formula in *Section 310-60(d)* under engineering supervision. *Appendix B* gives examples of the use of the formula under engineering supervision.

Article 328 provides specifications for the materials and installation for flat conductor cable to be placed under carpet squares. This is a technique for extending power to workstations in a large open room. Definitions are given in *Section 328-2*. Uses permitted and not permitted are covered in *Sections 328-4* and *328-5*. The maximum voltage permitted for flat conductor cable is 300 volts between ungrounded conductors and a maximum of 150 volts from any ungrounded conductor to the grounded conductor. General use circuits are not permitted to be rated in excess of 20 amperes. These voltage and current requirements are stated in *Section 328-6*. Flat conductor cable is permitted to be installed only under carpet squares defined in *Section 328-10* as having a maximum dimension of 36 inches on each side.

Article 330 covers the construction and installation of type MI mineral-insulated, metal-sheathed cable. Type MI cable has the insulated conductors contained within a liquidtight and gastight continuous sheath with a densely packed mineral insulation filling the space between the conductors and the copper or alloy steel sheath. The cable can be cut to length with special fittings applied at the terminations. Type MI cable has a wide variety of applications as described in *Section 330-3*, and it is permitted to be fished into existing building spaces, *Section 330-12(1) Exception*.

Article 333 covers type AC armored cable. Type AC cable generally consists of two or three insulated conductors and an equipment grounding conductor within a flexible metallic covering. Cables manufactured to be flame-retardant and limited-smoke producing will be marked with the letters **LS.** The uses permitted and not permitted are covered in *Sections 333-3* and *333-4*. The cable shall be supported within 12 inches of the end of each run and at intervals not to exceed 4 1/2 feet. Exceptions to this support rule are given in *Section 333-7*. The cable is subject to damage similar to other cables; therefore, it is important to follow the installation requirements of *Sections 333-10, 333-11,* and *333-12*.

Article 334 provides information on the uses permitted, installation, and construction of metal-clad type MC cable. The conductors are contained within a flexible metallic sheath. It is permitted to be used for services, feeders, and branch circuits. Permitted means of installation are covered in *Section 334-10*. *Section 334-10(a)* requires that type MC cable be supported at

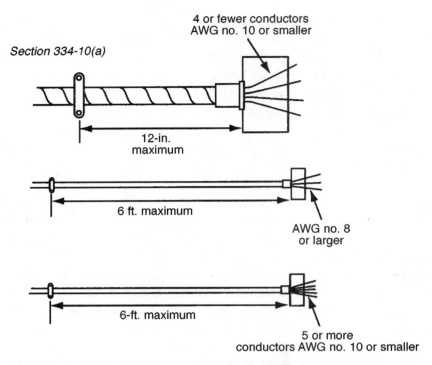

Figure 1.16   Type MC cable is required to be secured within 12 inches of a termination if the cable contains four or fewer conductors size AWG number 10 or smaller. For other cases, the cable is required to be supported within 6 feet of terminations.

intervals not to exceed 6 feet and within 12 inches of a termination when four or fewer conductors size AWG number 10 or smaller are used, as shown in Figure 1.16. For conductor sizes larger than AWG number 10, or more than four conductors size AWG number 10 and smaller, the cable shall be supported within 6 feet of the termination.

Article 336 deals with nonmetallic-sheathed cable, usually type NM. *Section 336-30(b)* specifies that the wires within the nonmetallic sheath shall have insulation rated at 90°C, but the wires shall be considered to be 60°C for the purpose of determining the ampere rating as stated in *Section 336-26.* Cables rated as flame-retardant and limited-smoke producing are permitted to be marked **LS.** Uses permitted are stated in *Section 336-4,* and uses not permitted are stated in *Section 336-5.* For one- and two-family dwellings, nonmetallic-sheathed cable is permitted to be used regardless of the height of the dwelling or the number of floors. For multifamily dwellings and other structures, nonmetallic-sheathed cable is permitted to be used only if the structure does not exceed three habitable floors above grade level. Determining the number of floors of height of a dwelling or structure is given in *Section 336-5(a)(1).* Installation requirements are provided in *Part B* of *Article 336.* It is important to protect nonmetallic-sheathed cable from damage. One form of damage is bending the cable in a small radius. The minimum radius of bend is five times the diameter of the cable, as stated in *Section 336-16.*

The type NM cable is required to be supported at intervals not exceeding 4 1/2 feet and within 12 inches of boxes, cabinets, and fittings. The exception to *Section 370-17(c)* requires the cable to be supported within 8 inches of a single-gang nonmetallic box, where the cable is not secured directly to the box. Nonmetallic-sheathed cable is permitted to be installed through bored holes or in notches, as specified in *Section 300-4.* It is permitted to be attached to the underside of joists in basements for AWG number 6 two-wire and AWG number 8 three-wire. These rules are covered in *Section 336-6(c).* Smaller sizes of cables are permitted to be

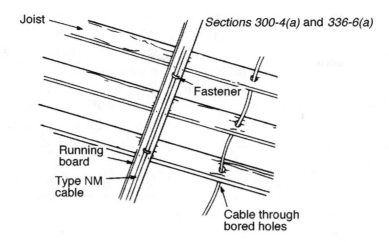

Figure 1.17   Small sizes of type NM cable are required to be attached to running boards or run through bored holds in structural members.

installed directly to the surface of joists and rafters in attics and crawl spaces, but protection is required near access openings, as specified in *Section 336-6*. This is illustrated in Figure 1.17.

Article 338 covers the use of type SE service entrance cable and type USE underground service entrance cable. For the purpose of Unit 1, the discussion will center around the use of type SE cable for circuits and feeders within a building. The common types of service entrance cable in use are type SE style U with two insulated conductors and a bare conductor within the nonmetallic covering, and type SE style R, which has three insulated conductors and a bare equipment grounding conductor within the nonmetallic outer covering. These cable types are shown in Figure 1.18. According to *Section 338-4*, type SE cable when used for interior wiring is installed according to the same rules as for type NM cable.

Article 339 covers type UF underground feeder and branch circuit cable for use underground and within buildings. The installation of this cable within buildings is required to follow the requirements of type NM nonmetallic-sheathed cable in *Article 336*. It is important to note that there will be two grades of this cable in use. When type UF cable is installed within a building, the individual insulated wires in the cable are required to have a 90°C insulation temperature rating. This requirement on the 90°C rated insulation is stated in *Section 339-3(a)(4)* where the conductor requirements for type UF cable are to be the same as for type NM cable. Cables are available with wires that have an insulation temperature rating of only 60°C. Both are permitted to be installed underground.

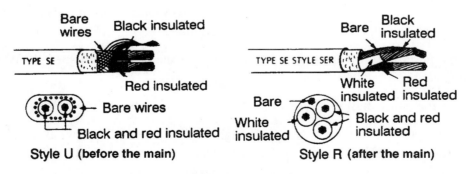

Figure 1.18   Type SE service entrance cable is available as style U or style R.

Article 340 covers the installation of type TC power and control tray cable. The uses permitted and not permitted are the main emphasis of this article. The ampacity of type TC cable is specified in *Section 340-7*. If the cable is smaller than AWG number 14, the ampacity is determined using the rules of *Section 402-5* for fixture wires. If the wire is AWG number 14 and larger, the ampacity is determined using the rules of *Section 318-11*. Determination of conductor allowable ampacity when installed in cable trays will be discussed in Unit 12. Type TC cable is available as a multiconductor cable and as single conductor cable. Most single conductor cables size AWG number 1/0 and larger have a type TC rating in addition to the normal conductor insulation rating.

Article 342 deals with nonmetallic extensions, typically a nonmetallic-sheathed cable installed on the surface of walls or ceilings. It is permitted to be attached as a nonmetallic extension to the exposed surface of walls and ceilings to extend power from an existing outlet of a 15- or 20 ampere circuit. This technique provides a means of extending wiring without having to enter the wall or ceiling cavity. This type of material is permitted to be messenger supported as an aerial cable for industrial buildings where the nature of the application requires a flexible means of connecting power to equipment.

Chapter 9, Tables 8 and *9* list data on bare and coated bare electrical conductors. *Table 8* gives AWG number and kcmil size, dimensions, and resistance on a 1,000-foot basis. *Table 9* gives the values of resistance and inductive reactance for the conductors in different types of installations, as well as the effective impedance of the conductors if the power factor of the load is 0.85. These data are useful in determining voltage drop on conductors or for cases where the dimensions of bare conductors are needed.

USING THE CODE TO ANSWER QUESTIONS

The numerous people who write the Code attempt the difficult task of choosing words that prohibit unsafe installations while not excluding any of the numerous acceptable installation methods. They try to achieve this task with words that we can understand. The Code is constantly changing, however, because of the introduction of new materials and techniques and because unacceptable confusion and loopholes are discovered.

One point that has led to confusion and misunderstanding in the field is the phrase **approved for the purpose. Approved** is defined in the Code as meaning approved by the authority having jurisdiction (the inspector). In many cases, it simply means that a judgment call must be made based on the inspector's knowledge and background and local conditions. The Code has made an attempt to place more responsibility on the manufacturer to indicate the suitability of various products for a specific purpose. The Code now uses the phrase **listed for the purpose** rather than **approved for the purpose.** The following points will help in understanding the meaning of the Code.

1. Read each section carefully and think about what the section is saying. Try not to let personal bias obscure the true meaning of the section.

2. Keep the purpose of the Code in mind as you read the Code, *Section 90-1*. All sections of the Code are directed to this purpose.

3. Look for the word **shall** in the sections.

4. Fine print notes (FPNs) are scattered throughout the Code to either act as an advisory or clarify the previous Code material. These FPNs are not considered a legally binding part of the Code.

5. Be sure to read the **scope** at the beginning of each article. The specific sections apply only under the conditions specified in the scope. Sometimes the scope is omitted and replaced with **uses permitted or not permitted,** as in *Sections 336-4* and *336-5.*

6. Definitions are frequently provided in an article to add important clarity. Read these definitions, as they are often necessary to the understanding of a particular section. Most definitions are grouped in *Article 100* if they are used in two or more articles. If used only in a specific article, the definitions will be given only in that article. If a definition is not given in the Code, then it is permitted to use the definition from a dictionary.

7. Do not confuse **wiring specifications** with Code requirements. Just because in your experience you have seen only one particular wiring method for an application, do not automatically assume this is a Code requirement. It is the option of the owner, architect, or another code to specify a particular method or material providing it is not in conflict with the Code.

8. Footnotes to the tables are considered part of the table and are a binding part of the Code.

9. Local jurisdictions have the right to adopt amendments to the Code. Sometimes the electric utility will have a requirement not covered in the Code. The utility standards apply to the installation of service equipment.

10. It is important to remember the limitations placed on manufacturers' materials and equipment by the testing process and the physical behavior of materials and equipment in a particular environment. For example, type THHN wire ampacity cannot be based on 90°C if the wire is connected to a circuit-breaker with a maximum temperature rating of 75°C. Some organizations providing testing services are Underwriters Laboratories (UL), Electrical Testing Laboratories (ETL), and Canadian Standards Association (CSA).

The following two example questions help illustrate the technique of finding information in the Code. Look for key terms that state the subject of the question. After using the Code for a period of time, you will remember many of the article numbers.

Example 1.11   A nonmetallic extension consisting of an assembly of two insulated conductors and an equipment grounding conductor within a nonmetallic jacket is permitted to be run on the surface of the outside concrete block wall of an unfinished basement of a dwelling from an existing outlet to a receptacle outlet further down the wall. (True or False)

Answer: You must learn to use the Code *Index.* The subject is **nonmetallic extensions.** Look up nonmetallic extensions in the index, and at the same time look for any listing that will further identify the specific question.

"As surface extensions" is listed and refers to *Section 342-7(a).* A check of this section does not obtain information on the specific question, "Is it permitted to mount this material on the concrete block wall of an unfinished basement of a dwelling?"

Another listing is "uses not permitted," which refers to *Section 342-4.* Examination of this section reveals that *Section 342-4(b)* states that this material is not permitted to be installed in an unfinished basement. Still another listing, "uses permitted," reveals in *Section 342-3(b)* that exposed and in a dry location area requirements apply to the installation of this material. A basement generally would be considered a wet or damp location with respect to placing wiring materials on an outside wall where earth is on the other side of the wall. The answer is False, and the Code reference is *Section 342-3(b)* or *342-4(b).*

Example 1.12   A branch circuit consisting of two or more ungrounded wires having a voltage between them, which share a common neutral conductor where the voltage between each ungrounded conductor and the neutral is the same, is defined as a _____ branch circuit.

Answer: The question is relating to a definition. Therefore, it is likely that *Article 100* will contain the answer, but it is possible that it may be a definition in a particular article. The key terms in the question are **branch circuit** and **common neutral.** Look up branch circuit in the index, or go immediately to *Article 100* and check to see if there is a definition of branch circuit that matches the question. This definition is found to be that of a multiwire branch circuit. The answer to the question is **multiwire,** found in *Article 100.*

STUDENT CODE PRACTICE

Answer the following wiring questions and give the Code reference where the answer is found. The answer will be found in the Code articles listed in the objectives at the beginning of this unit.

1. An additional outlet is to be added to an existing circuit of a home with wall construction of plaster on concrete block. Is it permitted to chip away the old plaster and run type AC cable from an existing box to a new outlet and then plaster over the cable as shown in Figure 1.19? (Yes or No)

    Code reference _____

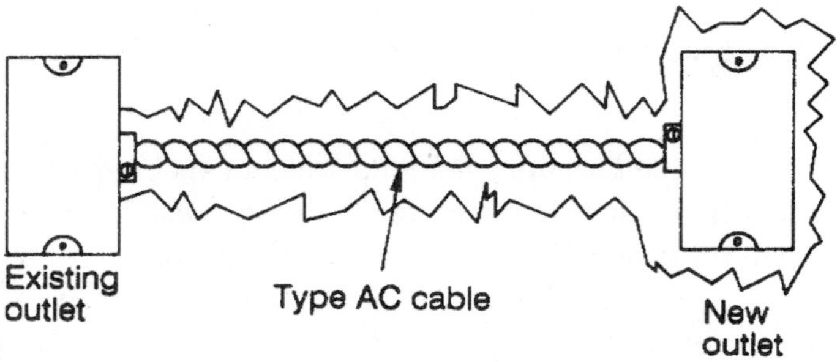

Figure 1.19   An underplaster extension using type AC cable supplies power to a new outlet from an existing outlet.

2. You are called to make a repair on a residential circuit wired concealed knob and tube. The problem is a damaged knob and a length of wire that has lost its insulation. Are you permitted to make this simple repair, or must you replace the entire circuit with cable or raceway wiring? (Permitted to make the repair, or Must replace with cable or raceway)

    Code reference _____

3. A feeder must be run through an area rated Class I Division 2 (hazardous). The electrician cannot get to the area without removing portions of the building at considerable time and expense. Is it permitted to fish type MI cable through this area? (Yes or No)

    Code reference _____

4. An office space is to be constructed in a commercial building with an existing concrete floor. Is it permitted to run flat conductor cable under carpet that is installed in one piece? The padding under the carpet is nonflammable. Outlets will be installed at each desk location. (Yes or No)

Code reference _____

5. Type SE cable with two insulated conductors and one uninsulated conductor within a nonmetallic outer covering is used to supply single-phase power from the service entrance to a subpanel in the same building as a 120/240-volt 3-wire feeder. Is this installation permitted as shown in Figure 1.20? (Yes or No)

Code reference _____

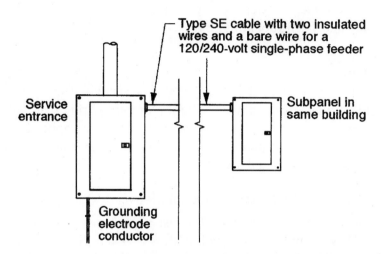

Figure 1.20   Type SE cable with two insulated wires and the bare wire within the outer covering is used to supply a subpanel in the same building from the service entrance panel.

MAJOR CHANGES TO THE 1999 CODE

These are the changes to the 1999 *National Electrical Code®* that correspond to the Code sections studied in this unit. The following analysis explains the significance of the changes from the 1996 to the 1999 Code only, and this analysis is not intended to be used in place of the Code. Refer to the actual section of the 1999 Code for the exact wording and meaning of each section discussed. Changes are indicated in the Code with a vertical line in the margin. If material has been deleted or moved to another part of the Code, the location of the deletion is indicated with a dark dot in the margin.

Article 90   **Introduction**

90-5: This section explains the protocol for understanding the Code. The section was expanded from one short statement in the previous edition of the Code to an explanation of what constitutes a **mandatory rule,** a **permissive rule,** and what is considered to be **explanatory material.** In mandatory rules, the words **shall** or **shall not** are used. For a permissive rule, usually for the purpose of giving options or alternative methods, the words **shall be permitted** or **shall not be required** are used. Explanatory material is contained in fine print notes. A new fine print note to this section explains that the Code follows the

"Style Manual" and a copy can be obtained from the National Fire Protection Association.

Article 100 **Definitions**

Cross-references to other definitions in this article were deleted in many locations throughout the article.

Feeder: In this definition, the words **or other power source** were added to make it clear that the definition of a feeder applies in all situations, even when there is an alternate source such as a generator.

Listed: This definition was revised to include **electrical services** that can be listed.

Multioutlet Assembly: This definition was revised to add the words **free standing** so that it is clear that a power pole designed to extend from the floor up through a suspended ceiling is included. The product is listed as a multioutlet assembly, but the definition only recognized surface and flush mounting, not free standing which is the case with a power pole.

Nonincendive Circuit: This definition was revised by adding the words **other than field wiring** to explain the criterion necessary to qualify as a nonincendive circuit.

Nonincendive Field Wiring: This is a new definition of field wiring that under normal operation will not create a condition that can ignite a flammable vapor or dust-air mixture. It further explains that normal operation includes the arcing during contact or switch opening, shorting of wires, or ground faulting.

Article 110 **Requirements for Electrical Installations**

This article was reorganized into four parts. The last part deals with requirements for installations of over 600 volts in tunnels, which in the previous edition of the Code was *Part F* of *Article 710*. *Article 710* was deleted and its contents were moved primarily to *Article 110* and *Article 300.*

110-14(c)(1)(d): This section was revised to eliminate the exceptions. The change has to do with electric motor connections when the conductors are not larger than AWG number 1. Even though terminations at electric motors are not necessarily marked when they are rated for 75°C terminations, they are indeed tested and listed for 75°C terminations. This new paragraph makes it clear that if conductors with 75°C insulation are being used for a motor circuit, the conductor ampacity is permitted to be determined at 75°C if the motor is a NEMA Design B, C, D, or E. These motor design designations are, however, 3-phase motors only. Single-phase induction motors have NEMA design designations L, M, N, and O. If the single-phase terminals are not marked 75°C, then the 60°C column of *Table 310-16* is used for determining conductor ampacity.

110-26: A new second sentence was added making it clear that when electrical equipment is installed in an enclosure controlled by lock and key, that electrical equipment is considered to be accessible to qualified personnel.

110-26(a)(2): There is a new requirement that the clear working space in front of equipment shall not be less than the width of the equipment when the equipment is wider than 30 inches, as shown in Figure 1.21. The previous edition of the Code only required the width to be 30 inches.

110-26(a)(3): Within the work space in front of equipment, other electrical installations within that work space mounted above or below the equipment are now permitted to extend

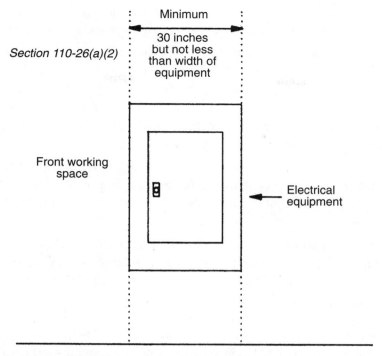

Figure 1.21   When electrical equipment is wider than 30 inches, now the clear working space in front of the equipment is not permitted to be less than the width of the equipment.

out not more than 6 inches from the front of the equipment, as shown in Figure 1.22. The previous edition of the Code required that other equipment or installations be of equal depth.

110-26(f)(1)(a): The space equal to the width and depth of electrical equipment up to a distance of 6 feet above electrical equipment, or to the structural ceiling, whichever is lower, is considered to be space dedicated only to equipment associated with the electrical

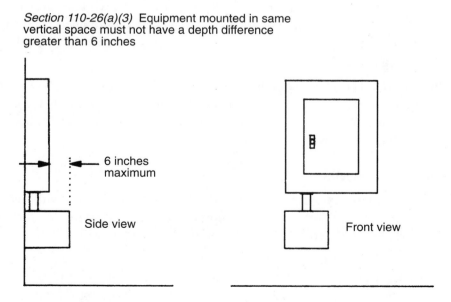

Figure 1.22   When electrical equipment is mounted in the same vertical space, the difference in depth from the mounting surface shall not exceed 6 inches.

power system such as conduits and electrical enclosures. There is an exception where special protection is provided for ducts and pipes passing through this area.

110-26(f)(1)(b): This section prohibits ducts, pipe, and other equipment not associated with the electrical power system from being installed in the space directly above electrical equipment up to the structural ceiling. If ducts or pipes pass through this area, it is necessary to install guards or shields to prevent condensation or leaks from damaging electrical equipment. This rule actually applies only when the structural ceiling is more than 6 feet above the electrical equipment.

110-26(f)(2): The previous edition of the Code did not specify clearance requirements for outdoor installations of switchboards, panelboards, and motor control centers. A new last paragraph was added that requires outdoor installations to meet the clearance requirements in *Section 110-26(a)*.

Part D: This is a new part to the article that provides installation requirements for portable and mobile high-voltage equipment in tunnels. It relates to digging, dredging, and drilling equipment and the associated power distribution equipment. *Article 710* in the previous edition of the Code covered requirements relating to high-voltage installations as well as parts of several articles. Specific requirements relating to high-voltage equipment in tunnels were located in *Article 110,* and the balance of the requirements was moved to a new *Article 490* covering installation of equipment in general operating at over 600 volts. *Part D* of *Article 490* also contains several requirements relating to high-voltage installations in tunnels.

Article 200   Use and Identification of Grounded Conductors

200-6(b): This section permits a grounded conductor AWG number 4 and larger to be identified at each termination with a distinctive white marking. The change is that this white marking is required to completely **encircle** the conductor. Also, AWG number 4 and larger conductors are now permitted to be identified by placing three white stripes on the conductor. This technique of identifying the grounded conductor is permitted provided the conductor insulation is not green, and the striping occurs along its entire length.

200-7(c): The previous edition of the Code permitted a white or gray conductor of any size to be reidentified at the time of installation as an ungrounded conductor. Now this practice is permitted only for white or gray conductors that are part of a **cable assembly,** as shown in Figure 1.23.

200-7(c)(2): The previous edition of the Code permitted a white or gray conductor in a cable to serve as an ungrounded conductor for switch loops. This practice is still permitted, but the Code now requires that the white or gray conductor be reidentified at all terminations.

Article 300   Wiring Methods

300-5(d): A new paragraph was added to this section dealing with underground conductors, which in the case of a direct burial service lateral requires that a warning ribbon be installed in the trench 12 inches above the conductors.

300-11(a): This section deals with the support or wiring above a suspended ceiling. A new sentence was added that permits wires installed in addition to the ceiling support wires to support raceways or cables in either a fire-rated or nonfire-rated ceiling system, as illustrated in Figure 1.24.

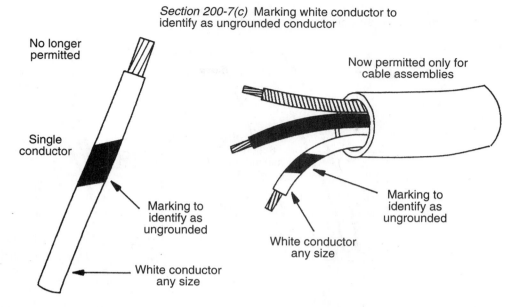

Section 200-7(c) Marking white conductor to identify as ungrounded conductor

No longer permitted

Single conductor

Marking to identify as ungrounded

White conductor any size

Now permitted only for cable assemblies

Marking to identify as ungrounded

White conductor any size

Figure 1.23   A white or gray conductor is no longer permitted to be reidentified as an ungrounded conductor unless it is a part of a flexible cord or cable assembly, or part of a cable assembly where used in a switch loop.

300-11(a)(1): When independent support wires are used as part of a fire-rated ceiling, the support wires are now required to be **distinguishable.** Color, tagging, or other means are permitted methods for identifying these support wires from other fire-rated support wires.

300-14: The rule for determining the minimum permitted length of free conductor at a box was changed. The 6-inch length is measured from the point at which the conductors emerge from the raceway or from the cable sheath, as shown in Figure 1.25. There must also be a minimum of 3 inches of free conductor that extends outside of the box opening where any dimension of the opening is less than 8 inches.

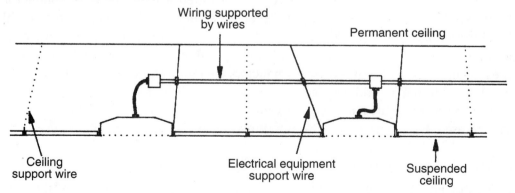

Section 300-11(a) Electrical equipment support wires are permitted, but must be identified

Wiring supported by wires

Permanent ceiling

Ceiling support wire

Electrical equipment support wire

Suspended ceiling

Figure 1.24   For a suspended ceiling, support wires are permitted to be installed in addition to the support wires required for the ceiling, and these additional wires are permitted to be secured on the lower end to the ceiling grid and then used to support wiring above the ceiling.

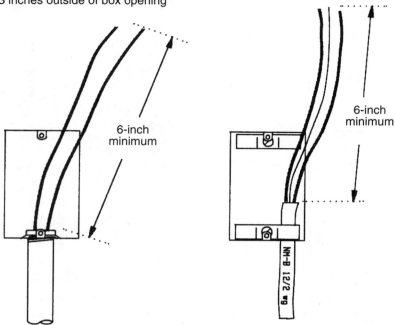

Section 300-14 Minimum length of free conductor; minimum of 3 inches outside of box opening

Figure 1.25   The 6-inch minimum length of free conductor is measured from the point where the conductor emerges from the raceway or cable sheath, and there must be a minimum of 3 inches of conductor extending outside of the box.

300-18(a): This section requires a raceway system to be completely installed before the conductors are installed. There are wiring assemblies available with wires already in the raceway, and this section recognizes these products and allows the material to be installed with the wires already in place provided they are permitted elsewhere in this Code. In the previous edition of the Code, only flexible metallic tubing and flexible metal conduit were recognized. Now for trades sizes 1/2 inch through 1 inch, the Code specifically permits the use of electrical nonmetallic tubing and liquidtight flexible nonmetallic conduit as a prewired assembly, as covered in *Section 331-3* and *Section 351-23*.

300-37: This section was *Section 710-4(a)* in the previous edition of the Code. Above-ground runs of high-voltage cable are now permitted to be run in electrical metallic tubing (EMT). There was a sentence added that permits above-ground high-voltage busbars to be either copper or aluminum.

300-50(a): This section, which was *Section 710-4(b)* in the previous edition of the Code, specifies the protection of high-voltage conductors emerging from the ground and running up a pole or entering a building. The raceway is now required to be **listed.** Also, where the cables and raceway are subject to movement, such as the heaving of frozen ground, provisions are now required to be taken to prevent damage to the cable.

Article 305   **Temporary Wiring**

305-4: For the purpose of supplying temporary power such as during construction, type NM and type NMC cables are permitted to be installed in buildings that have more than three floors above grade level, as shown in Figure 1.26. This change applies to both temporary feeders and branch circuits, in items (b) and (c) respectively.

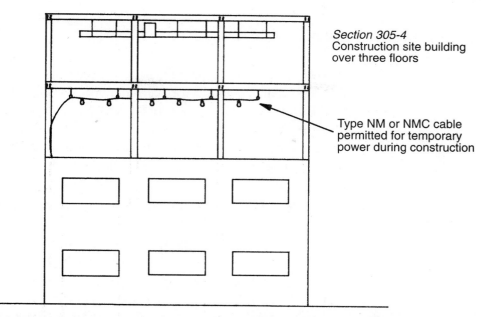

Section 305-4
Construction site building
over three floors

Type NM or NMC cable
permitted for temporary
power during construction

Figure 1.26   Type NM and NMC cable is permitted to be used for temporary power in buildings over three floors during construction.

305-4(b) Exception: The exception now specifically states that **single insulated** conductors are permitted to be installed as temporary wiring in case of emergencies or for special testing purposes as stated in *Section 305-3(c).* The previous edition of the Code left this application open to a variety of temporary wiring installation. It now also states that the wiring is to be under the supervision of qualified personnel.

305-4(c): This section permits single conductors to be run for branch circuits. The previous edition of the Code permitted this practice only for emergency purposes. Now branch circuits are permitted to be installed as single conductors for any temporary purpose up to 90 days. The restrictions are that the voltage to ground is limited to 150 volts, the conductors are not permitted to be subject to physical damage, and the conductors are to be supported on insulators with a spacing of not more than 10 feet. In the case of festoon lighting, the wiring is to be arranged so that there will not be strain on the lampholders.

305-4(j): This is a new section that requires cable assemblies, cords, and cables to be supported in an appropriate manner for the conditions.

Article 310   **Conductors for General Wiring**

310-15(a)(1): This section authorizes the use of the tables to determine allowable ampacity of conductors or for the ampacity to be determined by calculations or alternate tables under engineering supervision.

310-15(b): Most of the *Notes* that followed *Tables 310-16* through *310-19* are now contained in this section. *Note 9* in the previous edition of the Code was deleted because it is a repeat of *Section 240-3(b),* which allows rounding up to the next standard rating overcurrent device when the ampacity does not correspond with a standard rating.

Table 310-15(b)(2)(a): This table contains the conductor ampacity adjustment factors for more than three conductors in cable or raceway or bundled together for a distance of more than 24 inches. This was *Note 8* in the previous edition of the Code.

310-15(b)(4): This subsection specifies when the neutral is to be counted as a current-carrying conductor for the purpose of adjusting the allowable ampacity of the conductors. In the previous edition of the Code, this was *Note 10*. The rules did not change.

Table 310-15(b)(6): This section was *Note 3* in the previous edition of the Code, and it specifies the minimum size ungrounded conductors permitted for service entrance and feeder conductors supplying a dwelling served by a 120/240-volt 3-wire system. Some language was added that makes it clear that this section applies to the conductors supplying the lighting and appliance branch circuit panelboard of the dwelling. Also, additional insulation types for cables and conductors have been added to this table. The table that was in *Note 3* is now *Table 310-15(b)(6)*.

Table 310-16 footnote: The footnote to this and the other tables was moved to *Section 240-3(d)*. This is the footnote that limited the overcurrent protection to not more than 15 amperes for an AWG number 14 copper wire and also put similar limitations on other small sizes.

Table 310-19: The table now only gives conductor ampacities up to AWG number 4/0. The bare or covered conductor ampacity column was deleted, but a new *Table 310-21* for bare and covered conductors was added.

Table 310-20: This is not a new table; it was *Table B-310-2* in the previous edition of the Code. The table gives the allowable ampacity of multiplex cables in air such as duplex, triplex, and quadruplex, as shown in Figure 1.27. Because this table was a part of *Appendix B* in the previous edition of the Code, it could be used only under engineering supervision. There was no table available for electricians to use to determine size of overhead multiplex cables.

Table 310-21: This table was *Table B-310-4* in the previous edition of the Code. This table gives the allowable ampacity for bare or covered copper and aluminum conductors. The abbreviation AAC means all-aluminum conductor as opposed to ACSR, which means aluminum conductor, steel-reinforced. These values of ampacity are for individual conductors with a wind speed of 1.36 miles per hour across the conductor. At first it may seem strange that the covered conductor has a higher ampere rating than a bare conductor. But the covered conductor has a larger circumference and, therefore, cools faster, thus allowing a higher current.

310-60: This is a new section that deals with conductors with insulation capable of operating from 2,001 to 35,000 volts. There is a definition of electrical duct that is the same

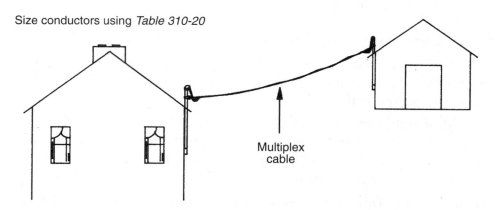

Size conductors using *Table 310-20*

Multiplex cable

Figure 1.27 Now there is a table that can be used to determine the allowable ampacity of overhead multiplex cable such as duplex, triplex, and quadruplex.

definition as was in *Section 310-15(d)* in the previous edition of the Code. There is also a definition of thermal resistivity, Rho, which is not new but was contained in *Appendix B* and as *Note 4* to the tables of the previous edition of the Code. There were no significant changes to the tables in this section.

Article 324   Concealed Knob-and-Tube Wiring

No significant changes were made to this article.

Article 326   Medium-Voltage Cable

No significant changes were made to this article.

Article 328   Flat Conductor Cable

No changes were made to this article.

Article 330   Mineral-Insulated, Metal-Sheathed Cable

No significant changes were made to this article.

Article 333   Armored Cable

333-3: Under uses permitted, the words **or are below grade line** were deleted. This section permits type AC cable to be run in voids of concrete block or tile where the walls are not subject to excessive moisture. The intent is that exterior walls exposed to earth are subject to moisture. It may have been interpreted as meaning that interior walls would be considered damp locations when below grade line. This change clears up this question.

Article 334   Metal-Clad Cable

No significant changes were made to this article.

Article 336   Nonmetallic-Sheathed Cable

336-5(a)(1): Type NM cable can now be installed in one- and two-family dwellings that exceed three floors. This new provision does not apply to multifamily dwellings or other structures.

336-6(b): Nonmetallic-sheathed cable run through floors is now permitted to be protected from damage by use of listed surface metal or nonmetallic raceway.

336-18 Exception 3: A new exception was added to the support requirements for nonmetallic-sheathed cable. Where a ceiling space is accessible, nonmetallic-sheathed cable is permitted to terminate at lighting fixtures or other equipment with the cable secured to the lighting fixture or equipment and the free length to the next support not to exceed 4 1/2 feet, as shown in Figure 1.28. The previous edition of the Code limited the free length of cable to 12 inches to the first support from the fixture or equipment. This requirement made it impractical to use nonmetallic-sheathed cable for terminations to lighting fixtures in most cases.

Article 338   Service Entrance Cable

338-4: There are now support requirements when service entrance cable is installed as **exterior** wiring. The support requirements for service entrance cable are the same as those for nonmetallic-sheathed cable, 12 inches from every box or enclosure and at intervals not

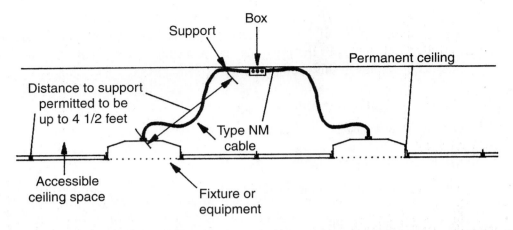

Figure 1.28   For equipment or lighting fixtures mounted in an accessible ceiling space, the distance from the fixture or equipment to the first support is now permitted to be 4 1/2 feet when nonmetallic-sheathed cable is used.

exceeding 4 1/2 feet, unless the cable is supported by a messenger. The requirements of *Article 339,* for type UF cable, apply to the installations where type USE cable is installed underground.

Article 339   Underground Feeder and Branch Circuit Cable

339-3(b)(10): This section covers uses not permitted for type UF cable. This is a new requirement that does not permit type UF cable to be installed where it may be subject to physical damage.

Article 340   Power and Control Tray Cable

No significant changes were made to this article.

Article 342   Nonmetallic Extensions

No significant changes were made to this article.

Chapter 9, Tables 8 and 9   Conductor Properties

No changes were made to these tables.

WORKSHEET NO. 1
GENERAL WIRING AND FUNDAMENTALS

These questions are considered important to understanding the application of the *National Electrical Code®* to electrical wiring, and they are questions frequently asked by electricians and electrical inspectors. People working in the electrical trade must continue to study the Code to improve their understanding and ability to apply the Code properly.

DIRECTIONS: Answer the questions and provide the Code reference or references where the necessary information leading to the answer is found. Space is provided for brief notes and calculations. An electronic calculator will be helpful in making calculations. You will keep this worksheet; therefore, you must put the answer and Code reference on the answer sheet as well.

1. Determine the current in amperes drawn by a 300-watt incandescent lamp operating at 120 volts.

   _____ amperes

2. A length of wire has a resistance of 0.7 ohm. If there are 12 amperes flowing in this wire, determine the voltage drop along the wire.

   _____ volts

3. A 120/208-volt electrical panelboard is mounted on a concrete block wall, and the panelboard extends out a distance of 10 inches. Metal equipment contacting the concrete floor is located in front of the panelboard and is 4 feet from the same concrete block wall as shown in Figure 1.29. The working clearance in front of the panelboard is: (adequate, not adequate).

   Code reference _____

4. An office building consists of five occupied floors above grade, and the building is wired with metal raceway. An extension to a circuit results in a load of less than 16 amperes. Is it permitted to install a nonmetallic surface extension, not penetrating a floor or partition, on the wall supplied from an existing outlet of a 20-ampere, 120-volt circuit? (Yes or No)

   Code reference _____

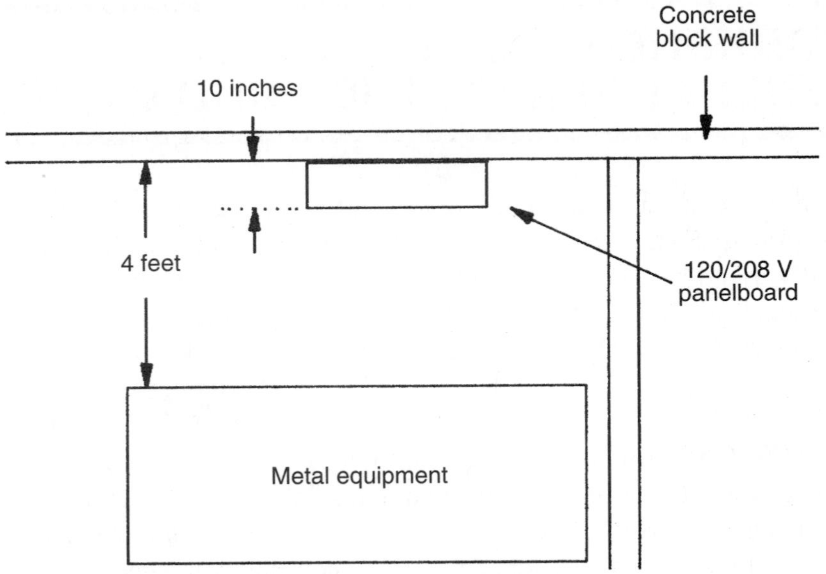

Figure 1.29   A 120/208-volt panelboard is mounted on a concrete block wall, and metal equipment, 4 feet from the same wall, is located in front of the panelboard.

5. The *National Electrical Code*® is considered law after it has been adopted by the **American National Standards Institute (ANSI).** (True or False)

Code reference _____

6. A single-phase electrical device in a building draws 44 amperes and is supplied with two type THWN copper wires in conduit where the conduit is run up through the cores of concrete blocks in a building. The wire ampacity required to supply the load is 55 amperes, and the circuit over-current protection is rated 60 amperes. The minimum size wire permitted is AWG number _____.

Code reference _____

7. If you would like to submit a proposal to amend a section of the 1999 *NEC*®, which would possibly be included in the 2002 *NEC*®, what is the final date the proposal is required to be received at the NFPA office?

Code reference _____

8. Three type THWN copper circuit conductors are run in raceway in a building, and the conductors are required to have an allowable ampacity of not less than 115 amperes. The circuit is protected with 125-ampere time-delay fuses and is illustrated in Figure 1.30. The minimum wire size permitted is AWG number _____.

Code reference _____

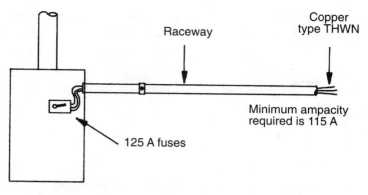

Figure 1.30   A circuit protected with 125-ampere fuses, and with a minimum requirement of 115 amperes, is run in raceway with three type THWN copper conductors.

9. Aluminum triplex conductors, with type XHHW insulation, are run overhead to supply an outbuilding from another building. The 120/240-volt, 3-wire feeder is protected at the source with a 100-ampere circuit-breaker, and the calculated load is 62 amperes. The minimum size ungrounded feeder conductors permitted are AWG number _____ .

    Code reference _____

10. Electrical equipment and materials (under 600 volts) installed by an electrician shall be UL approved. (True or False)

    Code reference _____

11. If you were asked to do the wiring on a floating store building, does the Code apply in this case? (Yes or No)

    Code reference _____

12. A 15-ampere, 120-volt branch circuit consisting of type UF cable is supplied from the service panel in a dwelling, and the entire circuit is protected with a ground-fault circuit-interrupter. If the direct burial type UF cable supplies lighting outlets in the yard, the minimum permitted depth of burial is _____ inches.

    Code reference _____

13. A temporary electrical service supplied by a utility source is established at a construction site to provide power and lights for the construction workers. Are the 15- and 20-ampere, 120-volt receptacle circuits required to be protected with a ground-fault circuit-interrupter? Maintenance records are not kept for electrical equipment on the job site. (Yes or No)

    Code reference _____

14. The minimum length of free conductor to be provided in a box to make up splices and connections measured from the point of entry of the conductors in the box is _____ inches provided that at least 3 inches extend out beyond the face of the box. Refer to Figure 1.31.

    Code reference _____

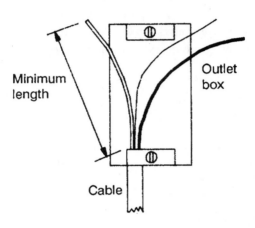

Minimum length

Outlet box

Cable

Figure 1.31   There is a minimum length of free conductor required at an outlet or junction box for making up splices and connections.

15. The cross-sectional area of an AWG number 2 bare seven-stranded copper wire is _____ Circular Mils.

    Code reference _____

16. According to *Section 339-3(a)(4),* type UF cable used for interior wiring in a building shall have the same conductor rating as type NM cable; therefore, the wire is required to have 90°C rated insulation. Determine the ampacity of an AWG number 10 copper 2-wire, type UF cable with an ambient temperature not exceeding 30°C.

    Code reference _____

17. Type MC cable contains three conductors size AWG number 12 copper. Is it permitted to install the cable so that the support for the cable is located 24 inches from the box or termination as shown in Figure 1.32? (Yes or No)

    Code reference _____

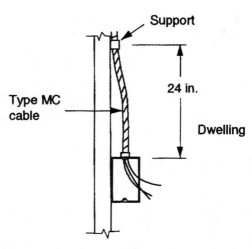

Figure 1.32   When type MC cable contains three AWG number 12 conductors, is the support permitted to be located 24 inches from the box or termination?

18. The greatest root-mean-square difference of potential between any two conductors of a circuit is called _____.

    Code reference _____

19. What is the maximum voltage permitted between conductors for a general-purpose single-receptacle outlet in a dwelling that does not supply cord-and-plug connected loads greater than 1,440 volt-amperes or greater than 1/4 horsepower?

    Code reference _____

20. Is it permitted to install AWG number 10 solid, type THWN copper wire in raceway? (Yes or No)

    Code reference _____

21. Two-wire type NM cable with an equipment grounding wire is used for a switch leg from a lighting outlet to a single-pole switch, as shown in Figure 1.33. Is the white wire between the lighting fixture and the switch required to be painted or taped to identify it as an ungrounded wire? (Yes or No)

    Code reference _____

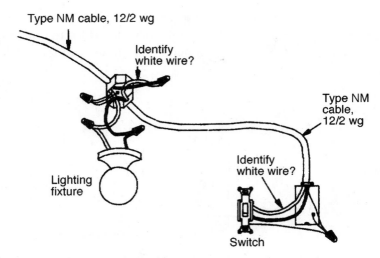

Figure 1.33   Nonmetallic-sheathed cable is used to wire a circuit containing a ceiling lighting fixture controlled by a single-pole wall switch.

22. Type SE cable style U with two insulated conductors and an uninsulated conductor is permitted to supply a 240-volt motor in a building supplied from a 120/240-volt 3-wire electrical system provided the uninsulated wire is used only for equipment grounding. (True or False)

Code reference _____

23. A duplex receptacle is installed on a multiwire branch circuit. (Neutral common to more than one ungrounded wire.) Are all of the wires connecting to the device (neutral and ungrounded) required to be pigtailed, as shown in Figure 1.34? (Yes or No)

Code reference _____

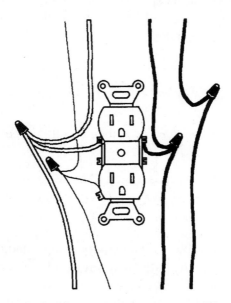

Figure 1.34   A duplex receptacle supplied by one wire and the neutral of a multiwire branch circuit.

24. Type NM cable, AWG number 12 with two insulated wires and a bare equipment grounding wire (type NM, 12-2 wg) has a flat dimension of 0.5 inches. What is the minimum radius of bend permitted for the cable, as shown in Figure 1.35? R = _____ inches.

    Code reference _____

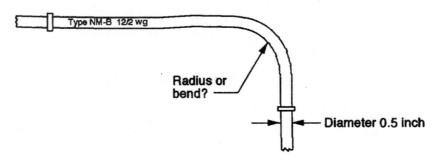

Figure 1.35   Nonmetallic-sheathed cable laid flat and attached to a surface is bent to form the minimum radius.

25. Nonmetallic-sheathed cable is used with single-gang metal boxes where both the cable and the boxes are accessible at the time of installation. If the cable is stapled to wood framing, what is the maximum distance permitted between the box and the first staple?

    Code reference _____

FORM FOR PROPOSALS FOR 2002 *NATIONAL ELECTRICAL CODE®*

Mail to: Secretary, Standards Council
National Fire Protection Association
1 Batterymarch Park, P.O. Box 9101
Quincy, Massachusetts 02269-9101
Fax to: 617-770-3500

FOR OFFICE USE ONLY
Log # _____
Date Rec'd _____

Notes: 1. All proposals must be received by 5:00 p.m. EST on Friday, November 5, 1999.
　　　　　Proposals received after 5:00 p.m. EST, Friday, November 5, 1999, will be returned to the submitter.
　　　　2. Type or print legibly in black ink. Limit each proposal to a SINGLE section. Use a separate copy for each proposal.
　　　　3. If supplementary material (photographs, diagrams, reports, etc.) is included, you may be required to submit sufficient copies for all
　　　　　members and alternates of the technical committee.

Please indicate in which format you wish to receive your ROP/ROC: ❑ electronic or ❑ paper

Date _____ Name _____ Tel. No. _____

Company _____

Street Address _____

Please Indicate Organization Represented (if any) _____

1. Section/Paragraph _____

2. Proposal Recommends: (Check one)　❑ new text　❑ revised text　❑ deleted text

3. Proposal (include proposed new or revised wording, or identification of wording to be deleted): (Note: Proposed
text should be in legislative format: i.e., use underscore to denote wording to be inserted (inserted wording) and strike-through to denote wording to
be deleted (~~deleted wording~~).

4. Statement of Problem and Substantiation for Proposal: (Note: State the problem that will be resolved by your recommendation; give the specific reason for your proposal including copies of tests, research papers, fire experience, etc. If more than 200 words, it may be
abstracted for publication.)

5. ❑ This Proposal is original material. (Note: Original material is considered to be the submitter's own idea based on or as a result of
his/her own experience, thought, or research and, to the best of his/her knowledge, is not copied from another source.)

❑ This Proposal is not original material; its source (if known) is as follows: _____

| If you need further information on the standards-making process, please contact the
Standards Administration Department at 617-984-7249.
For technical assistance, please call NFPA at 617-770-3000. |
| --- |

_____
Signature (Required)

PLEASE USE SEPARATE FORM FOR EACH PROPOSAL

ANSWER SHEET   Name _____

No. 1   GENERAL WIRING AND FUNDAMENTALS

Answer	**Code reference**
1. _____	_____
2. _____	_____
3. _____	_____
4. _____	_____
5. _____	_____
6. _____	_____
7. _____	_____
8. _____	_____
9. _____	_____
10. _____	_____
11. _____	_____
12. _____	_____
13. _____	_____
14. _____	_____
15. _____	_____
16. _____	_____
17. _____	_____
18. _____	_____
19. _____	_____
20. _____	_____
21. _____	_____
22. _____	_____
23. _____	_____
24. _____	_____
25. _____	_____

UNIT 2

Wire, Raceway, and Box Sizing

OBJECTIVES

Upon completion of this unit, the student will be able to:

- determine the size of a conductor for a circuit considering ambient temperature and more than three conductors in the raceway, cord, or cable.
- determine the minimum size conduit permitted when the conductors are all the same size and type of insulation.
- determine the minimum size conduit permitted when the conductors are different sizes and different types of insulation.
- determine the minimum size wireway and conduit nipples permitted for conductors.
- determine the minimum size junction box or device box permitted to take conductor fill into consideration.
- determine the minimum dimensions for pull boxes for straight pulls and angle pulls permitted.
- determine the minimum dimensions permitted for conduit bodies for various applications.
- answer wiring installation questions relating to *Articles 331, 343, 345, 346, 347, 348, 349, 350, 351, 352, 362, 370, 373, 374, Chapter 9, Tables 1, 4, 5, or 5A* and *Appendix C, Tables.*
- state at least five significant changes that occurred from the 1996 to the 1999 Code for *Articles 331, 343, 345, 346, 347, 348, 349, 350, 351, 352, 362, 370, 373, 374, Chapter 9, Tables 1, 4, 5, or 5A,* or *Appendix C, Tables.*

CODE DISCUSSION

The emphasis of this unit is to determine the minimum size of conductors for specific circuit and feeder applications if the actual or calculated load current is known. The ampacity tables in *Article 310* were discussed in Unit 1. Emphasis of this unit is on determination of the size and the installation of conductors, raceway systems, and boxes. A brief discussion of some key points made in the articles dealing with raceways, cabinets, and boxes follows, with example calculations later in this unit.

Article 331 deals with electrical nonmetallic tubing (ENT), which is a pliable corrugated raceway that can be bent by hand. It is available in 1/2-inch through 2-inch trade diameter. The uses permitted are discussed in the article; however, in general, it is intended for use as a concealed wiring method in walls, floors, and ceilings, and above suspended ceilings. If the building is more than three floors above grade, then walls, ceilings, floors, and suspended ceilings must have a 15-minute finish fire rating. An ENT installation is illustrated in Figure 2.1. Electrical nonmetallic tubing is permitted to be installed as surface wiring, provided it is not subjected to physical abuse and provided the building is not over three floors in height. Electrical nonmetallic tubing shall be supported at intervals not exceeding 3 feet, and it shall be supported within 3 feet of a termination at a box, fitting, or enclosure.

Article 343 describes the construction, use, and installation of a preassembled cable in nonmetallic underground conduit. This is a smooth outer surface nonmetallic conduit, which comes in continuous lengths usually on reels. The cable or conductors are already installed in the conduit. It is permitted to be used only for underground installations except for terminating in a building. The purpose of this product is the ease of installation of underground circuits to minimize the possibility of damage to the conductors during installation. This product is available in sizes from 1/2-inch to 4-inch trade size. The same fill requirements apply as for other conduit installations. It is permissible to remove the conductors at a later time and replace them with new conductors.

Article 345 is on intermediate metal conduit (IMC). It is a metal raceway that is permitted to be threaded. It is permitted to be used in most applications where rigid metal conduit is used. This type of conduit has a smaller wall thickness than rigid metal conduit. Intermediate metal conduit is available in sizes 1/2-inch through 4-inch trade diameter. It shall be supported within 3 feet of a box, fitting, or cabinet, and at intervals not more than 10 feet unless threaded couplings are used. The 3-foot spacing requirement for supports at IMC terminations is permitted to be increased not more than 5 feet when building structural supports do not permit supporting the IMC within 3 feet of the termination according to *Section 345-12*. When threaded couplings are used, IMC is permitted to be supported with the same maximum intervals as rigid metal conduit, which are given in *Table 346-12*. Exposed vertical risers with threaded couplings in industrial areas are permitted to be supported at intervals not to exceed 20 feet.

Article 346 covers rigid metal conduit (RMC) that has thicker walls than other types of metal conduit and tubing, and it is permitted to be threaded. Galvanized rigid steel conduit is generally used where high mechanical strength is needed, and rigid aluminum conduit is often used where weight is required to be minimized. Rigid metal conduit is available in sizes from

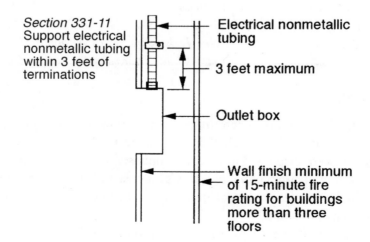

Figure 2.1   A minimum finish fire rating of 15 minutes is required when electrical nonmetallic tubing is installed as concealed wiring in buildings more than three floors above grade level.

1/2-inch through 6-inch trade diameter. There is a minimum radius of bend permitted for one-shot field bending. These minimums depend on the trade diameter of the conduit and are given in *Table 346-10 Exception*. Rigid metal conduit shall be supported within 3 feet of a box, fitting, or cabinet, and at intervals of not more than 10 feet unless threaded couplings are used. This is illustrated in Figure 2.2. The 3-foot spacing requirement for supports at conduit terminations is permitted to be increased not more than 5 feet when building structural supports do not permit supporting the conduit within 3 feet of the termination according to *Section 346-12*. When threaded couplings are used, they are permitted to be supported at intervals greater than 10 feet. Maximum support spacings are given in *Table 346-12*. Exposed vertical risers with threaded couplings in industrial areas are permitted to be supported at intervals not to exceed 20 feet.

Article 347 concerns rigid nonmetallic conduit (RNC), which is resistant to corrosion from moisture and most chemicals. It is available made from polyvinyl chloride (PVC) or from a fiberglass-reinforced nonmetallic material. The fiberglass-reinforced type is stiffer than PVC and is permitted to have support spacings increased from those listed in *Table 347-8*. It is available as Schedule 40, which is the standard wall thickness, and as Schedule 80, which has a thicker wall and thus will generally withstand greater impact before damage will occur. It is available in trade diameters of 1/2-inch through 6-inch. Standard lengths are 10 feet, although it is available as a continuous length from reels. The conduit is not threaded. It is joined to fittings by brushing an adhesive solvent on the conduit and then placing the conduit into the fitting. If done properly, this will form a watertight seal at the fitting. The conduit is bent by applying heat to the area to be bent until the conduit softens. Then the conduit is placed in a form of some type to make the desired bend and allowed to cool and harden.

Rigid nonmetallic conduit changes length when it is exposed to changes in temperature. If an installation will be subject to a large temperature variation during normal use, then it may be necessary to install expansion fittings to allow for the thermal expansion and contraction. It is important to remember thermal expansion and contraction when installing rigid nonmetallic conduit, especially when installing conduit supports. *Table 347-9* gives the number of inches of expansion or contraction for 100 feet of PVC rigid nonmetallic conduit for different temperature variations. Limited-smoke producing, type LS, high-density polyethylene rigid nonmetallic conduit actually has a greater coefficient of thermal expansion than for PVC rigid nonmetallic conduit. Manufacturer literature should be consulted to find values of expansion and contraction of type LS rigid nonmetallic conduit with change in temperature.

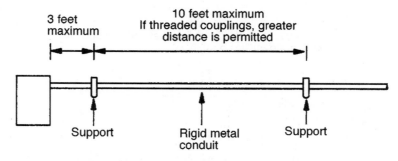

Section 346-12 Maximum distance between supports for rigid metal conduit is given in *Table 346-12* if threaded couplings are used

Figure 2.2   Maximum support spacing for rigid metal conduit is 10 feet unless threaded couplings are used; then maximum spacing is found in *Table 346-12*.

Fiberglass-reinforced rigid nonmetallic conduit has a lower coefficient of thermal expansion than PVC. *Table 347-9(b)* gives the length of change per 100 feet for various temperature changes for the fiberglass-reinforced rigid nonmetallic conduit. If the supports are installed in the correct locations, the conduit can move without damage. If possible, avoid confining the conduit so expansion or contraction will not apply stress to boxes, fittings, cabinets, and supports, as shown in Figure 2.3. If a straight run of rigid nonmetallic conduit is installed so it cannot expand and contract with a change in temperature, then an expansion fitting is required if the length due to temperature change will be more than 1/4 inch.

A disadvantage to this type of conduit is that the conduit tends to sag between supports; therefore, supports are required to be closer together than for comparable sizes of metal conduit and tubing. The maximum permitted support spacing for different trade diameters is given in *Table 347-8*. The maximum support spacing permitted for rigid nonmetallic conduit is shown in Figure 2.4. For trade diameter 1/2-inch through 1-inch, the maximum permitted spacing is 3 feet. Rigid nonmetallic conduit of all trade diameters shall be supported within 3 feet of a box, fitting, or cabinet except support spacing for fiberglass type may be greater.

Several types of rigid nonmetallic conduit are intended only for underground installations. Type A is a thin-walled rigid nonmetallic conduit that must be installed underground embedded in concrete. Type EB is also a thin-walled rigid nonmetallic conduit that has a stiffer wall thickness than type A. It too must be installed underground encased in concrete. Type HDPE schedule 40 rigid nonmetallic conduit is permitted to be installed without concrete encasement, but only for underground direct burial applications. It is also important to note the designation markings on fiberglass-reinforced rigid nonmetallic conduit. If it is marked **underground,** it is only permitted to be installed underground.

Article 348 deals with electrical metallic tubing (EMT), and it is frequently used where the raceway is not exposed to physical damage. The tubing is not permitted to be threaded. This type of raceway is popular because it is easy to cut, bend, and install. Electrical metallic tubing is available in trade diameters 1/2-inch through 4-inch. Minimum bending radius to the inside surface of the tubing is the same as for rigid metal conduit. If the bends are made in the field, the minimum bending radius is found in *Table 346-10 Exception.*

The maximum support spacing for electrical metallic tubing is 10 feet. The tubing shall also be supported within 3 feet of a box, fitting, or enclosure. There is an exception that permits the support to be up to 5 feet from the box, fitting, or enclosure if a practical means of support is not available at a lesser distance.

Section 347-9 Allow for thermal expansion and contraction of rigid nonmetallic conduit

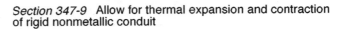

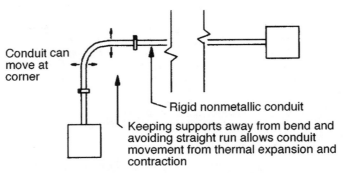

Conduit can move at corner

Rigid nonmetallic conduit

Keeping supports away from bend and avoiding straight run allows conduit movement from thermal expansion and contraction

Figure 2.3   Install rigid nonmetallic conduit so it can move due to thermal expansion and contraction if it will be exposed to a large temperature variation.

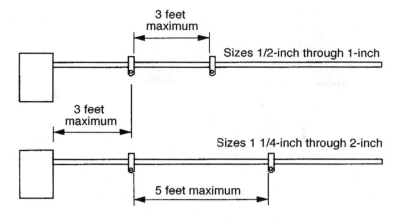

Section 347-8 Support spacing of rigid nonmetallic conduit
depends on the trade diameter

**Figure 2.4   Support spacing for rigid nonmetallic conduit depends on the
trade diameter of the conduit.**

Article 349 deals with a flexible metallic tubing (FMT) that is liquidtight. It is important
to read the list of uses permitted and uses not permitted covered in *Section 349-3* and *Section
349-4.* This material is not permitted to be used in lengths to exceed 6 feet, as stated in *Section
349-5(6).* The use of flexible metallic tubing as an equipment grounding conductor is covered
in *Section 250-118(8).* An equipment grounding conductor is required to be installed unless the
circuit conductors in the tubing are protected from overcurrent at not more than 20 amperes,
the fittings are listed for grounding, and the total length of flexible metallic tubing in any
grounding path is not more than 6 feet. Radius of bends is given in *Section 349-20,* and the
radius depends on whether the bend is fixed or may be infrequently flexed after installation.

Article 350 is on flexible metal conduit (FMC). It is of a spiral metal construction to provide flexibility and mechanical strength. This type of raceway is permitted for use in dry locations, and it is popular for use where flexibility is needed to connect raceway wiring systems
to lighting fixtures and equipment. The minimum trade diameter permitted to be used is
1/2-inch; however, there are several applications given in *Section 350-10(a)* where smaller
than 1/2-inch trade diameter is permitted.

Flexible metal conduit shall be supported within 12 inches of a box, fitting, or enclosure
with some exceptions. Where flexibility is necessary, as shown in Figure 2.5, lengths up to
3 feet are permitted to be supported only at the ends. Flexible metal conduit taps to lighting

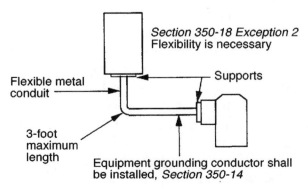

**Figure 2.5   Flexible metal conduit in lengths up to 3 feet is permitted to
be supported only at the ends when flexibility is necessary.**

fixtures are permitted in lengths up to 6 feet, supported only at the terminals. An equipment grounding conductor is generally required to be installed because the flexible metal conduit is usually not considered to be an acceptable equipment grounding conductor. The only condition is that, where the fittings are listed for grounding, circuit conductors within the listed flexible metal conduit, not more than 6 feet in length, are protected from overcurrent at not more than 20 amperes as provided in *Section 250-118*.

Article 351 describes liquidtight flexible metal conduit (LFMC) and liquidtight flexible nonmetallic conduit (LFNC) as having a nonmetallic liquidtight outer covering. It is available in trade diameters up to 4-inch. Both liquidtight flexible metal conduit and liquidtight flexible nonmetallic conduit are permitted for use for exposed and concealed wiring when the conditions of installation, operation, or maintenance require flexibility or require protection from liquids, vapors, or solid materials. Liquidtight flexible nonmetallic conduit shall be listed and marked for the purpose when installed outdoors or for direct burial. Liquidtight flexible nonmetallic conduit is permitted to be installed in lengths greater than 6 feet.

Liquidtight flexible metal conduit installed as a fixed wiring system is required to be supported at a distance of not more than 12 inches from a box, fitting, or enclosure, and at intervals not to exceed 4 1/2 feet. It is permitted to be used as a tap to a lighting fixture in lengths up to 6 feet supported only at the terminations. For other applications where flexible connections are to be made, such as a motor, liquidtight flexible metal conduit is permitted to be installed in lengths up to 3 feet that are supported only at the terminations.

An equipment grounding conductor is required to be installed through the liquidtight flexible nonmetallic conduit if the equipment or circuit supplied is required to be grounded. In the case of liquidtight flexible metal conduit, it is not permitted to be used as an equipment grounding conductor unless both the conduit and the fittings are listed for equipment grounding and the length does not exceed 6 feet. If these requirements are met, then trade diameters of 3/8-inch and 1/2-inch are not required to have a supplemental equipment grounding conductor, provided the circuit conductors are protected from overcurrent at not more than 20 amperes. Trade diameters of 3/4-inch through 1 1/4-inch are not required to have an equipment grounding conductor if the circuit overcurrent protection is not more than 60 amperes. Where necessary because flexibility is required, an equipment grounding conductor must be installed. These rules are found in *Section 250-118*.

Article 352 is about surface metal raceways and surface nonmetallic raceways. These are raceways attached to the surface of walls or ceilings. A typical application is where the raceway is run on the surface to extend from an existing outlet to a new outlet location. It is obvious that surface nonmetallic raceway requires an equipment grounding conductor. Equipment grounding for surface metal raceway is covered in *Section 352-9*. In the case of strut-type channel raceways, the grounding requirements are found in *Section 352-50*. Equipment grounding of surface metal raceway and strut-type channel is also covered in *Section 250-118*. It is not permitted to be used as an equipment grounding conductor unless the surface metal raceway is specifically listed for the purpose. Raceway support requirements are not stated in this article, but they are required to meet the general support requirements of *Section 300-11*. *Section 300-11* simply states that raceway shall be securely fastened in place, and there are no specific Code requirements for surface raceway support.

Article 362 covers wireways, which are raceways of square cross section with a removable cover along one side. These are used as a raceway for conductors from one location to another. Metallic and nonmetallic wireway is available. Change in length due to change in temperature must be considered when nonmetallic wireway is installed in locations where it will be exposed to a change in temperature. The cross-sectional area of the wire is not permitted to exceed 20 percent of the cross-sectional area of the wireway. The derating factors of *Section 310-15(b)(2)* do not apply if there are not more than 30 current-carrying wires in the metal wireway and the fill is not over 20 percent. For nonmetallic wireway, there is no 30 current-carrying conductor limit, but the derating factors of *Section 310-15(b)(2)* must be applied whenever there are more than three current-carrying conductors at any one cross section. Wireways are not permitted to serve as equipment grounding conductors unless they are listed for the purpose. When equipment grounding is not covered in an article, it is necessary to look in *Article 250* for specific requirements. Wireway would fall under *Section 250-118*.

Metallic wireway mounted horizontally is required to be supported at each end and at intervals not to exceed 5 feet unless listed for greater support spacings. If the wireway is manufactured as one solid length of more than 5 feet, the support spacing is permitted to be at the ends but at intervals not to exceed 10 feet. Vertical runs of wireway are permitted to be supported at intervals of not more than 15 feet with not more than one joint between supports. Nonmetallic wireway is required to be supported at terminations and at intervals not to exceed 3 feet unless listed for greater support intervals. For a vertical run, the minimum support spacing is every 4 feet.

Wire bending space requirements apply to wireways. If a wireway makes a change in direction of an angle of more than 30°, the wire deflection space requirements of *Table 373-6(a)* shall apply. The space requirements of this table also apply to locations where wire enters or leaves the wireway by means of raceway, cable, or flexible cord.

Article 370 applies to device and junction boxes, pull boxes, and conduit bodies. Fittings permitted to contain splices or devices as permitted elsewhere in the Code are required to meet the requirements of this article. Boxes, conduit bodies, and fittings shall be installed such that, after installation, they are accessible without removing any part of the building, or excavating, according to *Section 370-29*. Boxes are permitted to be installed behind easily removable panels such as ceiling tiles in suspended ceilings where the ceiling tiles are easily removed.

Minimum requirements for cubic inch capacity for the wires, devices, and fittings are given in *Section 370-16*. In the case of standard device boxes, the maximum number of wires permitted in a box is given in *Table 370-16(a)*. The issue is that there is adequate physical space to prevent damage to conductors, prevent unnecessary pressure on splices and terminations, and prevent excessive heat produced within the box from current flowing in the wires and devices. If different sizes of wire enter a box or if a standard box is not used, then the minimum permitted cubic inch capacity of the box is determined using *Table 370-16(b)*. When wire sizes AWG number 4 and larger are contained in a box, the minimum size permitted is determined on the basis of the physical length of the box size, according to the rules of *Section 370-28*. Now the box is known as a pull box.

The rules for supporting boxes are given in *Section 370-23*. Cable or raceway is required to be secured to metal boxes according to *Section 370-17(b)*. Nonmetallic-sheathed cable is permitted to enter a nonmetallic box with dimensions not exceeding 2 1/4 inches by 4 inches without being secured to the box, according to the *Exception* to *Section 370-17(c)*. The cable is required to be fastened at a distance along the cable of not more than 8 inches, and the cable sheath is required to extend into the box opening a distance of not less than 1/4 inch, as shown in Figure 2.6.

Article 373 is on cabinets and cutout boxes used to enclose electrical equipment. Wire bending space and space requirements for wires in gutters within the enclosures are covered in *Section 373-6*. When a wire or wires leave a lug or terminal and leave the enclosure through the wall opposite the lug, the distance from the lug to that enclosure wall is determined from *Table 373-6(b)*. When the conductors leave an enclosure wall adjacent to the lug, the distance

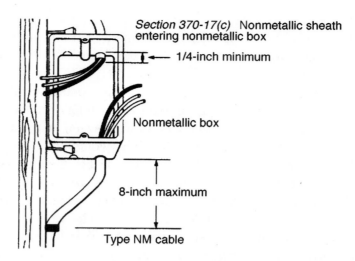

Figure 2.6   Nonmetallic-sheathed cable is not required to be secured to a single-gang nonmetallic box if it is secured within 8 inches of the box and extends into the box at least 1/4 inch.

from the lug or terminal to the opposite wall is found in *Table 373-6(a)*. These wire bending space requirements are illustrated in Figure 2.7.

Article 374 is about auxiliary gutters, which are the same material as wireway except the purpose is different. Auxiliary gutters are limited to 30 feet in length, and their purpose is to contain wiring between enclosures and devices such as at a motor control center or to connect the wiring and make taps for a group of enclosures making up a service entrance to a building. Auxiliary gutters are to be supported at intervals not to exceed 5 feet. *Section 374-5* permits up to 30 current-carrying wires in a metal auxiliary gutter without applying the derating factors of *Section 310-15(b)(2)*, provided that the conductor cross section does not exceed 20 percent of the cross-sectional area of the metal auxiliary gutter. For nonmetallic auxiliary gutter, there is no 30 current-carrying conductor limit, but the derating factors of *Section 310-15(b)(2)* must be applied whenever there are more than 3 current-carrying conductors at any one cross section. *Section 374-9(a)* requires adequate electrical and mechanical continuity of the complete auxiliary gutter system. This requirement would indicate that the auxiliary gutter would be considered to be acceptable to serve as an equipment grounding conductor. *Section 374-9(d)* requires that the same wire bending space as for wireways be provided following the provisions of *Section 373-6*.

Chapter 9, Notes to the tables provide information necessary for the use of the tables. *Notes 3* and *4* are of particular interest. *Note 3* states that equipment grounding conductors, if present, are to be counted when determining conduit or tubing fill. *Note 4* covers conduit and

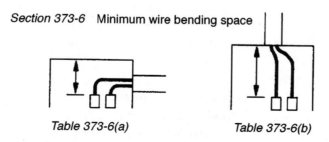

Figure 2.7   Minimum wire bending space is required from a lug or terminal to the opposite wall of the enclosure.

tubing nipples. As illustrated in Figure 2.8, conduit or tubing not more than 24 inches in length is considered to be a nipple. The wire fill is permitted to be 60 percent of the conduit or tubing total cross-sectional area. Also, the derating factors of *Section 310-15(b)(2)* do not apply in the case of a conduit or tubing nipple.

Tables in *Chapter 9* or *Appendix C* are used to determine the maximum number of wires and cables permitted to be installed in raceway. The total cross-sectional area of the conductors including the insulation is not permitted to exceed a maximum percentage of the cross-sectional area of the conduit or tubing. For most applications, the maximum is 40 percent fill. *Table 1* gives the maximum permitted percentage of cross-sectional area that the wires are permitted to fill in conduit and tubing. *Table 4* gives the internal diameter and cross-sectional area for the common types of conduit and tubing. The available trade diameters for the different types of conduit and tubing are provided. The table also gives the maximum cross-sectional area permitted for each size and type of conduit or tubing for one, two, and three or more conductors. Specific directions are provided in the notes of this table that tell how to determine the minimum permitted conduit or tubing diameter for specific wires or cable. *Note 9* tells how to determine the cross-sectional area of conduit or tubing for a multiconductor cable with an elliptical cross section.

Example 2.1   A type NM cable with two insulated AWG number 14 conductors and one bare conductor has a maximum dimension of 3/8 inch (0.375 inch). Determine the cross-sectional area of the cable to find the minimum trade diameter Schedule 80 PVC conduit permitted to be installed, as shown in Figure 2.9.

Answer: *Note 9* of *Table 1, Chapter 9* requires that for the purpose of determining conduit or tubing fill, the maximum dimension of the cable is to be used as though it was the diameter of a circular cable. The minimum trade diameter Schedule 80 PVC conduit required is 1/2 inch. Look up the area in the one conductor column of *Table 4*. The area of a circle is as follows:

$$\text{Area of a circle} = \frac{3.14 \times \text{Diameter} \times \text{Diameter}}{4}$$

$$= \frac{3.14 \times 0.375 \times 0.375}{4} = 0.110 \text{ in.}^2$$

When the wires are all of the same size and type of insulation, the cross-sectional area of the wires will be identical. In this case, *Note 1, Chapter 9,* specifies that the appropriate table

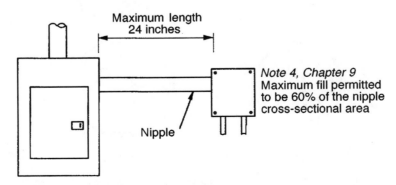

Figure 2.8   Conduit or tubing between enclosures that is not more than 24 inches in length is considered to be a nipple.

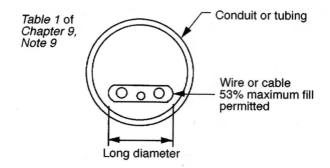

Figure 2.9 A type NM cable is installed in conduit.

in *Appendix C* is permitted to be used to look up the minimum size of conduit or tubing for the wires. There are separate tables in *Appendix C* for each type of conduit or tubing. Figure 2.10 compares conductors with compact strands and round strands. The conductors with compact strands have a smaller cross-sectional area than conductors of the same wire gauge with round strands. The tables in *Appendix C* for compact wires are designated with the letter A such as *Table C1A* through *Table C12A*. A summary of *Appendix C* is given in Table 2.1.

If the wires are different sizes and different insulation types, the cross-sectional areas of each size or type will be different. In this case, it will be necessary to calculate the total cross-sectional area of the conductors. *Tables 5* and *5A* give the diameter and cross-sectional area of conductors and their insulation. For aluminum conductors with compact conductor configuration, the values are found in *Table 5A*. Then *Table 4* is used to determine the minimum permitted size of conduit or tubing for the wires. *Table 4* provides data on trade sizes of conduit and tubing, such as internal diameter and internal cross-sectional area. The table gives values of area that are percentages of the total area, for example, 1.342 square inches is 40 percent of the total cross-sectional area of 2-inch trade size electrical metallic tubing.

SAMPLE CALCULATIONS

Methods for making the calculations to select the minimum size permitted for conductors, conduit, and boxes for specific installations are discussed. These methods are the same as those used in the Code, but, as the Code often uses the trial and error method for minimum

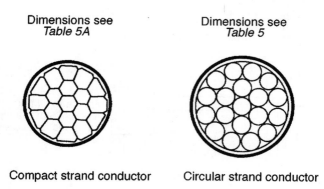

Figure 2.10 A conductor with compact strands will have a smaller cross-sectional area than a conductor of the same AWG size with round strands.

Table 2.1 Tables in Appendix C are permitted to be used to size raceway when all conductors are of the same size and insulation type.

Abbreviation	Raceway type	Appendix C table number		
(Tubing)				
EMT	Electrical metallic tubing	C1	&	C1A
ENT	Electrical nonmetallic tubing	C2	&	C2A
(Conduit)				
IMC	Intermediate metallic conduit	C4	&	C4A
RMC	Rigid metal conduit	C8	&	C8A
RNC-80	Rigid nonmetallic conduit, Schedule 80	C9	&	C9A
RNC-40	Rigid nonmetallic conduit, Schedule 40	C10	&	C10A
(Flexible conduit)				
FMC	Flexible metallic conduit	C3	&	C3A
LFNC-B	Liquidtight flexible nonmetallic conduit	C5	&	C5A
LFNC-A	Liquidtight flexible nonmetallic conduit	C6	&	C6A
LFMC	Liquidtight flexible metallic conduit	C7	&	C7A
(Underground encased in concrete)				
RNC-A	Rigid nonmetallic conduit, Type A thin wall	C11	&	C11A
RNC-EB	Rigid nonmetallic conduit, Type EB thin wall stiff	C12	&	C12A

size determination, the methods presented in this unit use direct calculations from which the minimum permitted size may be determined. It is suggested that the student copy the various formulas in the margin of the Code page where the size is to be determined for easy reference in the future.

Protecting Conductors

Electrical wires are required to be protected to prevent insulation damage. There are three common ways that insulation damage can occur, and insulated conductors are required to be protected from such damage.

- Protect insulation from excessive temperature (*Section 310-10*).
- Protect insulation and wire from physical damage (*Section 300-4*).
- Prevent deterioration of insulation by environmental factors such as chemicals and moisture (*Section 110-11*).

The ampere rating of a conductor will have an effect on the operating temperature of a conductor. A conductor heats as electrical current flows through the conductor. The amount of heat produced by current flow can be calculated using Equation 1.16. Figure 2.11 represents current flow through a conductor.

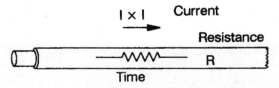

Figure 2.11 Heat produced in a conductor is proportional to the square of the current times the resistance of the conductor times the amount of time the current flows through the conductor.

$$\text{Heat} = I^2 \times R \times t$$

Where: I = current, in amperes
R = resistance, in ohms
If **t** is in hours, heat is in watt-hours
If **t** is in seconds, heat is in joules

Watt-hours can be converted to British thermal units (Btu) by multiplying by 3.413. As electrical current flows through a conductor, the temperature will rise. It takes approximately three hours of steady current flow for the temperature of the wiring system to reach a maximum operating temperature. This is why, in the Code, continuous load is considered a load operating for three hours or longer. The Code is concerned about heat produced under various conditions, such as 100 percent of the rated maximum conductor ampacity, 80 percent maximum conductor ampacity, and 50 percent conductor ampacity. Table 2.2 compares the joules of heat produced in 100 feet of AWG number 12 copper with a wire resistance of 0.16 ohm.

Compare the amounts of heat produced by the wire, as shown in Table 2.2. Conductors produce only one-quarter as much heat when operating at 50 percent load as they do at 100 percent of the rated maximum conductor ampacity. The conductors produce only 64 percent as much heat at 80 percent of conductor ampacity as at 100 percent. These percentages were determined based on the heating of 100 feet of AWG number 12 wire using Equation 1.16.

Ambient or surrounding temperature has an effect on the operating temperature of a wire. The ampacity of conductors for conditions specified in the tables of *Article 310* is determined at a specified ambient temperature. Temperature correction factors are provided at the bottom of the ampacity tables of *Article 310,* which are used to adjust the ampacity of the table for ambient temperatures other than those specified for the table. For example, consider a copper wire, AWG number 3, which is in conduit in free air with 60°C insulation with an ambient temperature of 120°F. The ampere rating of the wire is found in *Table 310-16* as 85 amperes. The ampacity correction factor at the bottom of the table for 120°F is 0.58, which results in an allowable ampere rating of the wire under these conditions of 49 amperes.

Allowable wire ampacity = 85 A × 0.58 = 49 A

The environment around a conductor affects the rate at which heat produced by current flow is removed from the conductor. This is the reason why several ampacity tables are in *Article 310* of the Code. Heat is removed from the conductors at different rates if the conductors are overhead in free air, in cable, or directly buried in the earth. If the conductors are wet or dry, the rate at which heat is removed from the conductor changes under certain conditions. For example, referring to *Table 310-16,* conductors with insulation type XHHW are permitted to have the ampacity determined as a 90°C rated conductor, but if the conditions are wet, it is considered a 75°C conductor.

When there is more than one conductor in cable or raceway carrying electrical current, the overall temperature of the conductors builds more rapidly because each conductor is

Table 2.2   **Approximate heat produced by current flow along 100 feet of AWG number 12 copper wire.**

Circuit rating	Amperes	Heat produced
50%	10	16 J/s
80%	16	41 J/s
100%	20	64 J/s

producing heat. The ampacity tables of *Article 310* were based on a maximum of three current-carrying conductors. When there are more than three current-carrying conductors in a cable or raceway, an adjustment factor is used to determine the allowable ampere rating for a conductor. These adjustment factors are found in *Section 310-15(b)(2)*. When there are ten current-carrying conductors in cable or raceway, the adjustment factor drops to 50 percent. Because of the significant drop in the adjustment factor from nine to ten conductors in a single run of cable or conduit, there is a real incentive to limit the number of wires to not more than nine.

There may be a situation where all of the conductors will not be energized at the same time. This is called load diversity. If the load diversity is not more than 50 percent, then other adjustment factors can be used. A 50 percent load diversity would mean that in any given cable or run of conduit or tubing, only 50 percent of the conductors would be energized at any time. The fine print note in *Section 310-15(b)(2)* calls attention to *Table B-310-11*, which is in *Appendix B* at the end of the Code. These adjustment factors are permitted to be used only when there is a load diversity of at least 50 percent and approved by the authority having jurisdiction.

Consider the case of more than three conductors in conduit, tubing, cord, or cable. The first step is to actually determine the number of current-carrying conductors. A neutral conductor, when serving as a common conductor to more than one ungrounded conductor, may carry only the unbalanced load between the ungrounded conductors. *Section 310-15(b)(4)* discusses this issue of when to count the neutral. Frequently, in the case of multiwire feeders and branch circuits, the neutral conductor is not considered a current-carrying conductor. Therefore, it is not counted for the purpose of derating. The following discussion explains when to count the neutral conductor for the purpose of derating for more than three conductors in raceway or cable.

The single-phase, 120/240-volt 3-wire feeder, or multiwire branch circuit, does not require the neutral to be counted as a current-carrying conductor. This is true even if the major portion of the load is electric discharge lighting or data processing equipment. This type of circuit or feeder is shown in Figure 2.12.

For the case of the 120-volt 2-wire circuit, the neutral is counted as a current-carrying conductor, as illustrated in Figure 2.13. It does not matter if the neutral is from a 3-wire 120/240-volt single-phase system, or from a 120/208-volt 3-phase system, the neutral is counted as a current-carrying conductor. The neutral also is counted as a current-carrying conductor of a 2-wire 277-volt circuit.

A 3-wire feeder, or multiwire branch circuit, can be obtained from a 3-phase, 4-wire wye electrical system with the feeder or branch circuit operating at 120/208 volts. Balancing the 120-volt loads will not reduce the current on the neutral *(Section 310-15(b)(4)(b))*. The

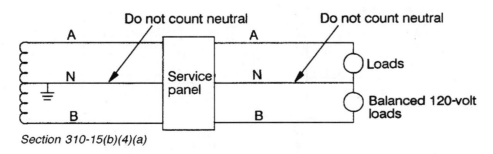

Section 310-15(b)(4)(a)

Figure 2.12 The neutral is not counted as a current-carrying conductor for a balanced single-phase 120/240-volt feeder or multiwire branch circuit.

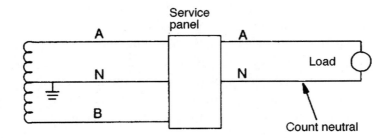

Figure 2.13   The neutral is counted as a current-carrying conductor for a 2-wire, single-phase 120-volt circuit or feeder.

neutral will always carry significant current, and the neutral is counted as a current-carrying conductor. Figure 2.14 shows this 3-wire circuit derived from a wye electrical system.

The 3-phase, 4-wire delta 120/240-volt feeder has a neutral conductor that serves as a common conductor for two of the ungrounded conductors, just like the 120/240-volt 3-wire single-phase system of Figure 2.12. The 4-wire, 120/240-volt delta system is shown in Figure 2.15. The neutral is not counted as a current-carrying conductor because the neutral only carries current due to the unbalance between the 120-volt loads on each of the ungrounded conductors (ungrounded conductors A and C). The 120-volt loads are not required to be balanced. The heat produced by current in the 120-volt circuit conductors is not greater in the unbalanced condition than when in the balanced condition.

For the 3-phase, 4-wire wye, 120/208-volt and 277/480-volt electrical systems, the neutral is not counted unless the major portion of the load does not consist of electronic computers, data processing equipment, or electric discharge lighting. In the case of a 4-wire set of branch circuits with a common neutral serving electric discharge lights in a room, the neutral is required to be counted. If less than half of the load in a building was electric discharge lighting, electronic computers, or data processing equipment, then the feeder neutral most likely would not be required to be counted as a current-carrying wire. The feeder and multiwire branch circuits of a 4-wire wye system are shown in Figure 2.16.

Sizing Conductors

The Code in *Article 310* requires that the allowable ampacity permitted for a conductor be determined based on the ampacity derating factors for ambient temperatures higher than

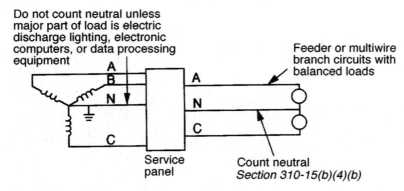

Figure 2.14   The neutral is counted as a current-carrying conductor for a 3-wire feeder or branch circuit derived from a wye system.

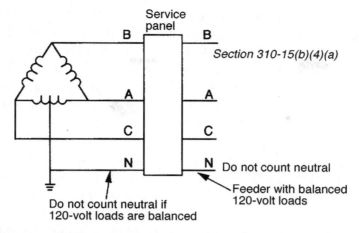

Figure 2.15   The neutral is not counted as a current-carrying conductor for a 4-wire feeder derived from a 120/240-volt delta electrical system with reasonably balanced 120-volt loads.

that for which the ampacity table was developed. When there are more than three current-carrying conductors in cable or raceway, the derating factors of *Section 310-15(b)(2)* shall also apply. A conductor is chosen from the proper ampacity table in the Code by choosing the wire that seems to be adequate, and then multiplying the allowable ampacity in the table by the temperature correction factor if it applies, and by the derating factor from *Section 310-15(b)(2)* if it applies. This derated allowable ampacity value is not permitted to be smaller than the rating of the circuit subject to the provisions of *Section 240-3*. If it is smaller than the circuit rating, then it will be necessary to choose a larger wire and repeat this calculation. The allowable ampacity of a conductor if a temperature derating factor and a derating factor for more than three conductors in a cable or conduit apply can be determined using Equation 2.1.

Adjusted Allowable Ampacity = Table Allowable Ampacity ×

Derating Factor (*Section 310-15(b)(2)*) ×

Temperature Factor   **Eq. 2.1**

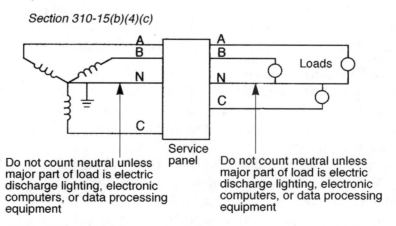

Figure 2.16   The neutral is not counted as a current-carrying conductor for a 4-wire feeder or multiwire branch circuit unless the major portion of the load is electric discharge lighting, electronic computers, or data processing equipment.

It is also necessary to know the temperature rating of splicing and terminating devices. Generally, if the temperature rating of a termination is lower than the temperature rating of the insulation, the column in the ampacity table that must be used is the one with the temperature equivalent to the termination. For example, assume a conductor has type THHN insulation rated at 90°C, but the termination is only rated at 75°C. If *Table 310-16* is used to determine conductor ampacity, the 75°C column must be used. If the termination temperature is not known, the rules for assuming termination temperature are found in *Section 110-14(c)*. If the overcurrent device is rated 100 amperes or less or the conductor is AWG number 1 or smaller, then the termination is considered to be rated 60°C unless otherwise indicated. If the overcurrent device is rated more than 100 amperes or the conductor is larger than AWG number 1, the termination is considered to have a 75°C rating unless otherwise indicated.

Continuous load, as discussed in *Sections 210-19(a), 215-3,* and *210-20(a),* applies to the circuit or feeder overcurrent device, and to the selection of the minimum permitted size of conductor for the circuit or feeder. The overcurrent device is required to have a rating not less than the noncontinuous load plus 125 percent of the continuous load. The minimum conductor size for a circuit is also determined on the same basis.

Conductor sizes AWG numbers 14, 12, and 10 have special overcurrent device limitations as given in *Section 240-3(d)*. The maximum overcurrent device rating permitted for copper conductors is 15 amperes for AWG number 14, 20 amperes for AWG number 12, and 30 amperes for AWG number 10, unless otherwise specifically permitted elsewhere in the Code. The maximum permitted overcurrent device rating for aluminum and copper-clad aluminum conductors for AWG sizes 12, 10, and 8 is also given in *Section 240-3(d)*.

Overcurrent devices are frequently not rated for operation at 100 percent load; therefore, they are permitted to be loaded only to 80 percent of their rating, according to *Section 384-16(d)*. The maximum load current for copper conductors for general branch circuits is 12 amperes for AWG number 14, 16 amperes for AWG number 12, and 24 amperes for AWG number 10.

Example 2.2 The home runs for eight 120-volt lighting circuits supplied from a 120/240-volt, 3-wire single-phase electrical system are contained in the same conduit, which runs for 100 feet through an area of a building where the ambient temperature typically runs to 120°F. The panelboard supplying the circuit is more than 10 feet from the room. A circuit supplies eight fluorescent fixtures, each of which draws 1.6 amperes. Multiwire branch circuits are used with a common neutral for two ungrounded conductors. The total conductors within the conduit are eight ungrounded conductors and four neutral conductors. This circuit is illustrated in Figure 2.17. Determine the minimum size of type THHN copper wire for these circuits, which are protected with 20-ampere circuit-breakers with 75°C terminations.

Answer: Use Equation 2.1 to determine the minimum wire allowable ampacity required for this installation. Determine the derating factor from *Table 310-15(b)(2)(a)* for eight current-carrying wires in the conduit. Only the ungrounded conductors are counted because these are multiwire branch circuits. Even though the lighting load is electric discharge, the neutral conductors are not counted because this is a single-phase, 120/240-volt 3-wire electrical system. In this case, the adjustment factor is 0.70. Find the temperature adjustment factor for a 120°F ambient temperature at the bottom of *Table 310-16,* which is 0.82 using the 90°C column (*Section 310-15(a)(2) Exception*).

Select a wire size and type, and then apply the ampacity reduction factors to see if it is adequate for the circuit rating. If it is not, then select the next larger wire size and do the calculation again. The ampacity of AWG number 12 copper wire with type THHN

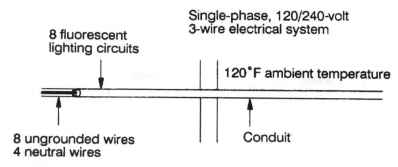

Figure 2.17   Eight circuits of a 120/240-volt 3-wire electrical system supply fluorescent fixtures using multiwire branch circuits that run through a room with an ambient temperature of 120°F.

insulation is 30 amperes. Multiply the 30 amperes by 0.70 and by 0.82 to get 17.2 amperes. This result is less than the rating of the 20-ampere circuit, but AWG number 12 copper is permitted for the circuit because of *Section 240-3(b)*.

An additional consideration for Example 2.2 is the rating of the overcurrent device. This circuit is protected by a 20-ampere circuit-breaker. Code *Section 384-16(d)* states that an overcurrent device is not permitted to carry load continuously in excess of 80 percent of its rating unless rated for continuous operation at 100 percent. In this example, the load current is 12.8 amperes, and the 20-ampere overcurrent device is permitted to carry 16 amperes continuously. It is the actual load current that is used to determine the minimum permitted size of overcurrent device for the circuit along with the adjusted allowable ampacity of the conductor if adjustment factors apply. Also keep in mind when using *Section 240-3(b)*, the conductor may be permitted to be protected with the next higher standard rating overcurrent device.

The key factor in *Section 240-3(b)* is that the allowable ampacity of the conductor is as large or larger than the load to be served. A major restriction is that this provision is not permitted to be applied where the circuit supplies multiple receptacle outlets for cord-and-plug connected portable loads. Also, the conductor allowable ampacity does not correspond to a standard rating of overcurrent device as listed in *Section 240-6*. If these conditions are met, and the circuit rating does not exceed 800 amperes, then the conductor is permitted to be protected using the next larger standard size overcurrent device. In the case of Example 2.2, the calculated allowable ampacity of a type THHN copper conductor is 17.2 amperes. The circuit supplies eight electric discharge lighting fixtures, each drawing 1.6 amperes for a total circuit load of 12.8 amperes. Because the allowable ampacity (17.2 amperes) is greater than the load (12.8 amperes), the next larger standard rating overcurrent device (20 amperes) is permitted to protect the circuit.

Another question arises concerning Example 2.2 with respect to the allowable ampacity of the conductor being computed based upon the allowable ampacities of *Table 310-16* from the 90°C column. It is likely that the overcurrent device in the supply panelboard is only rated for a 60°C termination. This would seem to limit the conductor to the 60°C column of *Table 310-16*. But the key factor is the *Exception* to *Section 310-15(a)(2)*. If the section of the circuit with the higher ambient temperature is more than 10 feet from a termination with a lower temperature rating (or 10 percent of the length of the section in the high ambient temperature area), then the allowable ampacity calculation is permitted to be based upon the higher conductor temperature.

Example 2.3   Fluorescent lighting, 120-volt circuits, each with a load current of 14.4 amperes, for a commercial building are supplied from a 120/208-volt, 4-wire, 3-phase wye system. A conduit from the lighting panel runs to an area of the building where the fixtures are to be installed. The conduit contains wires for six multiwire circuits. There are six phase wires and two neutrals. The 20-ampere circuit-breakers are only rated for 60°C terminations; therefore, even though the wires have type THHN insulation, the 60°C column of the ampacity table is required to be used for wire size selection. Determine the minimum copper wire size to supply the circuits.

Answer: This is a 3-phase wye, 4-wire electrical system supplying electric discharge lighting with a common neutral, and according to *Section 310-15(b)(4),* the neutrals are required to be counted as current-carrying conductors. The total number of current-carrying conductors for this conduit is eight; that requires a derating factor of 0.7 according to *Section 310-15(b)(2).*

Look up the allowable ampacity for an AWG number 12 copper wire in *Table 310-16* using the 60°C column. Use Equation 2.1 to determine the adjusted allowable ampacity of the wire, which is 17.5 amperes (25 A × 0.70 = 17.5 A). The load current is less than 16 amperes, which meets the continuous load derating factor of 80 percent for the 20-ampere circuit-breaker required by *Section 384-16(d).* The 20-ampere circuit-breaker satisfies *Section 240-3(d)* for overcurrent protection for an AWG number 12 copper wire. Using the provision of *Section 240-3(b),* this AWG number 12 wire with an adjusted allowable ampacity of 17.5 amperes is permitted to be protected with a 20-ampere circuit-breaker.

Example 2.4   Two 3-phase, 3-wire circuits that carry a continuous load current of 48 amperes are run in the same conduit. The total number of current-carrying conductors in the conduit is six. The wires are type THHN, copper, and the overcurrent protective devices used are rated for terminations at 75°C. At some distance from the supply panelboard, the conduit runs through an area with an ambient temperature of 125°F. Determine the minimum size wire for these circuits. Note in Figure 2.18 that the area of high ambient temperature is at a distance of more than 10 feet from the electrical panelboard. Also, the length of circuit run in the high ambient area is 30 feet.

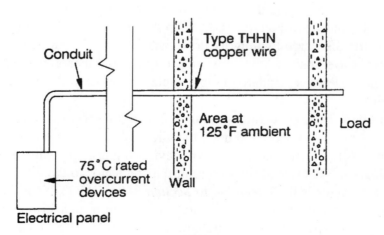

Figure 2.18   A run of conduit passes through a room with a high ambient temperature that is located more than 10 feet from the panelboard.

Answer: There are six wires in the conduit, each with a continuous load current of 48 amperes, and the temperature in one area is 125°F. First determine the minimum rating of overcurrent device permitted for the circuit. This is a continuous load of 48 amperes; therefore, unless rated for 100 percent rating, the load is not permitted to exceed 80 percent of the overcurrent device rating, which is equivalent to multiplying the load current by 1.25 (48 A × 1.25 = 60 A). The overcurrent device for this circuit will be 60 amperes.

If there were only three wires in the conduit and the 75°C column of *Table 310-16* were used, an AWG number 6 wire would be permitted. But there are six current-carrying wires in the conduit, and according to *Section 310-15(b)(2),* the allowable ampacity of the wire must be derated to 80 percent, which is 52 amperes (65 A × 0.80 = 52 A). Note that the 75°C column is used even though this is a type THHN wire because the conductor terminations are only rated at 75°C and the conditions of *Section 310-15(a)(2) Exception* are not met for the six wires. But the area of the building where the ambient temperature is 125°F also must be considered. Because this area is more than 10 feet from the terminations, it is permitted to size the wire based upon the 90°C column of *Table 310-16.* The adjusted allowable ampacity of the conductor for this area is determined using Equation 2.1, which gives a value of 45.6 amperes (75 A × 0.80 × 0.76 = 45.6 A). The AWG number 6, type THHN, copper conductor has too low an adjusted allowable ampacity for a 60-ampere overcurrent device. Now try the calculation again using the next size larger type THHN copper conductor, which is an AWG number 4. For this conductor, the allowable ampacity is 57.8 amperes (95 A × 0.80 × 0.76 = 57.8 A). Using the provision of *Section 240-3(b),* a copper, type THHN, wire is permitted for these circuits.

Sizing Conduit and Tubing

If all conductors in the conduit are the same size with the same type of insulation, the minimum trade diameter conduit or tubing permitted can be determined in accordance with *Note 1* of *Chapter 9. Note 1* makes reference to the tables in *Appendix C.* There are separate tables for each of twelve types of conduit or tubing. There is one set of tables for conductors with round strands and a separate set of tables for conductors with compact strands. If a type of conduit or tubing is used that is not included in *Appendix C,* the maximum number of conductors permitted in the conduit or tubing is determined by using the cross-sectional area provided by the manufacturer and calculating the conductor fill.

When the conductors in a conduit or tubing are of different sizes or types of insulation, the cross-sectional area of the conductors must be determined to calculate the minimum trade diameter of the conduit or tubing to be used. The diameter and cross-sectional area of common types of conductors are given in *Table 5* or *Table 5A* in *Chapter 9.* The total cross-sectional area of all conductors is first determined. Then, the minimum trade diameter conduit or tubing is found in *Table 4* for the type of conduit or tubing to be used. In the case of three or more conductors in conduit or tubing, the 40 percent fill column is used to determine the minimum trade diameter. Some examples will help illustrate how conduit or tubing is sized for a particular situation.

Example 2.5 Two feeders, each consisting of three AWG number 3/0, type THHN conductors, are run in the same rigid metal conduit, as shown in Figure 2.19. Determine the minimum trade diameter rigid metal conduit permitted for the six conductors.

Answer: Look up the minimum permitted rigid metal conduit trade diameter directly in *Appendix C, Table C8.* The minimum permitted trade diameter for wire with round strands is 2 1/2 inches. If the wire was of compact strand construction, the minimum permitted trade diameter conduit would be 2 inches from *Table C8A.*

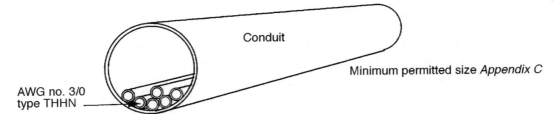

Figure 2.19   Six AWG number 3/0, type THHN wires are run in a rigid metal conduit.

Example 2.6   Determine the minimum trade diameter, Schedule 40 PVC rigid nonmetallic conduit permitted to contain three AWG number 2/0, type THW wires, six AWG number 12, type THWN wires, and one bare size AWG number 6 copper equipment grounding wire. This conduit is shown in Figure 2.20.

Answer: It is necessary to determine the total cross-sectional area of the wires. Wire cross-sectional area is not permitted to exceed 40 percent of the inside cross-sectional area of the conduit. Minimum size of 2 inches is found in *Table 4*.

AWG number 2/0, type THW, *Table 5*
$$3 \times 0.2624 = 0.7872 \text{ sq. in.}$$
AWG number 12, type THWN, *Table 5*
$$6 \times 0.0133 = 0.0798 \text{ sq. in.}$$
AWG number 6, bare, *Table 8*
$$1 \times 0.027 = \underline{0.0270 \text{ sq. in.}}$$
$$0.8940 \text{ sq. in.}$$

Sizing Conduit and Tubing Nipples

A conduit or tubing nipple is defined in *Note 4* of *Chapter 9* as a length of conduit or tubing that does not exceed 24 inches in length. Figure 2.8 illustrates an application of a tubing nipple. The cross-sectional area of the conductors is not permitted to exceed 60 percent of the cross-sectional area of the nipple. The nominal inside cross-sectional area of trade diameter conduit and tubing is given in the 100 percent column of *Table 4, Chapter 9*. For example, the total cross-sectional area of a 2-inch trade diameter rigid metal conduit nipple is 3.408 square inches.

The minimum trade diameter of a conduit or tubing nipple permitted is determined by first finding the total cross-sectional area of all conductors contained within the nipple. This

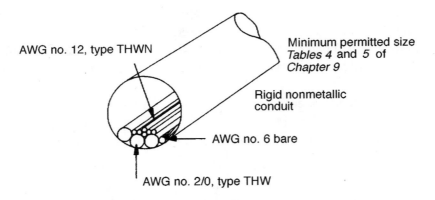

Figure 2.20   Insulated wires and bare copper equipment grounding wires are run in rigid nonmetallic conduit.

was done for the conductors in Figure 2.20, and the total cross-sectional area of the conductors was determined in Example 2.6 to be 0.8940 square inches. Assume the conduit of Example 2.6 is a nipple. Using the cross-sectional area, select a trade diameter from *Table 4* that is expected to be the correct size, and multiply the total cross-sectional area of that nipple by 0.6 (60%). The resulting area from this calculation is required to be equal to or larger than the cross-sectional area of the conductors. Consider a one-inch trade diameter Schedule 40 PVC conduit nipple. The conductor cross-sectional area is permitted to be 0.50 square inches.

$$0.832 \text{ sq. in.} \times 0.6 = 0.50 \text{ sq. in.}$$

This value is smaller than the cross-sectional area of the wire; therefore, the next size larger nipple is chosen, and the calculation is done a second time. This time the value determined is 0.87 square inches, which is still too small. When the calculation is done a third time for 1 1/2-inch trade diameter, the value determined is 1.19 square inches, which is adequate. This trial-and-error method can be time-consuming; therefore, Equation 2.2 may save time. The total conductor cross-sectional area is divided by 0.6 (60 percent) to determine the minimum permitted nipple cross-sectional area. Then, just proceed down the 100 percent column of *Table 4* until an area is found that is equal to or greater than the value calculated.

$$\textbf{Actual Inside Area} = \frac{\textbf{Wire Cross-sectional Area}}{\textbf{0.6}} \qquad \textbf{Eq. 2.2}$$

Example 2.7 Assume the conductors of Figure 2.20 with a total cross-sectional area of 0.8940 square inches are to be run through a nipple between two panels. Determine the minimum trade diameter conduit nipple permitted.

Answer: First, use Equation 2.2 to determine the total cross-sectional area of the nipple for which the wire cross-sectional area is 60 percent.

$$\text{Actual Inside Area} = \frac{0.8940 \text{ sq. in.}}{0.6} = 1.49 \text{ sq. in.}$$

Choose the minimum nipple diameter from the 100 percent column of *Table 4*. For this example, choose a 1 1/2-inch trade diameter nipple.

Wireway Sizing

The total cross-sectional area of the conductors is not permitted to exceed 20 percent of the cross-sectional area of the wireway, except at splices, *Section 362-19*. The derating rules of *Section 310-15(b)(2)* for more than three conductors in raceway do not apply to metal wireway when there are not more than 30 current-carrying wires in the metal wireway. Wireway is square in cross section; therefore, the cross-sectional area is determined by squaring the trade dimension of the wireway. For example, a 4-inch by 4-inch wireway has a cross-sectional area of 16 square inches. Equation 2.3 can be used to determine the minimum permitted dimension of wireway for electrical conductors. First determine the total cross-sectional area of the conductors, and then determine the minimum permitted cross-sectional area required with Equation 2.3.

$$\textbf{Actual Inside Area} = \frac{\textbf{Wire Cross-sectional Area}}{\textbf{0.2}} \qquad \textbf{Eq. 2.3}$$

Example 2.8   Determine the minimum wireway dimensions permitted to contain nine AWG number 2/0, type THW wires, and twelve AWG number 12, type THWN wires. This wireway is shown in Figure 2.21.

Answer: First determine the total cross-sectional area of the conductors in the wireway. Equation 2.3 can be used to determine the minimum permitted wireway cross-sectional area. A 4-inch by 4-inch wireway has a cross-sectional area of 16.0 square inches. Therefore, a 4-inch square wireway is adequate for the conductors of this example.

AWG number 2/0, type THW, *Table 5*
$$9 \times 0.2624 = 2.3616 \text{ sq. in.}$$
AWG number 12, type THWN, *Table 5*
$$12 \times 0.0133 = 0.1596 \text{ sq. in.}$$
$$\text{Total wire area} \quad 2.5212 \text{ sq. in.}$$

$$\text{Actual Inside Area} = \frac{2.5212 \text{ sq. in.}}{0.2} = 12.606 \text{ sq. in.}$$

Sizing Boxes

The easiest method to size standard boxes is to count the total conductors and conductor equivalents, then look up the size in *Table 370-16(a)*. This can be done only if the wires entering the box are all the same size. If they are not the same size, then the cubic inch capacity of the box must be determined using the cubic inch requirements of *Table 370-16(b)*. The rules for determining the minimum size device or junction box permitted are given in *Section 370-16*. The following steps are used when all wires in the box are the same size:

1.   Count each current-carrying wire entering the box. An unbroken wire only counts as one. (Pigtails do not count.)
2.   Where one or more internal cable clamps are used, one conductor equivalent shall be counted based upon the wire size secured by the clamp.
3.   Fittings such as fixture studs and hickeys count as one conductor equivalent for each type of fitting.
4.   Each device yoke counts as two wire equivalents based upon the wire size terminating at the device.

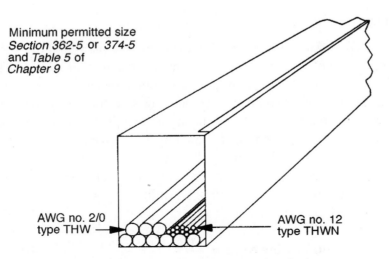

Minimum permitted size
Section 362-5 or *374-5*
and *Table 5* of
Chapter 9

AWG no. 2/0
type THW

AWG no. 12
type THWN

Figure 2.21   Electrical wires are run in wireway.

5. All grounding wires count as one wire equivalent unless the box contains equipment grounding wires for the circuit as well as for an isolated ground receptacle, such as may be used with computer equipment.

If wires of different sizes are in the same box, the equipment grounds are counted as a wire of the largest size. A device in the box is counted as a wire of the largest size connected to that device if more than one size of wire is in the box. Cable clamps are counted as a conductor equivalent to the size secured by the clamp.

Example 2.9 Determine the dimensions of the smallest size metal 2-inch by 3-inch standard device box required for the outlet shown in Figure 2.22. Both cables are AWG number 14 copper, with two insulated wires and a bare equipment grounding wire. There is a single-pole switch in this box with cable clamps.

Answer: Determine the number of conductors in the device box and the conductor equivalents for nonwire parts that take up space in the box. A 3-inch by 2-inch by 3 1/2-inch deep device box is found to be the minimum permitted from *Table 370-16(a)*.

Wires	4
Cable clamps	1
Other fittings	0
Device	2
Grounds	1
Total	8

Example 2.10 Consider the device box of Figure 2.22, but this time assume that one type NM cable entering the box is AWG number 14 with two insulated conductors and a bare equipment grounding wire, and the other cable is AWG number 12 with two insulated conductors and a bare equipment grounding wire. The AWG number 12 conductor connects to the device. Determine the minimum standard size device box permitted for this application.

Answer: To determine the minimum size box, it will be necessary to determine the cubic inch capacity of the box. First, do a conductor count and determine the number of

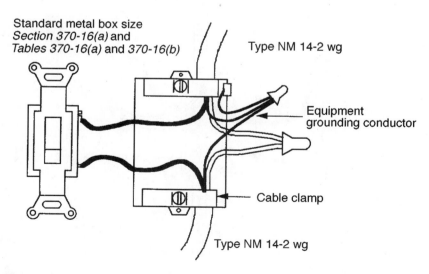

Figure 2.22 A switch is installed in a device box with AWG number 14 electrical cable with an equipment grounding wire.

conductors and conductor equivalents for each wire size. This is summarized below. Next, look up the cubic inch box capacity required for each wire size and multiply that value by the number of each size of wire in the box. This will give a minimum permitted capacity for the device box of 17.50 cubic inches. Then go to the minimum cubic inch capacity column of *Table 370-16(a)* and find the minimum permitted size of a device box with dimensions of 2 inches by 3 inches by 3 1/2 inches in depth.

	AWG no. 14	AWG no. 12
Wires	2	2
Grounds	0	1
Device	0	2
Cable clamps	0	1
Other fittings	0	0
Totals	2	6

AWG number 14	2×2.0 cu. in.	=	4.0 cu. in.
AWG number 12	6×2.25 cu. in.	=	13.50 cu. in.
	Total		17.50 cu. in.

Nonmetallic boxes are not listed in *Table 370-16(a)*. Instead, they are selected according to cubic inch capacity. The cubic inch capacity of a nonmetallic box is marked on the box.

Example 2.11   Determine the minimum cubic inch capacity permitted for a nonmetallic box with one type NM cable size AWG number 14 with two insulated conductors and a bare equipment grounding conductor entering the box and another leaving the box, as shown in Figure 2.23. The box is nonmetallic without cable clamps.

Answer: If the nonmetallic box does not contain cable clamps, then the equivalent conductor count is seven, as tabulated below. A nonmetallic device box will be marked with the

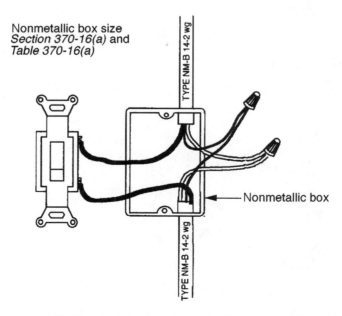

Figure 2.23   A nonmetallic box contains one device and has a type NM cable with two AWG number 14 insulated wires and a bare equipment grounding conductor entering the box.

cubic inch capacity. Look up the minimum cubic inch capacity requirement for each of the AWG number 14 wires in Code *Table 370-16(b)*. Then multiply the seven conductor count of this device box by 2.0 cubic inches for each wire in the box to get the minimum permitted box area of 14 cubic inches.

Wires	4
Grounds	1
Device	2
Clamps	0
Other fittings	0
Total	7

7 wires × 2.0 cu. in./wire = 14 cu. in.

Fixture wires, up to four including equipment grounding wire, entering a box from a canopy fixture are not counted for the purpose of determining the minimum size box. In some cases, fixture wires AWG number 14 and smaller may be required to be counted, *Section 370-16(b)(1) Exception*. Be sure to count fixture studs and similar devices, and if a fixture cover does not actually take up space inside a box, then it is not counted. Plaster rings and raised covers for devices add cubic inch capacity to a box. Trade literature for device covers will list the amount of cubic inch capacity added for the various types and sizes. The cubic inch capacity of a raised cover should be marked on the cover.

Pull Boxes

Minimum dimensions are required for pull boxes and conduit bodies when the wires are AWG number 4 and larger. There are straight pulls in which the conductors enter one end of the box and leave the opposite end. For a straight pull, the minimum distance between openings is eight times the largest trade diameter conduit entering the pull box. This rule is found in *Section 370-28(a)(1)*.

Example 2.12   Determine the minimum length permitted for a straight pull box with a 2-inch trade diameter conduit entering both ends, as shown in Figure 2.24.

Answer: The minimum permitted length is determined by multiplying the largest trade diameter conduit, which is 2 inches, by 8.

8 × 2 inches = 16 inches long

For an angle pull, the distance to the opposite wall is not permitted to be less than six times the largest diameter conduit plus the sum of the diameters of other conduits on the same side in the same row. What is being sized is the distance to the opposite wall. This is confusing, and it will be necessary to read *Section 370-28(a)(2)* carefully. An example of an angle pull box is shown in Figure 2.25. These required dimensions apply only when the wire size is

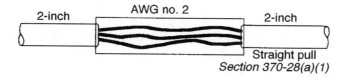

Figure 2.24   A straight pull box in a run of 2-inch trade size conduit where the conductors are AWG number 4 or larger.

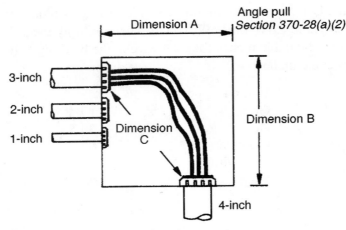

Figure 2.25 A pull box is used to make an angle pull for several conduits where the conductors are AWG number 4 or larger.

AWG number 4 and larger. If the wires are AWG number 6 and smaller, there is no minimum dimension requirement. In that case, the box is sized according to *Section 370-16*.

Example 2.13 Determine the minimum dimensions of the pull box shown in Figure 2.25 when all wires passing through the pull box are AWG number 4 and larger.

Answer: Dimension A is determined by multiplying the 3-inch diameter conduit by six and then adding to it the diameters of the other conduits on that same side in the same row. The minimum permitted length of dimension B is determined by multiplying the 4-inch diameter conduit by 6. The minimum size is 21 inches by 24 inches.

Minimum length of dimension B of the pull box:

$$6 \times 4\text{-inch} = 24 \text{ inches}$$

Minimum length of dimension A of the pull box:

$$
\begin{array}{r}
6 \times 3\text{-inch} = 18 \text{ inches} \\
+ \quad 2 \text{ inches} \\
+ \quad 1 \text{ inch} \\
\hline
21 \text{ inches}
\end{array}
$$

Also, when wire is AWG number 4 and larger, conduit openings are required to be installed at least six times the largest conduit diameter apart. For an example, see dimension C in Figure 2.25.

$$\text{Dimension C} = 6 \times 4\text{-inch} = 24 \text{ inches}$$

Conduit Bodies

The rules for determination of the minimum dimensions of pull boxes also apply to conduit bodies with one exception, which is the **LB**-type conduit body. The different situations considered are the straight run and the angle. These minimum dimension requirements apply only when the conduit body contains conductors size AWG number 4 and larger.

The rules of *Section 370-28(a)(1)* apply for a straight conduit body, and the length of the conduit body shall not be permitted to be less than eight times the trade diameter of the largest conduit entering the conduit body.

Example 2.14   A conduit body is used for wire pulling access in a straight run of conduit, as shown in Figure 2.26. There are four AWG number 4, type THHN wires installed in a 1-inch trade diameter conduit. Determine the minimum permitted length for this straight conduit body.

Answer: This is a straight run, and the conduit body shall have a length not less than eight times the trade diameter of the largest conduit entering the conduit body.

$$8 \times 1\text{-inch} = 8 \text{ inches}$$

The rule for an angle conduit body is found in *Section 370-28(a)(2)*. The dimension from a conduit entry to the opposite wall of the conduit body shall be required to be not less than six times the trade diameter of the conduit entering the conduit body. There is an exception that may be used for the dimension from the conduit entry to a removable cover. This same exception can be used when a conduit enters the back of a pull box opposite the removable cover. This exception allows the dimension to be determined by the one conductor per terminal column of *Table 373-6(a)*.

Example 2.15   A type LB conduit body is installed in a run of conduit to make a right angle access through a wall, as shown in Figure 2.27. If four AWG number 4 type THHN wires are run in 1-inch trade diameter conduit, determine the minimum dimensions of a type LB conduit body.

Answer: First, determine the minimum permitted length of dimension L2 that is required to be six times the trade diameter of the conduit to the opposite wall.

Dimension L2: $6 \times 1\text{-inch} = 6 \text{ inches}$

Dimension L1 is from the conduit entry to the removable cover. An exception permits this dimension to be determined using the one conductor per terminal column of *Table 373-6(a)*. Dimension L1 is a minimum of 2 inches.

If the wires in the conduit body are AWG number 6 and smaller, then the rules of *Section 370-28* do not apply. The conduit body is then required to have a minimum cross-sectional area of not less than twice the cross-sectional area of the largest conduit or tubing entering the conduit body. Also, if the conduit body has less than three conduit entries, it shall not contain splices, taps, or devices unless the cross-sectional area of the conduit body meets the minimum requirements for box fill, as given in *Section 370-16(c)(1)*.

Sometimes cable assemblies with conductors larger than AWG number 6 join in a box for splicing or terminating. The rules in *Section 370-28(a)* also apply, but there is no conduit or tubing size to use to determine the minimum dimensions of the box. The method to be used is explained in *Section 370-28(a)*. First, determine the minimum size conduit or tubing that would have been needed if the conductors had been run in conduit or tubing. Then use that minimum

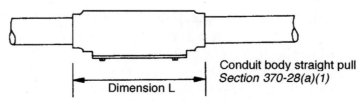

Conduit body straight pull
Section 370-28(a)(1)

Dimension L

Figure 2.26   A conduit body is used to provide access to a conduit to make a straight wire pull.

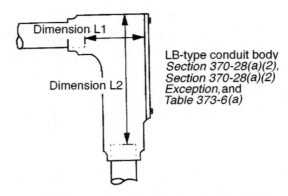

Figure 2.27   Minimum dimensions are required for a type LB conduit body when the wires are AWG number 4 and larger.

conduit or tubing size to determine the dimensions of the straight or angle box. The method is explained in Example 2.16.

Example 2.16   A cable assembly containing four conductors size AWG number 2 copper with type THWN insulation enter one side of a box and leave the adjacent side of the box. Determine the minimum size of box required for splicing these conductors.

Answer: Because the conductors are larger than AWG number 6, this is to be treated as an angle pull and sized according to *Section 370-28(a)(2)*. First determine the minimum trade diameter conduit or tubing that would have been required if the conductors had been run in conduit or tubing rather than as a cable assembly. Look up the cross-sectional area of the AWG number 2 type THWN conductors in *Table 5* and find the value 0.1158 square inches. Multiply this value by 4 conductors to get 0.4632 square inches ($0.1158\ \text{in}^2 \times 4 = 0.4632\ \text{in}^2$). Then look up the minimum size conduit or tubing from *Table 4* in the 40 percent column. The minimum size is 1 1/4 inch. If two different sizes are found, use whichever is smaller. Now use the rule of *Section 370-28(a)(2)* and multiply by 6 to get the minimum dimension of 7.5 inches (1.25 inch $\times$ 6 = 7.5 inches).

Sizing and Installing Parallel Conductor Sets

In the case of high-ampacity feeders, one conductor per phase may not be practical. Under these circumstances, two or more conductors per phase may be desirable or even necessary. The objective is to keep the resistance of each parallel path equal so the load current will divide equally on each wire of the set. The wires and terminations heat up when they carry current; therefore, the resistance changes when the temperature of the wire and terminations change. For these reasons, special requirements are necessary when installing parallel sets of wires. The following is a summary of some special installation rules of *Section 310-4* for sets of multiple conductors for each ungrounded conductor and the grounded conductor:

1. Power conductors are permitted to be paralleled only for size AWG number 1/0 and larger, except for special applications.
2. All parallel conductors of a phase or neutral set shall be the same length.
3. All conductors of the phase or neutral set shall be of the same material, cross-sectional area, and the same insulation type.
4. All conductors of a phase or neutral set shall be terminated in the same manner.
5. If run in more than one raceway, the raceways shall have the same physical properties, and the same length, and shall be installed in the same manner.

6. If more than one raceway is used, make sure each phase wire and neutral, if present, is placed in each of the raceways to prevent eddy currents, as shown in Figure 2.28.

7. If an equipment grounding wire is present and there is more than one raceway, an equipment grounding wire shall be in each raceway. The size of each equipment grounding conductor is determined according to *Section 250-122,* which means the overcurrent rating of the feeder determines the minimum permitted size of the equipment grounding conductor in each raceway.

These rules for the installation of parallel sets of conductors apply to conductors of the same phase or neutral. Phase A is not required to be identical to phase B, for example. But all of the wires of phase A are required to be identical, as discussed in the previous points.

Determining the minimum parallel conductor size for a particular application can be confusing. In the case where there is a single main disconnecting means, the minimum permitted conductor size is determined by dividing the rating of the feeder overcurrent device by the number of parallel sets of conductors. If there are more than three current-carrying conductors in a single raceway, then a derating factor according to *Section 310-15(b)(2)* must be applied. Equation 2.4 can be used to determine the minimum permitted wire size for a feeder consisting of parallel sets of conductors. If the feeder rating is not over 800 amperes, then the next wire size smaller may be permitted to be used by *Section 240-3(b).* For some services, calculated load may be used.

$$\textbf{Minimum Parallel Conductor Ampacity} = \frac{\textbf{Feeder Overcurrent Rating}}{\textbf{No. of sets} \times \textbf{Derate factor}} \qquad \textbf{Eq. 2.4}$$

Example 2.17   Determine the minimum permitted size of copper type THWN conductors when three parallel sets are used for a 3-phase, 3-wire feeder with a 500-ampere overcurrent device that is not rated for 100 percent continuous operation and when the actual load current does not exceed 80 percent of the rating of the overcurrent device. All nine conductors are run in the same raceway.

Answer: The result of the calculation of Equation 2.4 gives a minimum feeder conductor rating of 238 amperes when there are three parallel sets. The minimum type THWN copper wire size from *Table 310-16* is 250 kcmil with a rating of 255 amperes. This feeder has a rating of less than 800 amperes; therefore, *Section 240-3(b)* will apply. In this case, an AWG number 4/0 conductor rated at 230 amperes is adequate for the load and will satisfy the requirement of *Section 240-3(b).*

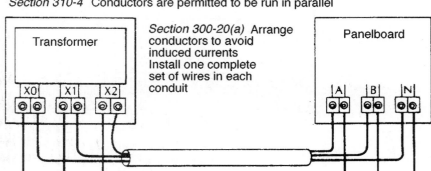

Figure 2.28   Care must be taken when paralleling sets of conductors, particularly when more than one raceway is used.

$$\text{Mimimum Parallel Conductor Ampacity} = \frac{500 \text{ A}}{3 \times 0.7} = 238 \text{ A}$$

Size 250-kcmil copper, type THWN:
$$255 \text{ A} \times 0.7 \times 3 = 536 \text{ A}$$
Size AWG number 4/0 copper, type THWN:
$$230 \text{ A} \times 0.7 \times 3 = 483 \text{ A}$$

MAJOR CHANGES TO THE 1999 CODE

These are the changes to the 1999 *National Electrical Code©* that correspond to the Code sections studied in this unit. The following analysis explains the significance of only the changes from the 1996 to the 1999 Code, and this analysis is not intended to be used in place of the Code. Refer to the actual section of the 1999 Code for the exact wording and meaning of each section discussed. Changes are indicated in the Code with a vertical line in the margin. If material has been deleted or moved to another part of the Code, the location of the deletion is indicated with a dark dot in the margin.

Article 331   **Electrical Nonmetallic Tubing**

331-3(6): Electrical nonmetallic tubing (ENT) is now permitted to be installed under grade-level concrete slabs if the ENT is placed on a sand base or similar approved material as illustrated in Figure 2.29.

331-3(8): Electrical nonmetallic tubing in sizes 1/2-inch through 1-inch trade diameter is permitted to be installed as a manufactured prewired assembly. It is permitted to be shipped as a continuous length in a coil, reel, or carton. A tag or marker is required at each end of the assembly giving the type and size of conductors and number of conductors in the assembly.

331-4(3): Electrical nonmetallic tubing of the polyvinyl chloride type is not permitted to be installed in an area with an ambient temperature above 50°C (122°F) unless a higher ambient temperature is listed.

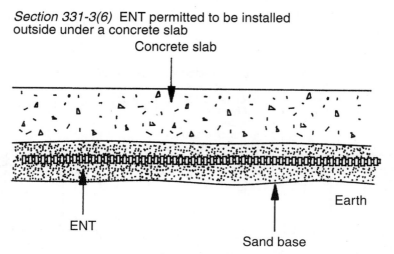

Section 331-3(6) ENT permitted to be installed outside under a concrete slab

Concrete slab

Earth

ENT

Sand base

Figure 2.29   Electrical nonmetallic tubing is permitted to be installed under a grade-level concrete slab where installed in a sand base.

Article 343   Nonmetallic Underground Conduit with Conductors

343-5: This new fine print note lists the metric equivalent dimensions for the different trade diameters of nonmetallic underground conduit with conductors. The metric numerical designations are approximately the inside diameter in millimeters. Table 2.3 lists the common trade diameters in inches and the metric equivalents.

343-10: A new provision was added to mandate a minimum bending radius. There is a new *Table 343-10* that lists the minimum bending radius for each trade diameter. The bending radii for nonmetallic underground conduit with conductors is more than twice as great as for rigid nonmetallic conduit.

There is also a new provision that the minimum radius of bend is now measured to the centerline of the conduit, not to the inner edge. The previous edition of the Code measured the minimum radius of bend for various types of conduit and tubing to the inner edge. This results in a significant reduction in radius of bend for the larger sizes of conduit and tubing.

Article 345   Intermediate Metal Conduit

345-16(a): This section specifies the standard length of intermediate metal conduit to be 10 feet. The words **for specific applications or use** were deleted from the second sentence. Lengths of intermediate metal conduit were permitted to be longer or shorter than 10 feet only if for special applications. Longer and shorter lengths are now permitted to be manufactured without threads. As a result of these changes, intermediate metal conduit can now be made available in lengths longer and shorter than 10 feet, with or without couplings, and with or without threads.

345-16(c): Intermediate metal conduit is now required to be marked as **IMC** at intervals not to exceed 5 feet. The previous edition of the Code required IMC to be marked at intervals not to exceed 2 1/2 feet as shown in Figure 2.30. Rigid metal conduit, electrical metallic tubing, and rigid nonmetallic conduit are only required to be marked at intervals of 10 feet. Presently with the IMC marking required to be every 2 1/2 feet, it is likely that any section of IMC will have an identification. With the spacing increased to 5 feet, there will be an increased number of installations where there is no marking on the IMC to distinguish it from rigid metal conduit. In some cases, the number of conductors in IMC as compared to the same size of rigid metal conduit is different.

Table 2.3   Nominal trade sizes of conduit and tubing compared to metric trade size numerical designation numbers.

Nominal trade size (inches)	Numerical designator (millimeters)
3/8	12
1/2	16
3/4	21
1	27
1 1/4	35
1 1/2	41
2	53
2 1/2	63
3	78
3 1/2	91
4	103

Section 345-16(a) New marking interval for IMC

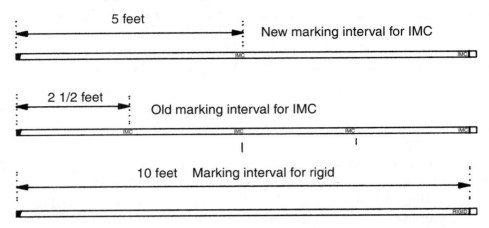

5 feet   New marking interval for IMC

2 1/2 feet   Old marking interval for IMC

10 feet   Marking interval for rigid

Figure 2.30   Intermediate metal conduit is only required to be marked at intervals of 5 feet so that it can be distinguished from rigid metal conduit.

Article 346   Rigid Metal Conduit

346-16(a): This section specifies the standard length of rigid metal conduit to be 10 feet. The words **for specific applications or use** were deleted from the second sentence. Lengths of rigid metal conduit were permitted to be longer or shorter than 10 feet only if for special applications. Longer and shorter lengths are now permitted to be manufactured without threads. As a result of these changes, rigid metal conduit can now be made available in lengths longer and shorter than 10 feet, with or without couplings, and with or without threads.

Article 347   Rigid Nonmetallic Conduit

347-3(d): The restrictions for use due to ambient temperatures have been revised. Unless listed otherwise, rigid nonmetallic conduit may be installed only in areas where the ambient temperatures are limited to 50°C.

Table 347-9(a): This was *Table 10* in *Chapter 9*. This table is used for determining expansion characteristics of rigid nonmetallic conduit. It has been moved forward in the Code for easier accessibility. No change of intent occurred.

Table 347-9(b): This is a new table in the Code that provides temperature expansion and contraction data for fiberglass-reinforced rigid nonmetallic conduit. The thermal expansion and contraction of fiberglass-reinforced rigid nonmetallic conduit is much less than for PVC rigid nonmetallic conduit. Compare the thermal expansion change in length of a 100-foot length of PVC and fiberglass-reinforced rigid nonmetallic conduit for a 110°F change in temperature. The 100-foot PVC changes in length 4.5 inches (*Table 347-9(a)*), while the fiberglass-reinforced conduit only changes in length 2 inches (*Table 347-9(b)*).

Article 348   Electrical Metallic Tubing

348-1: The Code now provides a definition of electrical metallic tubing. It is a listed metallic tubing with a circular cross section that when joined with listed fittings is approved for the installation of conductors.

Article 349 **Flexible Metallic Tubing**

349-2: This section provides the definition of flexible metallic tubing. The product is a listed tubing that has a circular cross section and is flexible, metallic, and liquidtight.

Article 350 **Flexible Metal Conduit**

350-14: The exception that permitted the use flexible metal conduits without the addition of an equipment grounding conductor, in lengths up to 6 feet, was moved to *Section 250-118.*

Article 351 **Liquidtight Flexible Metal and Liquidtight Flexible Nonmetallic Conduit**

351-22: The word **listed** was added to this section that defines liquidtight flexible nonmetallic conduit. The industry designators LFNC-A, -B, or -C were added to the definitions of the different types of liquidtight flexible nonmetallic conduits.

351-23(a): It is now permitted to use liquidtight flexible nonmetallic conduit as a listed manufacture prewired assembly. This provision is limited to 1/2-inch through 1-inch liquidtight flexible nonmetallic conduit that has a smooth inner surface with integral reinforcement in the wall of the conduit, type LFNC-B.

Article 352 **Surface Metal Raceway and Surface Nonmetallic Raceway**

352-41(2): Strut-type channel raceway is now permitted only in dry locations. The previous edition of the Code permitted it to be installed in damp locations.

Article 362 **Metal Wireways and Nonmetallic Wireways**

362-2: When wireway is installed in wet locations, it is now required to be listed for use in wet locations. The previous edition of the Code required wireway in wet locations to be of **raintight construction.**

362-6: A new provision was added to this section that requires raceways or cable entries into a wireway to be spaced apart at least six times the diameter of the largest raceways or cable when a set of wires enters one raceway or cable and leaves the other. This is a similar provision as *Section 370-28(b)* for pull boxes. This requirement for wireways only applies for wires AWG number 4 and larger. This is illustrated in Figure 2.31.

362-25: It is now required to use a **listed fitting** when closing dead ends of nonmetallic wireway.

362-26: This section requires a grounding means to be provided for any type of wiring method connecting to a nonmetallic wireway. The wording in the previous edition of the Code required an equipment grounding conductor to be installed for any type of extension method to a nonmetallic wireway. It was not intended that an equipment grounding conductor be installed when extending from a nonmetallic wireway using a method such as rigid metal conduit, intermediate metal conduit, or electrical metallic tubing that has been properly bonded to make it an adequate equipment grounding conductor. An example is shown in Figure 2.32. The change in wording now makes that clear.

Article 370 **Outlet, Device, Pull and Junction Boxes, Conduit Bodies and Fittings**

370-17(c) Exception: This exception permits nonmetallic sheathed cable to be installed in nonmetallic boxes without clamping the cable to the box if the cable is secured within

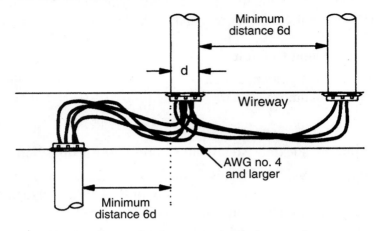

Section 362-6 Use largest diameter conduit or tubing for any run of conductors

Figure 2.31　Minimum spacing distances are required when conductors enter wireways.

8 inches of the box. The largest single-gang box that applies to this exception will have dimensions not exceeding 2 1/4 by 4 inches. The intent always was and still is that this method of installing nonmetallic sheathed cable in nonmetallic boxes applies only in the case of single-gang device boxes. However, now this exception also applies to underground feeder and branch circuit cables, type UF, as well as **nonmetallic-sheathed cables.**

370-23(c): When an enclosure such as a box is mounted in a finished surface, the enclosure is required to be securely fastened in place by clamps, anchors, or fittings. The change in this section is that the clamps, anchors, or fittings are now required to be listed for the purpose.

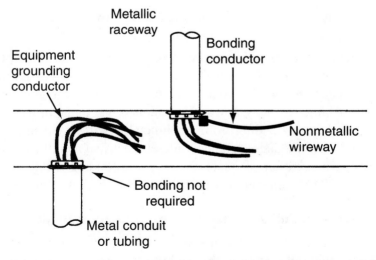

Figure 2.32　A bonding jumper is permitted to be run to a metal conduit or tubing or metal sheath of cable in place of running an equipment grounding conductor.

370-23(d): This is a new section that deals with the securing of enclosures to a suspended ceiling grid or support wires. *Subsection (2)* was taken from *Section 300-11(a)* in the previous edition of the Code. That section did not specify a maximum size of box, cabinet, or fitting. Now this new section refers only to **enclosures** supported by the ceiling support wires. The change is that this practice is limited to enclosures not more than 100 cubic inches in size. The rule is the same as in the previous edition of the Code. The ceiling support wires are permitted to support boxes only if tested as a part of the fire rating of the ceiling or floor assembly, and for a nonfire-rated ceiling assembly, boxes are only permitted to be supported according to the ceiling manufactureriś instructions.

370-23(e): The paragraph was rewritten resulting in a change in the meaning of the section. Enclosures that do not contain devices, equipment, or support fixtures and are supported by entering raceways, are not permitted to have a volume greater than 100 cubic inches.

370-23(e) Exception: The section deals with the support of enclosures by some types of raceway when the enclosure does not contain a device or support a fixture as shown in Figure 2.33. This exception permits RMC, IMC, RNC, and EMT to support conduit bodies. No support spacing is given; therefore, it is assumed to be the same as given in each article for terminations, which in some cases can be up to 5 feet. This exception now applies to a conduit body that has only one conduit entry, such as an E fitting. It also applies to conduit bodies of any volume. The volume of the conduit body is not limited to 100 cubic inches as is the case with other enclosures. That was necessary because for the larger trade sizes of conduit bodies, the volume was greater than 100 cubic inches.

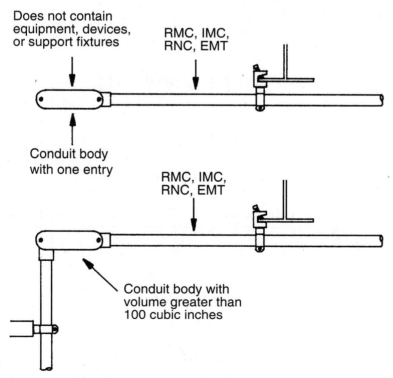

Section 370-23(e) Exception New conduit body support rules

Figure 2.33   Rigid metal conduit, intermediate metal conduit, rigid nonmetallic conduit, and electrical metallic tubing are now permitted to support conduit bodies.

370-23(f) Exception 1: This exception now applies to a conduit body that supports a fixture or contains a device that has only one conduit entry, such as an E fitting. It also applies to conduit bodies of any volume. The volume of the conduit body is not limited to 100 cubic inches as is the case with other enclosures. That was necessary because for the larger trade sizes of conduit bodies, the volume was greater than 100 cubic inches.

370-23(f) Exception 2: In item b. of the exception, it was required that when a box is cantilevered a distance not to exceed 3 feet, the length of unbroken conduit before the last support was required to be not less than 12 inches. Now it states that the conduit is required to be securely supported a distance of not less than 12 inches before the last support beyond which a box is cantilevered. Figure 2.34 illustrates this point.

370-27(a) Exception: This exception deals with the support of a fixture to a box. This was *Section 410-15(a)* in the previous edition of the Code. According to the previous edition of the Code, a fixture weighing not over 6 pounds and having a dimension not over 16 inches was permitted to be supported by the screw shell of a lampholder. A fixture weighing more than 6 pounds and not more than 50 pounds was permitted to be supported by the box, but specifications for support were not provided. Now only a wall-mounted lighting fixture weighing not over 6 pounds and having a dimension not more than 16 inches is permitted to be supported to a box by at least two number 6 or larger screws. It is no longer permitted to support a fixture by a screw shell of a lampholder.

370-27(b) Exception: This exception permits a floor box to be installed in an elevated show window if approved by the authority having jurisdiction, and if the location is judged to be free from physical damage, moisture, and dirt. The change is that now the box and cover are required to be listed for such use in show windows.

Part D: A new part was added to this article giving installation specifications for manholes and other electrical enclosures intended for personnel entry. Generally these installations are under the jurisdiction of a utility, but with the trend toward primary underground installation under private ownership, it was felt necessary to place requirements in the Code

Section 370-23(f) Exception 2   From box to second support must be an unbroken length

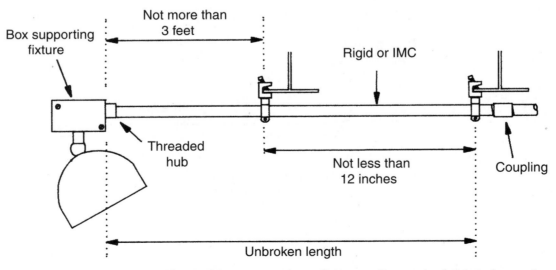

Figure 2.34   From the cantilevered box supporting a fixture to the required point of second support, the rigid metal conduit or intermediate metal conduit shall be an unbroken length.

dealing with manholes. The subject of manholes is not new in the Code. The previous edition of the Code contained two requirements for conductors installed in manholes. Those were *Section 300-32 Exception 2* and *Section 725-26(b) Exception 2.*

370-50: This is a general section stating the concern for installations of manholes under the jurisdiction of the Code. The concern stated is to **prevent damage to conductors** and the insulation during installation or withdrawal of the conductors. It is also concerned about **insuring a safe work space** about electrical energized equipment during examination, adjustment, servicing, and maintenance. It is not stated if other utilities such as steam, gas, water, or waste products are permitted to be transported through these same manholes or enclosures. Later in this section, vaults and tunnels are included even though they are not included in the title of *Part D.*

370-52: This section gives minimum horizontal clearances from conductors. If there are conductors on opposite sides of the vault, the clearance between the conductors is 3 feet. If there are conductors only on one side, the minimum clearance to the opposite wall is 2 1/2 feet. Figure 2.35 is a view of the clearances from above the vault. The minimum headroom is given as 6 feet, although there is an allowance for a lesser headroom depending upon the location of the personnel access opening.

Article 373   Cabinets, Cutout Boxes, and Meter Socket Enclosures

373-5(c) Exception: Multiple **nonmetallic-sheathed** cables are permitted to enter the top of a cabinet or cutout box through a raceway such as rigid metal conduit, intermediate metal conduit, electrical metallic tubing, or rigid nonmetallic conduit. The raceway is required to be at least 18 inches in length but no longer than 10 feet. The raceway is required to extend directly above the enclosure, but is not permitted to penetrate a structural ceiling. A fitting is placed on the end of the raceway to protect cables from abrasion. The cable

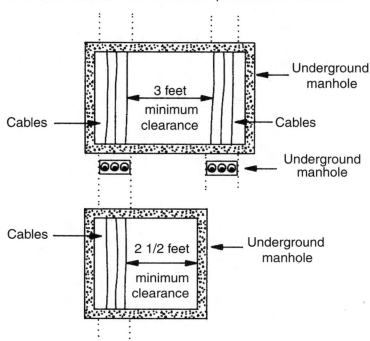

Section 370-52 Minimum clearances for personnel in manholes

Figure 2.35   Minimum clearances for personnel in manholes from cables and conductors are required.

sheath must be continuous through the raceway. The end of the raceway where the cables enter is accessible, and the cables are securely fastened within 12 inches of the point where they enter the raceway measured along the cable. The raceway is required to be securely fastened at the outer end. This is illustrated in Figure 2.36.

The fill requirements of *Table 1, Chapter 9* are required to be met if the raceway is conduit or tubing. The method of dealing with cable fill is found in *Note 9* to *Table 1* of *Chapter 9*. If the conduit or tubing is not more than 24 inches in length, then the cable cross-sectional area is permitted to be up to 60 percent of the conduit cross-sectional area. This practice is limited to nonflexible raceway; therefore, cables would be permitted to enter a cabinet through wireway if the requirements are met.

Using the method of *Note 9* to *Table 1* in *Chapter 9* to determine the cable cross-sectional area, type NM-B nonmetallic-sheathed cable size number 12 with two insulated conductors and a bare equipment grounding conductor has an approximate major diameter of 0.41 inches, which gives a cross-sectional area of 0.13 square inches. Two-conductor number 14 type NM-B nonmetallic-sheathed cable has an approximate cross-sectional area of 0.11 square inches. Type NM-B cable with three insulated conductors and a bare equipment grounding conductor is manufactured with the conductors twisted rather than parallel, and the cable diameter for three-conductor cable is slightly less than for the two-conductor cable of the same size. Table 2.4 gives the maximum number of AWG number 12 type NM-B cables permitted to be installed in a raceway not more than 24 inches in length. If the raceway is longer than 24 inches, the number of cables will be less because the fill is only 40 percent. The ampacity adjustment factors of *Section 310-15(b)(2)* will also apply when the raceway is more than 24 inches long.

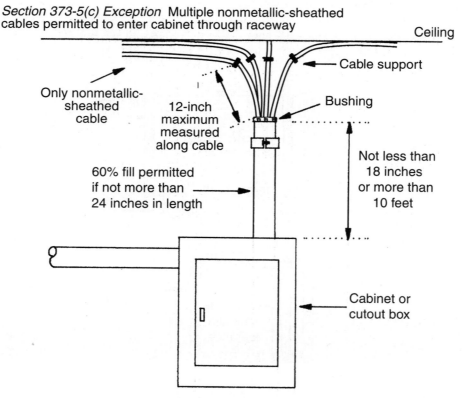

Section 373-5(c) Exception Multiple nonmetallic-sheathed cables permitted to enter cabinet through raceway

Ceiling

Cable support

Only nonmetallic-sheathed cable

12-inch maximum measured along cable

Bushing

60% fill permitted if not more than 24 inches in length

Not less than 18 inches or more than 10 feet

Cabinet or cutout box

Figure 2.36   A conduit or tubing not less than 18 inches in length is permitted to extend from the top of a panelboard to permit the entrance of multiple cables into the panelboard.

Table 2.4 Maximum number of type NM-B number 12-2 with ground nonmetallic-sheathed cables permitted in conduit or tubing *not more* than 24 inches in length.

	1/2-inch	3/4-inch	1-inch	1 1/4-inch	1 1/2-inch	2-inch
RMC	1	2	4	7	9	15
IMC	1	2	4	7	10	16
EMT	1	2	3	6	9	15
RNC-40	1	2	3	6	9	15
RNC-80	1	1	3	5	7	13

Article 374 Auxiliary Gutter

374-7: This section specifies the clearances for bare energized live parts. The term bare parts of **opposite polarity** was changed to **different potential.** A voltage difference does not necessarily mean there is an opposite polarity.

Chapter 9 Tables 1, 4, 5, and 5A

Table 10: This table provided temperature expansion and contraction data for PVC rigid non-metallic conduit and was moved to *Article 347.*

Appendix C Conduit and Tubing Fill Tables for Conductors of the Same Size

There were no significant changes made to the tables in the appendix.

WORKSHEET NO. 2
WIRE, RACEWAY, AND BOX SIZING

These questions are considered important to understanding the application of the *National Electrical Code*® to electrical wiring, and they are questions frequently asked by electricians and electrical inspectors. People working in the electrical trade must continue to study the Code to improve their understanding and ability to apply the Code properly.

DIRECTIONS: Answer the questions and provide the Code reference or references where the necessary information leading to the answer is found. Space is provided for brief notes and calculations. An electronic calculator will be helpful in making calculations. You will keep this worksheet; therefore, you must put the answer and Code reference on the answer sheet as well.

1. A metallic 2-inch by 3-inch device box contains one 3-way switch. The cables entering the box are one type NM size AWG number 14 with two insulated wires and a bare equipment grounding wire, and one type NM size AWG number 14 with three insulated wires and a bare equipment grounding wire, as shown in Figure 2.37. The box contains cable clamps. The minimum depth of device box permitted is _____ inch.

   Code reference _____

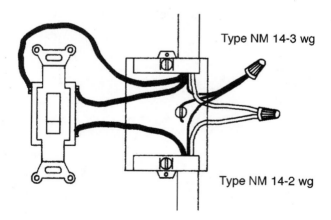

Figure 2.37   A 3-way switch is installed in a metal device box with AWG number 14 cable entering and leaving the box.

2. If a single-gang nonmetallic box without cable clamps is used instead of the metallic box of Figure 2.37, the minimum box cubic inch capacity permitted is _____ cubic inches.

   Code reference _____

3. Are aluminum fittings and enclosures permitted to be used with rigid steel conduit and intermediate metal conduit? (Yes or No)

    Code reference _____

4. A straight horizontal run of 1 1/2-inch intermediate metal conduit (IMC) with threaded couplings is required to be supported within 3 feet of a box or termination, and supported at intervals not exceeding _____ feet, as shown in Figure 2.38.

    Code reference _____

Figure 2.38   Intermediate metal conduit is run exposed on the ceiling of a room.

5. The minimum trade diameter Schedule 40 rigid PVC nonmetallic conduit permitted to contain four AWG number 4/0 copper, type THHW conductors is _____ inch.

    Code reference _____

6. Schedule 40 rigid PVC nonmetallic conduit contains four AWG number 3/0 copper, type THHN wires, six AWG number 10 copper, type THHW wires, and one bare copper AWG number 6 equipment grounding wire. The minimum trade diameter conduit permitted to contain these wires is _____ inch.

    Code reference _____

7. A rigid metal conduit nipple connects two enclosures. New wires are being run for a rewire job. Is it permitted to place four AWG number 1 type THWN wires, where the total cross-sectional area of the wires is 0.625 square inches, in the 1-inch trade diameter rigid metal conduit nipple? (Yes or No)

    Code reference _____

8. If a metal wireway contains not more than _____ wires and the cross-sectional area of the wires does not exceed 20 percent of the cross-sectional area of the wireway, the derating factors of *Section 310-15(b)(2)* do not apply.

    Code reference _____

9. A sheet metal auxiliary gutter is installed as part of a service entrance to make taps to several disconnect switches. The total cross-sectional area of the wires, not considering splices, is not permitted to exceed _____ percent of the cross-sectional area of the auxiliary gutter.

    Code reference _____

10. Splices are permitted in accessible metal surface raceway, provided the metal surface raceway has a removable cover and the cross-sectional area of the splices and conductors do not exceed 75 percent of the raceway area. (True or False)

    Code reference _____

11. Nonmetallic single-gang 2-inch by 3-inch device boxes are used in construction of a new dwelling. The nonmetallic-sheathed cable is not required to be secured to the box provided it is stapled or secured within _____ inches of the box measured along the cable, and at least 1/4 inch of cable sheath extends into the box.

    Code reference _____

12. A metal bushing is used to secure a metal conduit entry tightly to the surface of a cabinet as shown in Figure 2.39. This conduit is for branch circuits not in a mobile home or a recreational vehicle. The circuits operate at less than 250 volts to ground inside a building not exposed to adverse or hazardous environments. Is a locknut required on the inside as well as on the outside? (Yes or No)

    Code reference _____

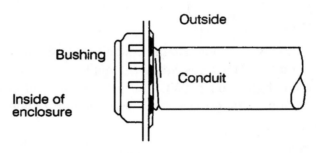

Figure 2.39   A conduit is secured to a cabinet with a locknut on the outside and a metal bushing on the inside.

13. A liquidtight flexible nonmetallic conduit, type B, that is 5 feet in length connects a motor controller to an electric motor on a machine where liquids are present in the area and flexibility of the wiring connection is required. There is no practical means of supporting this flexible nonmetallic conduit except at the terminations. Is this means of support permitted by the Code? (Yes or No)

    Code reference _____

14. Type UF cable emerges from the ground and runs up a pole to a mercury vapor lamp, as shown in Figure 2.40. The cable shall be protected from physical damage from 18 inches below the ground to a point at least _____ feet above the ground.

    Code reference _____

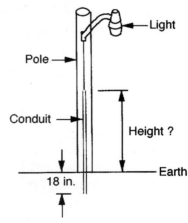

Figure 2.40   Conduit is used to protect type UF cable on a light pole in the yard.

15. Twenty current-carrying circuit wires are contained in a single conduit run. If the wires are AWG number 8, type THHN copper, the maximum allowable ampacity of the wires determined using *Section 310-15(b)* is permitted to be _____ amperes.

    Code reference _____

16. An office building has lighting circuits supplied from a 120/208-volt 3-phase, 4-wire electrical system. Assume the incandescent lighting in a room is supplied using 20-ampere rated multiwire branch circuits. The maximum number of circuits that can be run in the same EMT and not exceed the 0.7 derating factor of *Table 310-15(b)(2)* is _____.

    Code reference _____

17. Electrical nonmetallic tubing is permitted to be installed as surface wiring in dwelling and office building areas where not subjected to physical abuse and where the building is not more than three floors in height. (True or False)

    Code reference _____

18. The typical temperature at the top of a boiler room is 140°F. Four type THHN copper wires run in the same conduit supply two lighting circuits and run for 80 feet in the boiler room. Each circuit is protected with a 20-ampere circuit-breaker and carries a load of 14 amperes. The circuit-breaker with a maximum rating of 75°C is 40 feet from the boiler room. The minimum permitted size of wire for these circuits is AWG number

    _____.

    Code reference _____

19. A horizontal run of 2-inch trade diameter intermediate metal conduit is supported within 3 feet of boxes and terminations, and all couplings are of the threadless type. The maximum permitted spacing between supports for this horizontal run is _____ feet.

    Code reference _____

20. A 2-inch EMT enters the top of an auxiliary gutter, and out the bottom of the gutter is a 2-inch rigid metal conduit connecting to a panelboard. Four AWG number 1 copper, type THHN conductors enter the auxiliary gutter from the EMT and pass directly to the rigid metal conduit as shown in Figure 2.41. The minimum permitted offset distance between the two raceways containing the same conductors is _____ inches.

    Code reference _____

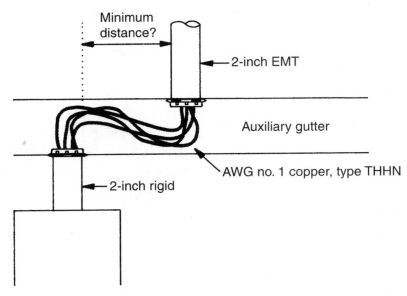

Figure 2.41   Determine the minimum permitted offset for the raceways when the AWG number 1 conductors pass directly from one raceway to the other.

21. A pull box is installed with a 4-inch conduit entering one end and leaving the opposite end. Wires in the conduit are larger than AWG number 6. The minimum permitted length of the pull box is _____ inches.

    Code reference _____

22. Fluorescent lay-in lighting fixtures are mounted in a suspended ceiling. Flexible metal conduit is run from the fixture to a junction box, which is mounted directly to the permanent ceiling of the building. This is a field-installed flexible metal conduit, and it is not a manufactured wiring system. Is flexible metal conduit permitted to be installed between the junction box and the fixture as a 5-foot length supported only at the termination to the box and at the fixture? (Yes or No)

    Code reference _____

23. Electrical metallic tubing (EMT) of the 1-inch trade diameter is to be run perpendicular to roof trusses of a building as illustrated in Figure 2.42. The roof trusses are spaced 5 feet on center, and a lighting fixture is secured to the underside of every second truss. Is the EMT permitted to be supported only to the underside of the truss between the light fixtures plus being supported to the lighting fixture boxes with a connector? (Yes or No)

    Code reference _____

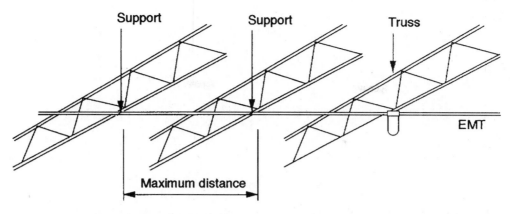

Figure 2.42   Electrical metallic tubing of the 1-inch trade size is attached to the underside of ceiling trusses at the boxes and at the truss between the lighting fixtures.

24. A 3-phase feeder consisting of two parallel sets of wires must be capable of carrying a calculated load of 560 amperes. The overcurrent device for the feeder is 700 amperes. If the 6-phase wires are copper, type THHW, all installed in the same rigid metal conduit, the minimum size phase wires permitted is _____.

     Code reference _____

25. A pull box has a 4-inch conduit entering one side and a 2-inch and a 3-inch conduit entering the adjacent side, as shown in Figure 2.43. The wires making the angle pull from the 4-inch to the 3-inch conduit are size AWG number 4/0. Other wires are AWG number 6 or smaller. Determine the minimum dimensions of the pull box.

     Code reference _____

     Dimension A _____ inches   Dimension B _____ inches

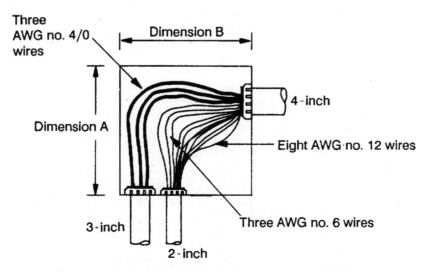

Figure 2.43   A pull box is installed for the purpose of making angle pulls.

ANSWER SHEET   Name _____

No. 2   WIRE, RACEWAY, AND BOX SIZING

Answer	**Code reference**
1. _____	_____
2. _____	_____
3. _____	_____
4. _____	_____
5. _____	_____
6. _____	_____
7. _____	_____
8. _____	_____
9. _____	_____
10. _____	_____
11. _____	_____
12. _____	_____
13. _____	_____
14. _____	_____
15. _____	_____
16. _____	_____
17. _____	_____
18. _____	_____
19. _____	_____
20. _____	_____
21. _____	_____
22. _____	_____
23. _____	_____
24. _____	_____
25. _____	_____

UNIT 3

Outlets, Lighting, Appliances, and Heating

OBJECTIVES

Upon completion of this unit, the student will be able to:

- determine the location of electrical outlets in a dwelling.
- determine the minimum number of general lighting, small appliance, and laundry circuits permitted in a dwelling.
- state the clearance requirements for outside aerial feeders and branch circuits.
- determine the minimum size flexible cord required for an application.
- state the installation requirements for lighting outlets in clothes closets.
- state the installation requirements for recessed lighting fixtures.
- determine the minimum size conductor for a storage-type electric water heater.
- determine the maximum number of baseboard electric heaters permitted to be installed on a 15- or 20-ampere branch circuit.
- state if it is permitted to power a room air conditioner from an existing general lighting branch circuit of a dwelling.
- state the cable type required to be installed for a push button of a door chime.
- answer wiring installation questions relating to *Articles 210, 220, Part A, 225, 320, 321, 380, 400, 402, 410, 411, 422, 424, 426, 720,* and *725.*
- state at least five significant changes that occurred from the 1996 to the 1999 Code for *Articles 210, 220, Part A, 225, 320, 321, 380, 400, 402, 410, 411, 422, 424, 426, 720,* or *725.*

CODE DISCUSSION

The emphasis of this unit is on circuits and installation of outlets, appliances, and equipment. Methods are covered for determining the minimum number of general illumination, receptacle, and other circuits required for various types of buildings. Outside feeders and branch circuit installations are also covered. Special circuits such as low-voltage, power-limited-control, and signaling circuit installation are covered in this unit.

Article 210, Part A contains specifications about branch circuits such as circuit ratings (*Section 210-3*) and voltage limitations for different types of occupancies and types of loads to be served (*Section 210-6*). For example, in a dwelling, the maximum permitted voltage between conductors for lighting fixtures and receptacles is 120 volts nominal. Multiwire branch circuits are discussed in *Section 210-4*. A multiwire branch circuit is actually two or more circuits that share a common neutral conductor. In the case of a single-phase, 120/240-volt system, three wires are used to supply two circuits with a common shared neutral and 240 volts between the ungrounded conductors. In the case of a 3-phase 120/208-volt system, three circuits can be supplied with only four conductors. Multiwire branch circuits are illustrated in Figure 3.1. Generally it is assumed that when the loads on the multiwire circuits are the same and all are operating at the same time, the current on the common neutral conductor will be zero or nearly zero. This is not necessarily the case with a multiwire circuit derived from a 3-phase wye electrical system. This is pointed out in the fine print note in *Section 210-4(a)*. In the case of dwellings, multiwire branch circuits are required to be supplied by circuit-breakers with a common trip (usually a 2-pole circuit-breaker) so that when one of the circuits trips off, the other will also trip off.

Ground-fault circuit-interrupter requirements for receptacles are covered in *Section 210-8*. There are two general requirements that apply to any type of facility. All 15- and 20-ampere, 125-volt receptacles in bathrooms and on rooftops of any type of facility are required to be ground-fault circuit-interrupter protected. This can be accomplished with a ground-fault detecting receptacle, or a ground-fault detecting circuit-breaker. Both the ungrounded conductor and the neutral pass through a current-sensing coil. If the current measured in those two wires is different by more than 0.006 amperes (6 milliamperes), the interrupter will trip. This means the current is finding a return path other than the neutral wire, which possibly could be a person. The dwelling ground-fault circuit-interrupter requirements for protection of 125-volt, 15- and 20-ampere receptacles are summarized below:

- Receptacles serving *kitchen counter top* surfaces
- *Bathroom* receptacles
- *Outside* receptacles (see exception for snow-melting equipment)
- *Crawlspace* receptacles
- *Garage* receptacles (see exceptions for dedicated equipment receptacles)
- Receptacles in *accessory buildings* at or below grade used for storage or work areas
- Receptacles in *unfinished basement* or unfinished portion of a basement (see exceptions)
- Receptacles serving counter top and within 6 feet of a *wet bar*

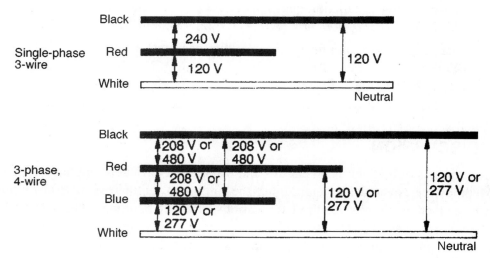

Figure 3.1 A common neutral wire is run with two or more ungrounded wires in multiwire circuits.

The minimum number of branch circuits required for a building are specified in *Section 210-11*. Calculating the minimum number of branch circuits is covered later under *Sample Calculation*. Four 20-ampere branch circuits are required as a minimum in a dwelling. Two 20-ampere branch circuits are required to supply the receptacles in the kitchen, dining room, and similar rooms. One 20-ampere branch circuit is required to supply laundry equipment such as an electric washer and possibly a gas dryer. Another 20-ampere branch circuit is required to supply the receptacle outlets in the bathroom or bathrooms of the dwelling. The minimum number of remaining general purpose 15- and 20-ampere circuits is calculated according to the method of *Section 210-11(a)*.

Article 210, Part B covers the ratings of branch circuits, including those in dwellings. *Section 210-19(a)* states that the branch circuit conductors shall have an ampacity of not less than the maximum load to be served. The rating of a branch circuit is the rating of the overcurrent device protecting the circuit. The actual rating of any one branch circuit is determined by the method in *Section 210-19(a)*. The overcurrent device protecting a circuit is required to have a rating not less than 1.25 times any continuous load served plus 1.0 times any noncontinuous load served. An example will help to illustrate the method.

Example 3.1   A circuit supplies 11.2 amperes of lighting load and four receptacle outlets, each assumed to have a load of 1.5 amperes. The receptacles in this case are not considered to be continuous loads, but the lighting is considered to be a continuous load. Can this load be served by a branch circuit rated at 20 amperes?

Answer: The continuous load is 11.2 amperes, and the noncontinuous load is 6 amperes (1.5 amperes × 4 = 6 amperes). The minimum permitted rating of the circuit is determined as follows:

Continuous load	11.2 A × 1.25 = 14 A
Noncontinuous load	6 A × 1.00 = 6 A
Minimum circuit rating	= 20 A

The load is permitted to be served with a 20-ampere rated circuit, but the circuit is fully loaded and there is no room for expansion in the future.

The minimum circuit rating of 40 amperes for a dwelling electric range is given in *Section 210-19(c)*. The neutral conductor of the range circuit is permitted to be of a smaller size than the ungrounded conductors, but it is not permitted to have an ampacity less than 70 percent of the branch circuit rating, and it shall be no smaller than AWG number 10 (*Section 210-19(c) Exception 2*).

A tap is a conductor that connects to a branch circuit or feeder conductor to serve a specific load. The tap conductor shall be of sufficient ampacity to supply the load. The tap conductor is permitted to be smaller than the branch circuit or feeder conductor, but the Code sets a minimum size based on the rating of the branch circuit or feeder. Additional discussion of taps is in *Units 6, 7,* and *8. Exception 1* to *Section 210-19(c)* permits a tap with sufficient ampacity to serve the load but not less than 20 amperes for an electric wall-mounted oven or counter-mounted cooking unit in a dwelling when the circuit is protected at not more than 50 amperes. This is illustrated in Figure 3.2. There is another tap rule in this article. *Section 210-19(d) Exception 1* permits taps to branch circuits provided the tap conductor is to individual lampholders or fixtures and is not more than 18 inches in length. For a branch circuit rated up to 30 amperes, the tap shall have an ampere rating sufficient for the load, and not less

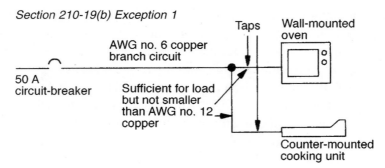

Figure 3.2   Taps in a dwelling range circuit shall have sufficient ampacity for the load, and shall not be less than 20 amperes.

than 15 amperes. For a branch circuit rated up to 50 amperes, the tap shall have an ampere rating sufficient for the load, and not less than 20 amperes.

Fine print note number 4 to *Section 210-19(a)* recommends a maximum of 3 percent voltage drop be permitted on branch circuits. *Section 210-21(b)(1)* states that when a single receptacle is installed on a branch circuit, the receptacle is not permitted to have a rating less than that of the branch circuit rating. A single receptacle is defined in *Article 100.* It has provisions for the connection of only one cord-connected device. A duplex receptacle outlet is considered two receptacles. Therefore, if a single receptacle on a yoke is installed on a 20-ampere circuit, the minimum receptacle rating permitted is 20 amperes. If a duplex receptacle outlet is installed on a 20-ampere branch circuit, the receptacle rating is permitted to be either 15 or 20 amperes. *Section 210-23* covers the permissible loads on branch circuits of various ratings. For example, *Section 210-23(a)* permits 15- and 20-ampere branch circuits to supply typical residential and commercial lighting outlets. It is necessary to read the remainder of this section to determine the restrictions on the type of outlets and equipment permitted to be supplied from branch circuits of various ratings.

The minimum requirements for providing outlets for lighting and receptacles are given in *Article 210, Part C.* The maximum permitted spacing of receptacle outlets in dwellings is given in *Section 210-52.* A receptacle outlet shall be installed in listed rooms of a dwelling such that any point measured along the wall is not greater than 6 feet from the outlet. Wall spaces 2 feet wide or wider shall have a receptacle outlet. These rules are illustrated in Figure 3.3. Receptacle outlets of the small appliance branch circuits serving the kitchen counter of a dwelling are required to be spaced such that no point along the wall line is more than 24 inches from a receptacle outlet. This requirement is in *Section 210-52(c)(1). Section 210-70(a)*

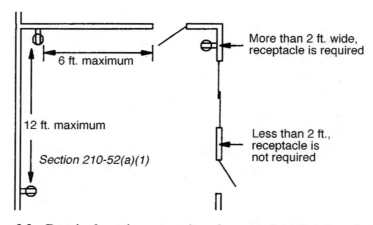

Figure 3.3   Required maximum spacing of receptacle outlets in a dwelling.

requires that every habitable room of a dwelling, including other listed areas, shall have a lighting outlet that is wall-switch-controlled. A switched receptacle outlet may be used in place of an actual lighting fixture in some rooms, as shown in Figure 3.4.

Article 220, Part A provides requirements for the determination of the number of outlets of various types permitted on a branch circuit. *Section 220-3* provides important information for the determination of the number of branch circuits required for receptacles and general lighting in buildings. *Table 220-3(a)* gives the minimum volt-amperes per square foot of building area considered in determining the minimum number of branch circuits for lighting and for determining the minimum lighting load for feeder calculation purposes. If the actual lighting load is known to be of a greater value, that load shall be considered the general lighting load.

Section 220-3(b)(9) permits the general use receptacles in a dwelling to be included in the general illumination calculation. Small appliance receptacle loads and laundry receptacle load are specified elsewhere in the Code and are not a part of this general illumination load calculation. This means that the 3 volt-amperes per square foot from *Table 220-3(a)* is used for a dwelling to determine the minimum number of general illumination branch circuits that will supply the receptacle outlets and lighting fixtures. The significance of this is that the 180-volt-amperes requirement of *Section 220-3(b)(9)* does not apply to dwelling general-use receptacles. The number of outlets permitted on a circuit depends upon the loads to be served, and according to *Section 220-4,* the total load is not permitted to exceed the rating of the circuit. In the case of general illumination in a dwelling, the load is determined based on the method described in *Section 220-3(a),* which is the area of the dwelling times 3 volt-amperes per square foot. This is converted into the number of circuits according to the method of *Section 210-11* where the total load in volt-amperes is divided by 120 volts and then divided into 15- or 20-ampere circuits.

Article 225 gives the requirements for the installation of branch circuit and feeder conductors outside. Minimum size of conductors for overhead spans, protection of conductors, and overhead conductor clearances is specified in this article. Clearances above ground of aerial conductors are covered in *Section 225-18.* Similar requirements for service conductors are covered in *Section 230-24.* The conductor clearance over areas accessible only to pedestrians is a minimum of 10 feet. For a residential driveway and other driveways not subject to truck traffic, the minimum clearance is 12 feet, provided the conductors do not exceed 300 volts to ground. This clearance is increased to 15 feet for conductors operating at more than 300 volts to ground. An 18-foot minimum conductor clearance is required when the driveway or road is subject to truck traffic. Branch circuit and feeder conductor clearances above roofs

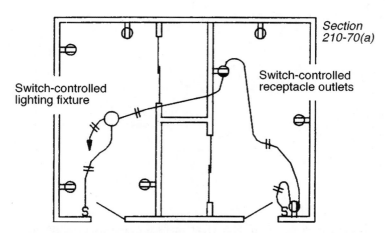

Figure 3.4   Wall-switch-controlled lighting is required in every habitable room of a dwelling.

are covered in *Section 225-19*. The minimum overhead conductor clearances for roofs accessible to only pedestrians are summarized in Figure 3.5.

Article 320 covers the situation in which conductors are run within a building or on the outside of a building where the individual conductors are supported by open wires on insulating devices. The wiring covered by this article is required to be exposed except where passing through a wall or floor.

Article 321 deals with messenger-supported wiring, such as aerial triplex or quadruplex cable where a bare conductor supports the insulated conductors. It is permitted to field construct a support messenger and suitably attach the conductors to the messenger. There are no minimum conductor size requirements for overhead spans using messenger-supported wiring, according to *Section 225-6(a)*. The ampacity of messenger-suported conductors is to be determined using the methods of *Section 310-15,* which basically means the ampacity is determined in most cases using *Table 310-16* or *Table 310-20.*

Article 380 covers the installation, rating, and use of switches of various types, including knife switches and circuit-breakers used as switches. All switches and circuit-breakers used as switches are not permitted to be installed such that the center of the handle is more than 6 feet 7 inches above the floor when in the "on" position, as stated in *Section 380-8(a)*. *Section 240-83(d)* specifies that when a circuit-breaker is used as a switch for 120- and 277-volt fluorescent lighting, the circuit-breaker shall be marked "SWD."

Article 400 provides information and requirements on the use of flexible cords and cables. *Table 400-4* gives information about the various types of cords and cables and states the uses permitted. *Table 400-5* gives the allowable ampacity and ampacity derating factors applicable when more than three current-carrying conductors are in the cable of flexible cord. Refer to Unit 2 for examples of how to determine the minimum size wire permitted when more than three current-carrying conductors are in flexible cord or cable. Markings on the flexible cord or cable with uses permitted and installation requirements are covered in this article.

Article 402 lists the markings on fixture wires, the types available and their uses, and general installation requirements. *Table 402-5* gives the permitted ampacity of the fixture wires of sizes AWG numbers 18 to 10.

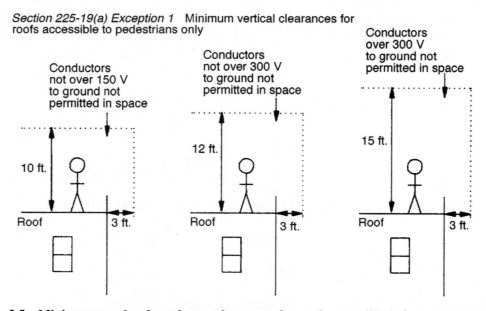

Figure 3.5   Minimum overhead conductor clearances for roofs accessible only to pedestrians.

Article 410 gives requirements on the installation, location, grounding, support, and wiring of lighting fixtures and associated auxiliary equipment. *Section 410-8* gives a definition of storage space in clothes closets. This is important because a lighting fixture is required to be installed such that a minimum clearance is maintained between the fixture and the storage space. *Figure 410-8* in the Code gives the dimensions of the storage space in a clothes closet. *Section 410-8(c)* states that incandescent lighting fixtures with exposed or partially exposed lamps are not permitted to be installed in a clothes closet. This means porcelain lamp receptacles with bare lamps are not permitted to be installed in clothes closets.

Fire is a danger if excessive heat is produced by improper installation or use of lighting fixtures. Incandescent fixtures require a higher wattage to obtain the same amount of light as electric discharge fixtures. Therefore, the heat produced by incandescent lamps is usually greater than for electric discharge fixtures. Electric discharge fixtures usually have a ballast that is a source of heat in addition to the lamp. Ballasts for fluorescent fixtures are required by *Section 410-73(e)* to be thermally protected, and ballasts for recessed high-intensity discharge fixtures are required by *Section 410-73(f)* to be thermally protected.

Recessed lighting fixtures, if not installed properly, can create a fire hazard. *Part M of Article 410* provides installation requirements for recessed fixtures. For most applications, recessed incandescent fixtures are required to be thermally protected. Overheating of the fixture will interrupt electrical power to the lamps. The lamps generally will light again when the fixture cools.

The installation of receptacles is covered in *Part L* of *Article 410. Section 410-56(b)* deals with the situation in which aluminum conductors are attached directly to a receptacle outlet. The receptacle outlet is required to be marked **CO/ALR** if it is suitable for use with aluminum terminations. Do not confuse this marking with the usual marking of cu/al, which is frequently used for other types of terminations suitable for both copper and aluminum conductors. *Section 410-56(c)* covers receptacles with the equipment grounding terminal isolated from the yoke. These receptacle types are permitted to supply electronic equipment where electrical noise may be a problem. These receptacles are identified by an orange triangle. Permitted means of identification of receptacles with isolated grounds are illustrated in Figure 3.6.

Article 411 provides specifications for lighting systems that operate at 30 volts or less. One example of such a system is low-voltage landscape lighting for gardens, walkways, decks, patios, and other building accent illumination. Lighting systems are required to be listed for the purpose. The uses not permitted are stated in *Section 411-4*. The low-voltage secondary circuit is required to be insulated from the supply branch circuit by an isolating transformer. Each secondary lighting circuit is not permitted to operate at more than 25 amperes. The lighting system isolating transformer is not permitted to be supplied from a branch circuit with a rating more than 20 amperes.

Article 422 provides information and requirements for electrical appliances in any type of occupancy. Branch circuit requirements, control of appliances, and disconnects are covered. Storage-type electric water heater wiring is covered in *Section 422-13*. When the capacity of the electric storage-type water heater is not more than 120 gallons, the branch circuit conductor shall be sized at not less than 1.25 times the nameplate rating of the water heater. If the water heater rating is given in watts, then a calculation must be done to determine the ampere rating of the water heater. An electric water heater with a rating of 3,500 watts at 240 volts will have an ampere rating of 14.6 amperes. This is determined by dividing 3,500 watts by 240 volts. The minimum branch circuit conductor rating permitted is 1.25 times the 14.6 amperes or 18.3 amperes. The minimum circuit conductor size would be AWG number 12 copper protected with a 20-ampere overcurrent device.

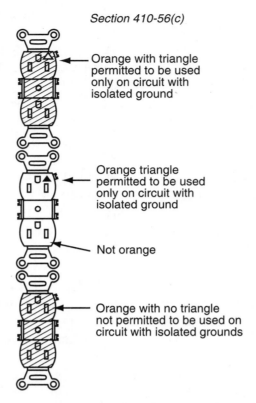

Section 410-56(c)

Orange with triangle permitted to be used only on circuit with isolated ground

Orange triangle permitted to be used only on circuit with isolated ground

Not orange

Orange with no triangle not permitted to be used on circuit with isolated grounds

Figure 3.6   Receptacles installed on a circuit with the equipment grounding terminals isolated from the other circuit equipment grounds, as permitted in *Section 250-146(d),* shall be identified with an orange triangle on the face of the receptacle.

Article 424 on fixed electric space heating equipment is covered including space heating cables. Branch circuit requirements, control, disconnection, grounding, location, and wiring are covered. The first part of the article gives general wiring requirements, while the later parts of the article relate to specific types of electric heating equipment or installations.

Article 426 deals with the installation and use of outdoor electric de-icing and snow-melting equipment, such as heating cable embedded in concrete. Definitions of different methods of using electricity for the production of heat are covered in *Section 426-2.* The minimum size of conductor permitted for branch circuits for electric outdoor snow-melting and de-icing equipment is covered in *Section 426-4* and shall be not less than 1.25 times the total load on the circuit.

Article 720 deals with electrical circuit installation, either alternating current or direct current, that operates at less than 50 volts. Minimum wire size and receptacle rating are covered. Some installation types are listed that are covered elsewhere in the Code. It is important to note that even though the voltage is low, it is electrical current flow that causes heating and can result in fire. Note the *Exception* to *Section 720-1* which states that there are specific requirements for installations operating at less than 50 volts in other articles.

Article 725 covers the installation of remote-control, signaling, and power-limited circuits that are not an integral part of a device or appliance. These circuit types are divided into Class 1, Class 2, and Class 3 circuits. Line voltage control circuits for motors and other equipment are Class 1 circuits. Thermostat circuits operating at 24 volts for furnace control, and door chime and similar circuits are considered Class 2 because they have supply transformers limiting the maximum current that can flow if the wires become shorted. Wiring methods for

Class 1 circuits are covered in *Section 725-25*. The wiring methods of *Chapter 3* of the Code shall be permitted. If the conductors are of sizes AWG numbers 18 and 16, the insulation type is specified in *Section 725-27(b)*. Wiring methods and materials for Class 2 and Class 3 wiring are specified in *Section 725-52*.

SAMPLE CALCULATIONS

Information found in *Articles 210* and *220* of the Code is used to determine the minimum number of circuits required in a dwelling or other building. The following are examples of how these calculations are used to determine the number of circuits required. The same information is used to determine the maximum number of outlets permitted on a particular circuit. An important issue is whether the loads are considered continuous loads. In many situations, it is a matter of judgment to the use of the loads on a circuit. Generally, lighting in other than dwellings is considered a continuous load. In the case of receptacles outlets, it would depend on the particular application. If the load is considered to be continuous, then the overcurrent device protecting the circuit is not permitted to be loaded to more than 80 percent of its rating. The exception is when the overcurrent device and the enclosure into which it is installed are rated for operation at 100 percent.

Circuit and Outlet Requirements

The minimum number of circuits for general illumination is based on the actual load when the load is known; however, the minimum permitted demand load for general illumination is specified in Code *Table 220-3(a)*. The demand load for receptacle outlets shall be considered a load of 180 volt-amperes (VA) per strap or yoke at an outlet. In the case of a dwelling, the receptacle outlets are considered loads for general illumination, and the load is determined based on the 3 volt-amperes per square foot from *Table 220-3(a)* and not from the 180 volt-amperes per outlet as stated in *Section 220-3(b)(9)*. This information on loads has three basic purposes: (1) it is used to determine the minimum number of branch circuits in a building or an area of a building; (2) it is used to determine the minimum number of outlets on a particular branch circuit; and (3) it is used to determine the demand load of a building or an area of a building for the purpose of sizing a feeder or electrical service or distribution panel. This latter function of feeders and panels will be the subject of Unit 4.

The minimum number of circuits for general illumination for a dwelling, which includes receptacle outlets and lighting outlets, is determined by multiplying the area of the building by 3 volt-amperes per square foot, as found in *Section 220-3* and *Section 210-11(a)*.

Example 3.2   A dwelling has 2,100 square feet of living area. Determine the minimum number of circuits required in this dwelling for general illumination.

Answer: The area does not include the unfinished basement, which is not considered in this calculation. The total volt-amperes determined is divided by 120 volts to find the total ampere load for general illumination. This ampere load is then divided into 15- and/or 20-ampere circuits. Four 15-ampere circuits or three 20-ampere circuits would be the minimum circuit requirement for general illumination loads in the dwelling of the example.

$$2{,}100 \text{ sq. ft.} \times 3 \text{ VA/sq. ft.} = 6{,}300 \text{ VA}$$

$$\frac{6{,}300 \text{ VA}}{120 \text{ V}} = 52.5 \text{ A} \quad \text{round up to 53 A}$$

$$\frac{53 \text{ A}}{15 \text{ A/circuits}} = 3.5 \text{ circuits}$$

There is seldom justification to consider dwelling loads for general illumination to be continuous loads. There are both lighting loads and receptacle outlets on the circuits, and generally this type of load combination will not apply a heavy load to the circuit on a continuous basis.

Buildings other than dwellings have the general illumination load generally limited to fixed lighting. If the actual ampere rating of the lighting fixtures is known, then this load is required to be used if it is greater than the load determined using the volt-ampere per square foot from *Table 220-3(a)*. To determine the actual number of lighting fixtures on the circuit, these lighting loads usually are considered to be continuous loads. Continuous load means the circuit is not permitted to be loaded more than 80 percent of the circuit rating. For a 20-ampere lighting circuit, the maximum permitted continuous load on the circuit would be 16 amperes.

Example 3.3　Electric discharge lighting fixtures to be installed in a building each are rated at 1.9 amperes at 120 volts. Determine the maximum number of these lighting fixtures permitted to be installed on a 20-ampere circuit in a commercial building.

Answer:　The 20-ampere circuit is only permitted to be loaded to 80 percent of the circuit rating, which is 16 amperes. Divide the 16 amperes by the rated current of each fixture to determine the maximum number of fixtures permitted to be installed on the circuit. It is necessary to round a fraction down to the next integer or the circuit will carry more than 16 amperes. The maximum number of fixtures permitted to be installed on the circuit is eight.

$$\frac{16 \text{ A per circuit}}{1.9 \text{ A per fixture}} = 8.4 \text{ fixtures}$$

When to consider loads to be continuous is frequently a matter of judgment. In a commercial building, for example, receptacle loads may be operated with a great amount of diversity. Therefore, the times will be infrequent when the circuit would be operated near the circuit rating, especially for three hours or longer. In this case, the receptacle circuit is not to be considered a continuous load. In another case, the receptacle circuit could be considered a continuous load.

In *Appendix D* of the Code, several examples illustrate how to determine the minimum number of branch circuits for different building types. *Example D1(a)* is an example for a single-family dwelling, and *Example D4(a)* shows how to determine the minimum number of branch circuits for each dwelling unit of a multifamily dwelling. *Example D3* shows how to determine the minimum number of branch circuits for a store building.

Electric Range for a Dwelling

An electric range for a dwelling seldom operates at full nameplate rating, and when it does, the load is on only for a short time. Damage to conductors requires heat-producing current over a time period. If the time period is known to be limited, the wires will not be damaged. The range circuit rating is, therefore, based on a range demand factor. The minimum rating circuit permitted is 40 amperes for ranges rated not less than 8 3/4 kilowatts, as specified in *Section 210-19(c)*. The range demand load is found in *Table 220-19*. Column A of *Table 220-19* gives the demand load of an electric range or ranges that have a rating of not less than 8 3/4 kilowatts nor more than 12 kilowatts. For example, one 12-kilowatt electric range can be taken at a demand load of 8 kilowatts.

Example 3.4   A 10-kilowatt electric range in a dwelling operates from a 120/240-volt circuit. Determine the minimum ampere rating of the circuit permitted to supply this electric range.

Answer: The demand load for a 10-kilowatt electric range is considered to be 8 kilowatts, from *Table 220-19*, column A. Divide this demand load by the circuit voltage to determine the minimum permitted circuit current rating, which in this case is 33.3 amperes. The minimum permitted circuit rating is 40 amperes.

$$\frac{8\ kW \times 1,000}{240\ V} = 33.3\ A$$

Remember that the minimum is 40 amperes; therefore, a 40-ampere circuit using AWG number 8 copper wire is required for the range circuit. The wiring method is permitted to be wire in conduit, type SE cable, type NM cable, or other suitable cable. For new branch circuit installations, the grounded circuit conductor and the equipment grounding conductor are required to be separate conductors according to *Section 250-140* and *Section 250-138(a)*. *Table 310-16* is used to determine the minimum size wire for the range circuit. When an electric range has a nameplate rating higher than 12 kilowatts, the notes to *Table 220-19* describe the method to determine the minimum permitted demand load in the case of dwellings. Example 3.5 illustrates how the method works for one range in a dwelling with a rating of more than 12 kilowatts.

Example 3.5   A 17-kilowatt electric range is installed in a dwelling. Determine the minimum circuit rating permitted for this electric range.

Answer: The demand load of this range is determined by the method described in *Note 1* to *Table 220-19*. The first 12 kilowatts is taken as a demand of 8 kilowatts. For each 1 kilowatt the actual size exceeds 12 kilowatts, the 8-kilowatt demand is increased by 5 percent. For this range, a 45- or 50-ampere circuit would be installed. *Example D6* in Appendix D of the Code shows how to make a range calculation.

First 12 kW	=	8 kW
Remainder above 12 kW		
17 kW – 12 kW = 5 kW		
Increase the 8 kW by 5% for each		
of these remainder kWs		
0.05 × 8 kW × 5	=	2 kW
Demand	=	10 kW

$$\frac{10\ kW \times 1,000}{240\ V} = 41.7\ A$$

An oven and cooking unit can be separate units and built into the kitchen, as shown in Figure 3.7. The range conductor is permitted to extend from the supply panel to a junction box, and then taps connect to the counter-mounted cooking unit and to the wall-mounted oven. The rules for determining the minimum size of conductors for a wall-mounted oven and a counter-mounted cooking unit supplied from the same branch circuit are found in *Note 4* to *Table 220-19*. The minimum conductor size for the counter-mounted cooking unit and the wall-mounted oven is determined based on the nameplate rating of the unit.

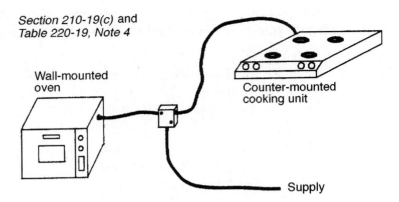

Section 210-19(c) and
Table 220-19, Note 4

Wall-mounted
oven

Counter-mounted
cooking unit

Supply

Figure 3.7   A wall-mounted oven and a counter-mounted cooking unit are installed on a single branch circuit.

Example 3.6   A dwelling electric range consists of a wall-mounted oven with a nameplate rating of 5 kilowatts and a counter-mounted cooking unit with a rating of 6.6 kilowatts. Determine the minimum size copper wires permitted for this branch circuit.

Answer: First, determine the minimum size of conductor permitted for the tap to the wall-mounted oven and the counter-mounted cooking unit. This tap rating is based on the nameplate rating of the unit. If the copper conductors were selected from the 60°C column of *Table 310-16*, then these tap conductors shall be permitted to be AWG number 10.

Wall-mounted oven:
$$\frac{5{,}000 \text{ W}}{240 \text{ V}} = 21 \text{ A}$$

Counter-mounted cooking unit:
$$\frac{6{,}600 \text{ W}}{240 \text{ V}} = 28 \text{ A}$$

Note 4 to *Table 220-19* permits the branch circuit conductor supplying both the counter-mounted cooking unit and the wall-mounted oven to be sized at the minimum permitted by adding the nameplate ratings of the separate units and treating it as if it were a single range. In this case, the total range nameplate rating would be 11.6 kilowatts, and according to column A of *Table 220-19*, the minimum branch circuit rating is permitted to be determined using an 8-kilowatt demand. The minimum permitted copper wire based on the 60°C column of *Table 310-16* would be AWG number 8, but the minimum size wire is not permitted to be smaller than specified in *Section 210-19(c)*.

$$\frac{8{,}000 \text{ kW}}{240 \text{ V}} = 33 \text{ A}$$

MAJOR CHANGES TO THE 1999 CODE

These are the changes to the 1999 *National Electrical Code*® that correspond to the Code sections studied in this unit. The following analysis explains the significance of only the changes from the 1996 to the 1999 Code, and this analysis is not intended to be used in place of the Code. Refer to the actual section of the 1999 Code for the exact wording and meaning of each section discussed. Changes are indicated in the Code with a vertical line in the margin. If material has been deleted or moved to another part of the Code, the location of the deletion is indicated with a dark dot in the margin.

Article 210   **Branch Circuits**

210-8(a)(2): The change deals with ground-fault circuit-interrupter protection for receptacles in accessory buildings associated with a dwelling. If the building is used for storage or as a work area, the receptacles are required to be ground-fault circuit-interrupter protected. There is no longer the requirement that the building be unfinished and at grade level, as illustrated in Figure 3.8. This requirement also applies to accessory buildings that are below grade level.

210-8(b) Exception: This is a new exception that applies to buildings other than dwellings. Ground-fault circuit-interrupter protection is not required for receptacles on rooftops when the receptacle is on a dedicated circuit for snow-melting or de-icing equipment.

210-11(c)(3) Exception: This is a new exception that permits the lighting and utilization equipment in a single bathroom to be on the same 20-ampere circuit with the bathroom receptacles if the circuit serves only that bathroom. This is illustrated in Figure 3.9.

210-12: This is a new section of the Code that will require the installation of a new type of circuit protection device called an arc-fault circuit interrupter. This device uses a small processor housed inside of the device to recognize the current and voltage characteristics of arcing faults. Arc-fault circuit interrupters are designed to replace the circuit's overcurrent device while providing an extra level of safety from arcing faults. Beginning in January, 2001, these devices will be required to protected all 125-volt, 15- and 20-ampere receptacle outlets installed in dwelling bedrooms.

210-52(b)(3): Small-appliance circuits are now permitted to serve only one kitchen. This new provision would require a dwelling with two kitchens to have two small-appliance circuits for each kitchen.

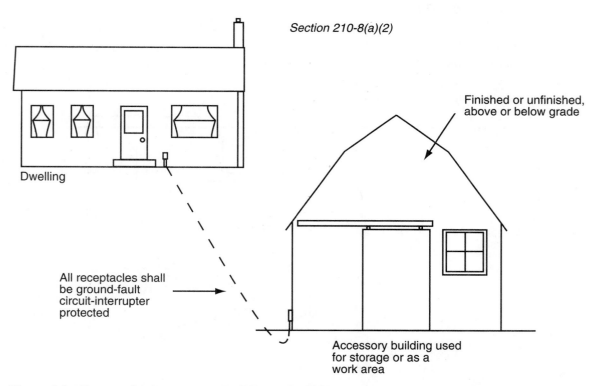

Section 210-8(a)(2)

Finished or unfinished, above or below grade

Dwelling

All receptacles shall be ground-fault circuit-interrupter protected

Accessory building used for storage or as a work area

Figure 3.8   Receptacles in accessory buildings of dwellings, whether finished or unfinished, or above grade or below grade, if used for storage or as work areas, are required to be ground-fault circuit-interrupter protected.

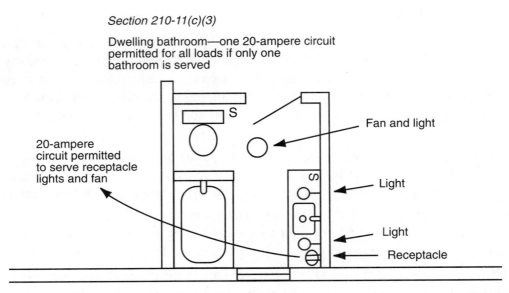

Section 210-11(c)(3)

Dwelling bathroom—one 20-ampere circuit
permitted for all loads if only one
bathroom is served

20-ampere
circuit permitted
to serve receptacle
lights and fan

Fan and light

Light

Light

Receptacle

Figure 3.9   If a dedicated 20-ampere, 120-volt circuit serves only one bathroom, the lighting and utilization equipment in that bathroom are permitted to be on the same circuit with the receptacle outlets.

210-52(c)(5) Exception: Receptacles serving a peninsular or island counter space in a dwelling kitchen are permitted to be mounted not more than 12 inches below the edge of the counter top when the counter top is flat and there is no practical means to mount the receptacle above the surface. If there is a backplate, or some other raised portion of the counter top, then this exception does not apply and the receptacles must be mounted above the surface. This is illustrated in Figure 3.10.

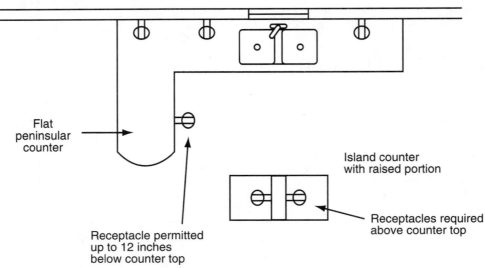

Section 210-52(c)(5)

Dwelling kitchen counter with no cabinets
above island or peninsular counter

Flat
peninsular
counter

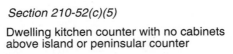

Island counter
with raised portion

Receptacles required
above counter top

Receptacle permitted
up to 12 inches
below counter top

Figure 3.10.   When a dwelling kitchen counter top is flat with no raised portions, and there is no other practical means to mount the receptacles within 18 inches above the surface, the receptacles are then permitted to be mounted not less than 12 inches below the counter top surface.

210-52(d): For a dwelling bathroom adjacent to the sink is now defined as on an adjacent wall and within 36 inches of the edge of the sink.

210-52(g): The required receptacle in the basement of a one-family dwelling is required to be installed in the unfinished portion of the basement when some of the basement is finished. A receptacle in the finished portion of the basement does not satisfy the requirement for a receptacle in the basement.

210-60(b): In the previous edition of the Code, all receptacle outlets installed in guest rooms were required to be readily accessible. Now only two receptacle outlets are required to meet this provision. In addition when receptacle outlets are placed behind beds, a guard is required to be installed or the outlet must be placed to prevent an attachment plug from touching the bed.

210-70(a)(3): Wall switches need only be installed at the usual place of entry, not every possible means of entry, when they control lighting outlets in attics, basements, underfloor spaces, or utility rooms that are used for storage or contain equipment requiring servicing.

210-70(c): When equipment requiring servicing is placed in either an attic or underfloor space, the lighting outlet control point need only be installed at the usual place of entry.

Article 220, Part A   **Branch Circuit Calculations**

There was a major revision of this part of the article to eliminate as many of the exceptions as possible. This article only deals with branch circuit and feeder calculations. Some of the material in this article dealing with subjects other than calculations was moved to other articles. Likewise, material in other articles dealing with calculations was moved to this article.

220-2(b): This is a new paragraph that permits the rounding to the nearest whole number of amperes when calculations are performed. There was a note to calculations in *Chapter 9* of the previous edition of the Code, but that could not be taken as a rule. Now rounding to the nearest whole number is a rule. For example, if a calculation results in 20.4 amperes, the number can be rounded down to 20 amperes. If the calculation results in 20.5 amperes, the number must be rounded up to 21 amperes.

Table 220-3(a): This was *Table 220-3(b)* in the previous edition of the Code, and it gives the minimum general illumination load to be used for different occupancies. There were no changes in the values of the table. A footnote applying to dwelling lighting loads in the previous edition of the Code was deleted and language was added to *Section 220-3(b)(9)* to explain the same issue. What it said in the past and still states in the present Code is that general-use receptacle outlets in a dwelling are to be included in the 3 volt-amperes per square foot calculation and, therefore, are not figured at 180 volt-amperes each, as is the case with receptacles in other occupancies. This means the number of receptacle outlets permitted to be installed on a dwelling circuit is determined based on the load to be served, and not on the basis of 180 volt-amperes per outlet.

Article 225   **Outside Branch Circuits and Feeders**

225-4: In the previous edition of the Code, open individual overhead conductors were required to be insulated or covered when within 10 feet of any structure or building. In the case of structures such as poles and towers, open individual overhead conductors on insulators are not required to be insulated or covered within 10 feet of the structure.

225-19(d): A new sentence was added to the first paragraph. The vertical clearances as given in *Section 225-18* are required to be maintained for final spans of overhead feeder and branch circuit conductors that are above and within a horizontal distance of 3 feet from any platform or other projection from which the conductors can be reached.

Part B: This is a new part providing specific requirements for the selection, location, sizing, and installation of the disconnecting means for a building or structure supplied from elsewhere on the property. The subject was covered by *Section 225-8* in the previous edition of the Code. There were some references to *Article 230* for specific requirements. It cannot be assumed that requirements for disconnecting means for services apply to feeders to other buildings unless those requirements are specifically referenced. The disconnecting means requirements for feeders to other buildings are illustrated in Figure 3.11.

225-32: The disconnecting means for an outbuilding is required to be located closest to the point of entry of the conductors to the building either inside or outside. There are several exceptions where the disconnecting means can be located elsewhere on the same property.

225-33: The disconnecting means is permitted to be up to six individual switches or circuit-breakers.

225-34: When the disconnect consists of more than one switch or circuit-breaker, they are required to be grouped at one location. An emergency disconnect is permitted to be located away from the other disconnects.

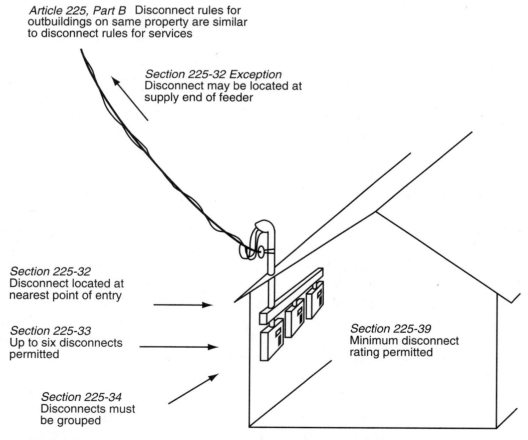

Article 225, Part B Disconnect rules for outbuildings on same property are similar to disconnect rules for services

Section 225-32 Exception Disconnect may be located at supply end of feeder

Section 225-32 Disconnect located at nearest point of entry

Section 225-33 Up to six disconnects permitted

Section 225-34 Disconnects must be grouped

Section 225-39 Minimum disconnect rating permitted

Figure 3.11   When one building is supplied power from another building or structure on the same property, the requirements for the disconnecting means for the building are generally the same as for installing a service in the building.

225-35: If the building is multiple occupancy, each occupant is required to have access to the disconnect serving that occupant's area. There is an exception where electrical maintenance personnel are provided for the building.

225-36: The disconnecting means for an outbuilding served from another point on the same property is required to be marked as suitable for use as service equipment. There is an exception for a single circuit on residential property.

225-39: The minimum rating of the disconnecting means for a building or structure is new and is the same as the requirements in *Section 230-79*. For example, the minimum rating of disconnecting means for a dwelling is 100 amperes.

Part C: This is a new part providing specific requirements for the placement of warning signs for, location of, and the types of the disconnecting means for a feeder that is over 600 volts. These subjects were covered by *Part H* of *Article 230* in the previous edition of the Code but did not necessarily apply in the case of branch circuits and feeders.

Article 320   Open Wiring on Insulators

No significant changes were made to this article.

Article 321   Messenger-Supported Wiring

No changes were made to this article.

Article 380   Switches

380-3(b): This new provision requires that an enclosure for a switch or circuit-breaker meet the 40 percent cross section fill requirement of *Section 373-8* if the enclosure is used as a raceway or junction box for routing or taping conductors to other switches or over-current devices. This means that when a switch is installed in a box and one or more conductors are feeding through the box, the cross-sectional area of the conductors are not permitted to occupy more than 40 percent of the remaining cross-sectional area of the box after the cross-sectional area of the switch has been deducted. This will be difficult to accomplish, since this cross-sectional area information is not available. For a typical switch installed in a 2-inch by 3-inch by 2 1/2-inch box, the cross-sectional area of the conductors and splicing devices is not a problem if the fill requirements of *Section 370-16* have been satisfied. This is illustrated in Figure 3.12. Assuming that a type NM-B, AWG number 12 copper cable with two insulated conductors and a ground enter and leave a box with a switch installed and assuming some conductors double back within the box so they are counted twice, the available cross-sectional area of the box less the area of a typical switch is 2.5 square inches. Counting six type THWN number 12 conductors, three number 12 bare conductors, and two splice caps at any one cross section, the cross-sectional area is much less than 75 percent of the available space.

380-6(c): This section was modified to include switches with "butt contacts," molded case switches, and circuit-breakers when they are used as switches. These switches as well as knife switches are required to be wired so that terminals connected to loads are not energized when the switch is in the open position. There is an exception where the switch is designed for a back feed source of power.

380-9(b) Exception: Now only in the case of existing installation where no grounding means is available is a snap switch permitted to be installed as a replacement without an equipment grounding connection. This is illustrated in Figure 3.13. If replacement snap

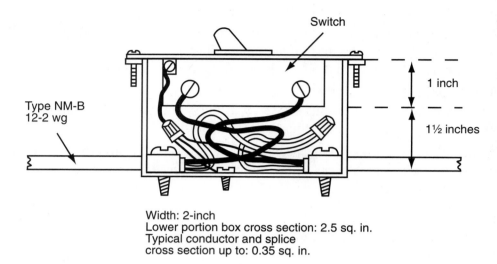

Section 380-3(b) If wires feed through, fill cross section must be observed

Switch

1 inch

1½ inches

Type NM-B
12-2 wg

Width: 2-inch
Lower portion box cross section: 2.5 sq. in.
Typical conductor and splice
cross section up to: 0.35 sq. in.

Figure 3.12   This rule requires the cross-sectional area of the conductors and splices to be considered when conductors feed through a switch box; however, when the fill requirements of *Section 370-16* are satisfied, there is plenty of cross-sectional area available.

switches are installed according to this exception, a nonconducting faceplate is required to be installed if the switch is located within reach of a conducting surface.

380-14(d): A minimum rating of 15 amperes is required for all 347-volt snap switches. In the previous edition of the Code, no minimum ampere rating was required. Also, a new provision for 347-volt flush-mounted switches is that their mounting method is required to be such that they are not readily interchangeable with other general-use snap switches.

Article 400   **Flexible Cords and Cables**

Table 400-4: Type G-GC portable power ground check cable is new in the Code. The cable is rated for portable use and is an extra hard usage cord. Circuit conductor sizes range from AWG number 8 to 500 thousand Circular Mils. Type G cord which was in the previous edition of the Code has from 2 to 6 circuit conductors with one or more equipment grounding conductors. Type G-GC is available only as a three-conductor cord with two equipment grounding conductors. This type is different in that it also has a low-voltage ground-check conductor for the purpose of monitoring the continuity of the equipment grounds. If continuity of the equipment grounding conductor is lost, the cable is de-energized. This cable is permitted to be used in commercial garages and on theater stages as well as in many other locations.

400-7(a): This section states uses permitted for flexible cords. Portable and mobile signs were added to this section to make it clear that flexible cords were to be used for this purpose. *Article 600* recognized flexible cords for portable and mobile signs, but *Article 400* did not recognize flexible cords for this purpose.

400-8: The list of uses not permitted for flexible cords was altered to clarify the meaning of ceilings. In the previous edition of the Code, flexible cords were not permitted to be run through holes in walls, ceilings, and floors. The meaning of the term **ceiling** as used in this section was changed to **structural ceiling, suspended ceiling, and dropped ceiling.**

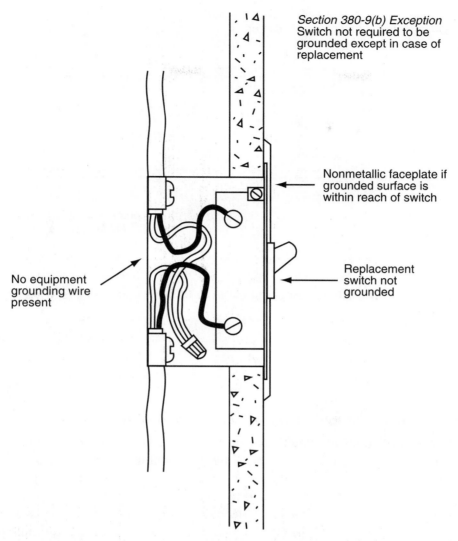

Section 380-9(b) Exception
Switch not required to be grounded except in case of replacement

Nonmetallic faceplate if grounded surface is within reach of switch

No equipment grounding wire present

Replacement switch not grounded

Figure 3.13 A switch is required to be grounded except in the case where the switch is a replacement for an existing circuit without an equipment ground available, and a nonmetallic faceplate is required if within reach of a grounded surface.

It is not permitted to make a hole in a suspended ceiling tile and pass a cord through that hole in a manner similar to Figure 3.14.

400-8 Exception: The wording in this exception has been deleted and replaced with a reference to *Section 364-8* for the provisions for attaching a flexible cord or cable to the surface of a building. A flexible cord or cable is permitted to be attached to the building with a strain relief device with a take up termination not more than 6 feet from the point of attachment of the strain relief. The previous edition of the Code did not limit the practice. Now this practice is only permitted for attachment to busways.

400-10 Exception: This section required some means of preventing strain to be transmitted directly to terminations of flexible cords. Some type of strain relief at the attachment point of the flexible cord is required. This new exception permits a single conductor flexible cable connection to a listed single-pole connector without the use of an additional strain relief connector. These listed devices are designed to compensate for strain to the terminal. An example is a single-pole pin and sleeve connector such as that which is often used for theater power.

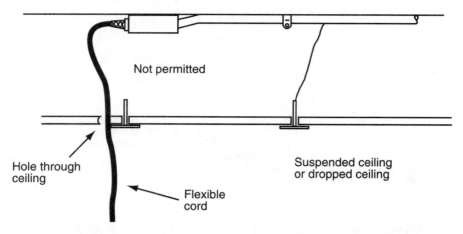

Section 400-8 Flexible cord is *not* permitted to be installed through suspended or dropped ceilings

Not permitted

Hole through ceiling

Flexible cord

Suspended ceiling or dropped ceiling

Figure 3.14   Flexible cords are not permitted to be installed through holes in suspended or dropped ceilings.

Article 402   **Fixture Wires**

No significant changes were made to this article.

Article 410   **Lighting Fixtures, Lampholders, Lamps, and Receptacles**

410-4(d): This section gives minimum spacing requirements for lighting fixtures, track lighting, and ceiling-suspended paddle fans from the bathtub rim and now the **shower stall** threshold. The fixture or fan is to be located at least 3 feet horizontally from the outside edge of the bathtub rim or shower stall threshold, or if within 3 feet, it shall be installed so that it is a minimum of 8 feet above the bathtub rim or shower stall threshold. The distances did not change from the previous edition of the Code, but it did not specifically include the shower stall rim as a point for making clearance measurements.

410-11: The previous edition of the Code required, in the case of incandescent fixtures, that wiring not supplying the fixture was only permitted to feed through an integral junction box if the fixture was identified for through wiring. Now this requirement applies to all types of fixtures, not just incandescent fixtures. A fine print note was added calling attention to a similar situation in *Section 410-31.*

410-15(b): Metal poles that support lighting fixtures are now considered to be raceways rather than just objects that **enclose** conductors. This would mean that if communication or other similar circuits are run inside of metal poles, the separation requirements found in other articles of the Code clearly apply to these installations.

410-16(a): Outlet boxes and fittings that are supported according to the methods of *Section 370-23* are permitted to support fixtures that weigh 50 pounds or less. The change in this section deals with fixtures that weigh more than 50 pounds. The previous edition of the Code required they be supported independent of the box or fitting. Now the Code permits the box to support a fixture over 50 pounds if the box or fitting is listed for such support.

410-16(c): When clips are used to fasten lighting fixtures to suspended-ceiling framing members, the clips were required to be identified for use with the type of fixture and type of framing member. Now the clips are also required to be **listed.**

410-18: This section deals with the grounding of exposed metal parts of lighting fixtures. In the previous edition of the Code, the rules for fixtures operating at over 150 volts were combined into one section so the rules were the same regardless of voltage. All fixtures, lamp-tie wires, mounting screws, clips, and other similar equipment are not required to be grounded if spaced at least 1 1/2 inches from the terminals of the lamp. This only applied to fixtures operating at more than 150 volts in the previous edition of the Code.

410-25: This section giving conductor types permitted to be used for fixtures with mogul base lampholders and those with other than mogul base lampholders was deleted from the Code because the section did not cover any installation requirements, and was duplicate material already covered in *Tables 310-13, 400-5(A), and 402-3.*

410-35(a): This marking requirement for fixtures now applies to all fixtures; in the previous edition of the Code, it only applied to electric discharge fixtures, not incandescent. The circuit conductor insulation temperature marking on the fixture was changed. Now if the fixture requires circuit conductors rated more than 60°C, the minimum permitted temperature must be marked on the fixture. The previous edition of the Code only required insulation temperature markings when the minimum required was greater than 90°C.

410-56(c): Now receptacles with the grounding terminal isolated from the metal receptacle strap (identified with an orange triangle) are permitted to be installed in nonmetallic boxes and have metal faceplates if the box has a feature that provides for the effective grounding of the faceplate. This is illustrated in Figure 3.15.

410-56(f): Item (3) in this section was broadened to include all mounting of receptacles in covers. The previous edition of the Code covered only raised covers. Now any receptacle that is mounted and supported by a **cover** shall be either held in place by more than one screw or held in place by one screw of a cover listed for one-screw mounting.

410-56(g): The prongs, pins, or blades of attachment plugs can become energized only after they have been inserted in a receptacle. Using a male cord cap to supply a receptacle is

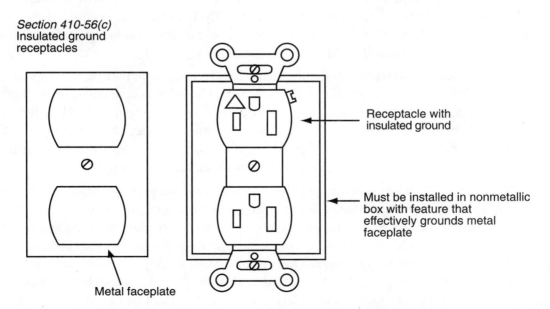

Figure 3.15   Receptacles with equipment grounding terminals insulated from the yoke of the receptacle installed in nonmetallic boxes and having metal faceplates are required to have provision for effectively grounding the faceplate.

not permitted. This was the intent in the previous edition of the Code, but there was no specific provision to prohibit the practice.

410-58(a): The previous edition of the Code required a fixed grounding pole on all grounding-type attachment plugs. Now the grounding pole is permitted to be movable and self-restoring provided the unit or cord assembly consists of a ground-fault circuit interrupter and the maximum voltage between any conductor and from any conductor to ground does not exceed 150 volts.

410-67: Tap conductors may be run from a fixture to a junction box to connect to the circuit conductors. That junction box was required to be at least 1 foot from the fixture, and the tap conductor was required to be **not less than** 4 feet long nor more than 6 feet long. Now the tap is permitted to be only 18 inches long.

410-73(f)(4): The previous edition of the Code required that all remote ballasts for high-intensity discharge fixtures be thermally protected. Now this requirement applies only if the remote ballast is recessed, and the thermal protection is required to be integral with the ballast. The previous edition of the Code did not require the thermal protection to be an integral part of the ballast.

410-101(c): A new item (9) was added to the list of locations where it is not permitted to install track lighting to be consistent with *Section 410-4(d)*. This is not a change.

Article 411 **Lighting Systems Operating at 30 Volts or Less**

No changes were made to this article.

Article 422 **Appliances**

This article was reorganized with many of the sections renumbered.

422-15: This is a new section in the Code that provides requirements for central vacuum outlet assemblies. These assemblies are permitted to be connected to 15- or 20-ampere branch circuits excluding small-appliance, laundry, and bathroom circuits. The connecting wires to the vacuum outlet must have an ampacity not less than that of the branch circuit from which the outlet is connected. Generally there are no exposed metal parts requiring grounding; therefore, the equipment grounding conductor is usually not required. A 2-pin polarized molded plug is generally used.

422-16(b): This section contains the requirements for flexible cords that are used to supply specific appliances. In the previous edition of the Code, flexible cords that supplied built-in dishwashers, trash compactors, and garbage disposers were required to be **identified for the purpose.** Some interpretations required the cord to be identified for use with these individual appliances. Now the cord supplying these appliances is required to be **suitable for the purpose** as indicated in the appliance manufacturer's instructions.

422-16(b)(2): The method of determining the length of cord supplying built-in dishwashers and trash compactors was changed. In the previous edition of the Code, the cord length was limited to a distance of 3 to 4 feet from the point of attachment to the appliance to the receptacle. This length of cord is now measured from the face of the cord cap to the plane of the back of the appliance. This distance shall be 3 to 4 feet, which means the actual cord length may be greater, as illustrated in Figure 3.16. This change in the rule will permit an appliance to be adequately moved so that it can be serviced.

422-18(b) Exception: This new exception permits the use of a listed box that has been identified for the purpose to support ceiling-suspended paddle fans that weigh not more than

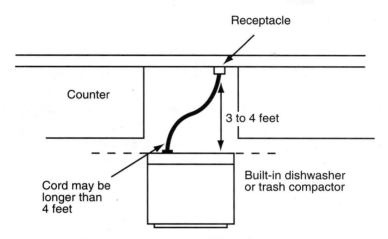

Section 422-16(b)(2) Three to 4 feet measured perpendicular to plane of back of appliance

Receptacle

Counter

3 to 4 feet

Cord may be longer than 4 feet

Built-in dishwasher or trash compactor

Figure 3.16   The 3- to 4-foot length of cord permitted for a dishwasher or trash compactor is measured from the receptacle perpendicular to the plane of the back of the appliance, not to the actual point of connection to the appliance.

70 pounds. The previous edition of the Code would have required support independent of the box for any ceiling fan that weighed more than 35 pounds.

Article 424   Fixed Electric Space Heating Equipment

424-12(a): Fixed electric space heating equipment is required to be protected in an approved manner when installed in locations where physical damage can occur. There are actually two changes. The protection of the heating equipment must be **approved** instead of being **adequately** protected. Protection is required if **physical damage** is likely where the previous edition of the Code specified only in the case of **severe** physical damage. The word **severe** was deleted.

424-44(g): This is a new ground-fault circuit-interrupter requirement for the installation of electric heating cables in concrete or poured masonry floors of bathrooms, spas, hot tubs, and hydromassage tubs. It is not required to protect the entire circuit, just that portion of the circuit that supplies the heating cable. This protection is required whether the finished floor surface is conductive or nonconductive.

Article 426   Fixed Outdoor Electric De-Icing and Snow-Melting Equipment

426-28: The phrase **branch circuit supplying** was deleted to make it clear that only the fixed outdoor electric de-icing and snow-melting equipment is required to have equipment ground-fault protection. It is no longer required to provide equipment ground-fault protection for mineral-insulated metal-sheathed de-icing and snow-melting cable that is embedded in a noncombustible material such as concrete.

Article 720   Circuits and Equipment Operating at Less than 50 Volts

No changes were made to this article.

Article 725   **Class 1, Class 2, and Class 3 Remote-Control, Signaling, and Power-Limited Circuits**

725-24 Exception 3: This new provision deals with listed power supplies for Class 1 circuits other than transformers. Overcurrent protection for electronic power supplies was not covered in the previous edition of the Code. In the case of single-voltage 2-wire power, an overcurrent device on the input circuit is permitted to protect the output if the following maximum rating is not exceeded. The overcurrent device rating on the input conductors is not greater than the Class 1 circuit conductor ampacity times the ratio of the Class 1 circuit voltage divided by the input circuit voltage. This is illustrated in Figure 3.17 where an electronic power supply is supplied by 240 volts and the output to the AWG number 14 copper Class 1 conductor is 120 volts. The primary fuses are permitted to protect the Class 1 conductors if they are not rated more than 10 amperes (20 A × 120 V ÷ 240 V = 10 A). If the power supply does not have a single-voltage 2-wire output, then overcurrent protection is required on both the input conductors and the Class 1 output conductors.

725-54: In the previous edition of the Code, this section provided the separation requirements for Class 2 and 3 circuits from electric light, power, Class 1, and nonpower-limited fire-alarm circuits. Now Class 2 and 3 circuit conductors must also be separated from medium- and high-powered broadband communication circuits. Low-powered broadband communication circuits, on the other hand, are permitted to occupy the same raceway or enclosure with Class 2 and 3 cables.

725-54(a)(1) Exception 6: This new exception permits the use of a **solid fixed** barrier to separate Class 2 and 3 circuits from electric light, power, or nonpower-limited fire-alarm circuits that are installed together in a cable tray. The barrier requirement can be avoided if the Class 2 and 3 circuits are run as type MC cable.

725-54(d): This section was revised to make it clear that power-limited fire-alarm circuit conductors may not be strapped, taped, tied, or attached in any fashion to the exterior of any raceway unless *Section 300-11(b)(2)* permits such attachment.

725-61: Some cable types for Class 3 power-limited circuits were deleted because they are not generally used for Class 3 applications.

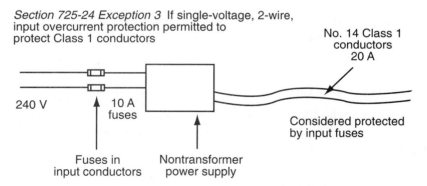

Figure 3.17   The primary overcurrent device is permitted to protect single-voltage, 2-wire, Class 1 conductors for a power supply without a transformer if the primary overcurrent device has a rating not greater than the Class 1 conductor ampacity times the ratio of the Class 1 voltage divided by the input voltage.

725-61(d) Exception 3: This new exception permits type PLTC to be used as open wiring from cable trays to utilization equipment in hazardous locations. Though the cable must be listed for the purpose and meet the same impact and crush requirements of type MC cable, it may not be installed unprotected in locations where it is exposed to physical damage. The overall length of the cable is limited to 50 feet and must be supported at intervals not exceeding 6 feet. This exception only applies to industrial locations where only **qualified persons** will provide service and maintenance.

WORKSHEET NO. 3
OUTLETS, LIGHTING,
APPLIANCES, AND HEATING

These questions are considered important to understanding the application of the *National Electrical Code®* to electrical wiring, and they are questions frequently asked by electricians and electrical inspectors. People working in the electrical trade must continue to study the Code to improve their understanding and ability to apply the Code properly.

DIRECTIONS: Answer the questions and provide the Code reference or references where the necessary information leading to the answer is found. Space is provided for brief notes and calculations. An electronic calculator will be helpful in making calculations. You will keep this worksheet; therefore, you must put the answer and Code reference on the answer sheet as well.

1. An aerial messenger-supported feeder is to be installed between a dwelling and an outbuilding. The feeder is protected from overcurrent at 100 amperes. The three insulated conductors are type XHHW AWG number 3 aluminum marked sunlight-resistant. Is it permitted to field assemble this aerial feeder at a dwelling using lashing material to secure the conductors to the messenger? (Yes or No)

   Code reference _____

2. A duplex dwelling (two-family dwelling), with both living units at grade level, has direct grade-level access from the front yard to the back yard. What is the minimum number of receptacle outlets required on the outside of this building?

   Code reference _____

3. An overhead feeder originates in a building where the feeder circuit-breaker and the service panel main circuit-breaker cannot be locked in the open position. An aluminum ladder is the typical relamping tool. Is an outside lighting fixture permitted to be installed on a pole above energized feeder conductors, as shown in Figure 3.18? (Yes or No)

   Code reference _____

Feeder protected by
circuit-breaker that
cannot be locked in the
open position

To building

Pole

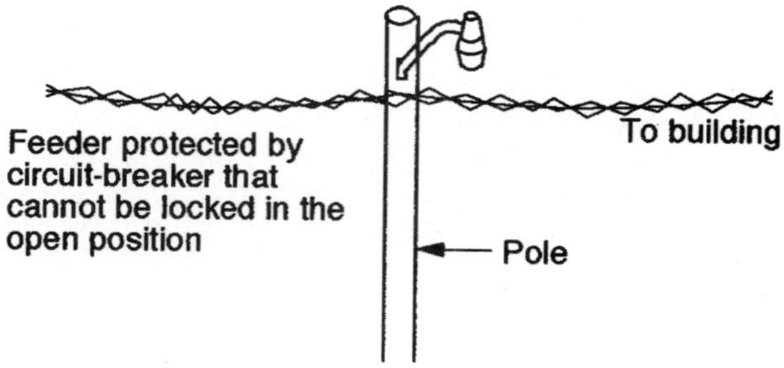

Figure 3.18   The energized feeder conductor overcurrent device cannot be locked in the open position, and the fixture is mounted above the conductors.

4. Are the two required 20-ampere dwelling kitchen small-appliance branch circuits permitted to also supply receptacle outlets in the dining room? (Yes or No)

   Code reference _____

5. Is a surface-mounted fluorescent fixture with an enclosed lamp permitted to be installed on the ceiling of a clothes closet in a dwelling? (Yes or No)

   Code reference _____

6. A living room in a dwelling has an unbroken wall section, as shown in Figure 3.19, with one side 4 feet, the long side 20 feet, and the remaining short side 7 feet. The minimum number of receptacle outlets permitted to be installed on this wall section is _____.

   Code reference _____

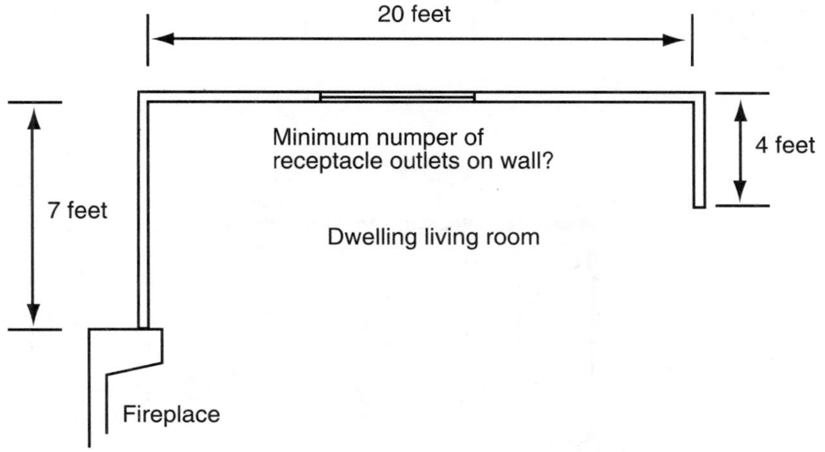

Figure 3.19   Determine the minimum number of receptacle outlets permitted to be installed on this length of unbroken wall in the living room of a dwelling.

7. A single-family dwelling has a living area of 3,860 square feet. What is the minimum number of 20-ampere general-purpose lighting circuits required for the dwelling, not including small-appliance, bathroom, and laundry circuits, if all general-illumination circuits are rated at 20 amperes and only bathroom receptacles are served by the bathroom circuit?

   Code reference _____

8. Class 2 power-limited wiring operating at 24 volts is installed for a dwelling thermostat circuit in a one-family dwelling. This circuit is to be wired with jacketed cable. There are several types of cable permitted to be used for this dwelling thermostat circuit. What cable type is limited only to this specific application and is not permitted to be installed in a commercial building as open cable without raceway protection?

   Code reference _____

9. In a single-family dwelling, what is the maximum number of lighting outlets and general-purpose receptacle outlets (not including small-appliance receptacles) permitted to be installed on a 15-ampere branch circuit?

   Code reference _____

10. The receptacle outlets of a dwelling required to be ground-fault circuit-interrupter protected are all supplied from a panelboard supplied by a feeder from the service entrance. Is it permitted to omit the individual circuit or outlet ground-fault protection if the feeder is personnel ground-fault protected? (Yes or No)

    Code reference _____

11. A 10-foot wall along one side of a room in a dwelling does not have a receptacle outlet, as shown in Figure 3.20. There is a receptacle outlet on each adjacent wall within 6 inches of turning the corner. If this 10-foot wall section does not have a receptacle outlet, is this a violation of the Code? (Yes or No)

    Code reference _____

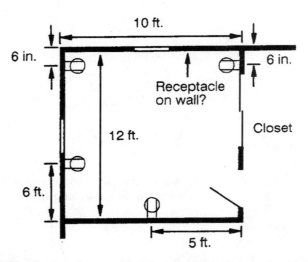

Figure 3.20   Is the absence of a receptacle outlet on the 10-foot wall section of the room of the dwelling a violation of the Code?

12. A hard service cord type SO, AWG number 12 supplies a portable device, and the cord was accidentally cut. Is the cord permitted to be spliced provided the wires are made mechanically secure and soldered and suitable tape is applied to restore equivalent original insulation and protective covering? (Yes or No)

    Code reference _____

13. A single-pole snap switch on a 15-ampere circuit fails in a dwelling, and during replacement, it is discovered that the wire is aluminum. If the aluminum wire is to be connected directly to the new switch, the switch is required to be marked cu/al. (True or False)

    Code reference _____

14. The box shown in Figure 3.21 has threaded hubs, and the conduit is rigid steel. The conduit is supported within one foot of the box with conduit supports that completely surround the conduit and are bolted together and bolted securely to clamps fastened to the structure of the building. Are the conduit supports permitted to serve as the support for the fixture, which weighs 10 pounds? (Yes or No)

Code reference _____

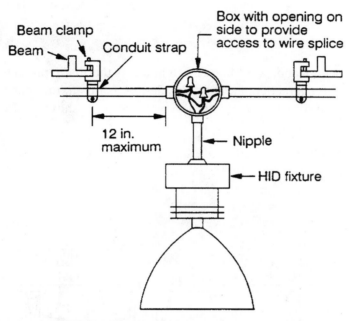

Figure 3.21 A 10-pound lighting fixture is mounted to a box with threaded hubs, which is supported by conduit hangers within one foot of the box.

15. Determine the minimum copper supply conductor size, type NM cable for an 80-gallon electric storage-type water heater with a rating of 4,500 watts at 240 volts. The 4,500 watts is the maximum load permitted by the water heater controls.

Code reference _____

16. What is the demand load in kilovolt-amperes for a 17.4-kilowatt electric range in a dwelling?

Code reference _____

17. An 80-gallon storage-type electric water heater is installed on the first floor of a single-family dwelling, and the service panel containing the overcurrent device for the water heater is in the basement. The individual circuit-breaker is not capable of being locked in the open position, and the lock on the panel cover is not considered an acceptable means of locking the circuit-breaker in the open position. Is a separate disconnect for the water heater required on the first floor within sight of the water heater? (Yes or No)

   Code reference _____

18. A recessed incandescent fixture is marked as thermally protected, but it is not rated for direct contact with insulation. Is this fixture permitted to be mounted in direct contact with the wood joist, as shown in Figure 3.22? (Yes or No)

   Code reference _____

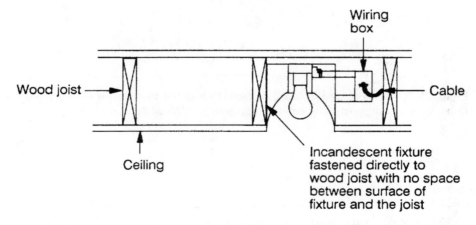

Figure 3.22   A recessed incandescent lighting fixture marked as thermally protected is mounted with the side of the enclosure in direct contact with a wood joist.

19. A 240-volt branch circuit supplies three baseboard electric heaters in a room of a dwelling. What is the maximum permitted rating of the branch circuit?

   Code reference _____

20. A line voltage thermostat (rated for operation at 240 volts) does not have an off position that would prevent the closing of the circuit with change in temperature. The thermostat controls 240-volt single-phase electric baseboard space heaters by directly interrupting current flow to the baseboard heaters. It is a 2-pole type thermostat that opens both ungrounded conductors. Is this thermostat permitted to serve as the disconnect for the baseboard heaters? (Yes or No)

    Code reference _____

21. A lighting fixture has provisions for twelve lamps with a maximum marked wattage of 100 watts per lamp. If one set of 120-volt fixture wires supplies all twelve sockets, is an AWG number 16 copper wire permitted to carry the total load within the fixture? (Yes or No)

    Code reference _____

22. When electric heating cable is installed to prevent icing of a sidewalk, for a particular installation, the heating cable is placed on a concrete masonry or asphalt base with a minimum thickness of _____ inches and then covered with a minimum 1.5-inch thick finish coat of concrete or asphalt.

    Code reference _____

23. A 120-volt cord-and-plug-connected window air conditioner in a dwelling draws 7.2 amperes. There is a receptacle outlet on a 15-ampere general-illumination branch circuit located on the wall below the window. It is determined the circuit normally supplies a small load of only a few amperes. Is it permitted to plug this air conditioner into this outlet as illustrated in Figure 3.23, assuming that this additional load would not overload the circuit? (Yes or No)

    Code reference _____

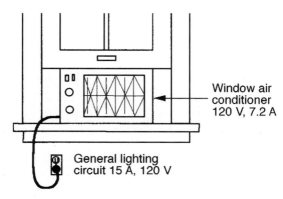

Window air
conditioner
120 V, 7.2 A

General lighting
circuit 15 A, 120 V

Figure 3.23   A room air conditioner in a dwelling is plugged into a 120-volt, 15-ampere general lighting branch circuit.

24. Is a thermostat cable, type CL2, permitted to be run in the same conduit with power wires of circuits not associated with the heating or cooling equipment? (Yes or No)

Code reference _____

25. A dwelling has three bathrooms, and the fan and lights in each bathroom are installed on the same circuit with the bathroom receptacles. The minimum number of 20-ampere, 125-volt branch circuits required for this dwelling to serve the bathrooms is _____.

Code reference _____

ANSWER SHEET

Name _____

No. 3   OUTLETS, LIGHTING, APPLIANCES, AND HEATING

Answer   **Code reference**

1. _____   _____

2. _____   _____

3. _____   _____

4. _____   _____

5. _____   _____

6. _____   _____

7. _____   _____

8. _____   _____

9. _____   _____

10. _____   _____

11. _____   _____

12. _____   _____

13. _____   _____

14. _____   _____

15. _____   _____

16. _____   _____

17. _____   _____

18. _____   _____

19. _____   _____

20. _____   _____

21. _____   _____

22. _____   _____

23. _____   _____

24. _____   _____

25. _____   _____

UNIT 4

Services and Feeder Calculations

OBJECTIVES

Upon completion of this unit, the student will be able to:

- determine the minimum permitted service entrance conductor size.
- determine the demand load to be included in a multifamily service calculation for more than two electric ranges in the building.
- determine the minimum permitted ampere rating for a single-family dwelling using the methods of *Article 220, Part B* or the optional calculation method.
- determine the maximum unbalance load for a single-family dwelling for the purpose of sizing the neutral conductor.
- determine the minimum ampere demand load for a small commercial building.
- determine the minimum permitted service entrance demand load for a multi-family dwelling.
- determine the minimum ampere rating of the service entrance for an existing dwelling where there has been an addition to the structure.
- determine the minimum demand load in amperes for a farm building where the loads operating with diversity and without diversity are known.
- look up the installation requirements for a service entrance from the Code.
- determine the minimum permitted ampere rating of central distribution equipment for a group of farm buildings, where the demand loads at each building are known.
- answer wiring installation questions relating to *Articles 215, 220, Parts B, C and D, 230, 600,* and *Appendix D, Examples.*
- state at least five significant changes that occurred from the 1996 to the 1999 Code from the previously stated articles.

CODE DISCUSSION

The installation of service equipment and calculations for the determination of the minimum rating of service equipment and feeders are covered in this unit. Grounding and bonding of service equipment are discussed in *Unit 5.*

Article 215 provides the minimum requirements for feeder conductors. Feeders are main conductors that are ultimately subdivided into smaller circuits. The minimum size and rating of conductors are given in *Section 215-2*. Maximum ratings of overcurrent protection for feeders are covered in *Section 215-3*. For feeders that operate with a voltage greater than 150 volts to ground and 1,000 amperes or more, equipment ground-fault protection is required, and this requirement is found in *Section 215-10*.

Article 220, Part B provides the information necessary to determine the load on a feeder conductor for a dwelling or any other type of building. Examples of how to use the information in *Article 220, Part B* to determine the minimum permitted size of service entrance for a building are included in the Sample Calculations section of this unit. Other examples are provided in *Appendix D* of the Code.

Article 220, Part B provides a method of determining the minimum permitted feeder load used to determine the minimum size of feeder conductors and electrical equipment. Service entrance conductors are feeders; therefore, this method can be used to determine the minimum size of service entrance equipment and service entrance conductors permitted for a building. Some optional methods are permitted to determine the minimum demand load for a feeder. These methods are discussed later. *Section 215-2(a)* describes the method to use to determine the load on a feeder where the load is continuous. A continuous load is one that, under normal conditions, is expected to operate for more than 3 hours. The load is required to be 1.25 times the continuous load. If the feeder load is a combination of continuous and noncontinuous load, then the load is the noncontinuous load plus 1.25 times the continuous load. Figure 4.1 illustrates how to determine the ampacity of a circuit or feeder supplying a combination of continuous and noncontinuous loads.

Experience has shown that all loads in a building may not operate at the nameplate rating or at the calculated value all at the same time. Demand factors are provided in the Code for certain loads to allow a reduction of the total connected load on a feeder when it is known that all loads will not operate at the same time. Demand factors for general illumination are given in *Section 220-11* and for receptacle loads in *Section 220-13*. In the case of four or more fixed-in-place appliances in a dwelling, other than the electric range, clothes dryer, space heating, or air conditioning, the nameplate rating is permitted to be multiplied by 0.75, as stated in *Section 220-17*. This demand reduction is permitted because it is not expected that all of these fixed-in-place appliances will operate at the same time. Sometimes it is not known exactly how much load will occur on certain circuits, but experience has shown how much load is reasonable. Examples are the loads on the small-appliance circuits for a dwelling kitchen and a

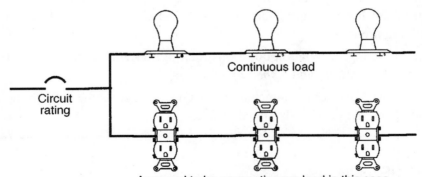

Continuous load

Circuit rating

Assumed to be noncontinuous load in this case

Section 215-2(a) Feeder rating shall not be less than the sum of:
1. noncontinuous load at 100%
2. 1.25 times the continuous load

Figure 4.1   The feeder rating is not permitted to be less than the noncontinuous load plus 1.25 times the continuous load.

dwelling laundry circuit. *Section 220-16(a)* requires that each of the two small-appliance circuits shall be considered to be 1,500 volt-amperes. For the dwelling laundry circuit, the load also is required to be not less than 1,500 volt-amperes.

The information in *Section 220-19* is used for dwellings and certain other buildings to determine the minimum demand load permitted for an electric range. Demand load for an electric range or for multiple electric ranges in a multifamily dwelling may be necessary for the determination of a minimum service entrance rating. The demand load for an electric clothes dryer is to be taken at 5,000 volt-amperes, unless the actual nameplate rating of the clothes dryer is higher. This requirement is found in *Section 220-18*. When there are more than four electric clothes dryers, such as would be the case for a multifamily dwelling, the load for the clothes dryers is permitted to be reduced by the amount of the demand factors in *Table 220-18*.

Article 220, Part D covers the methods of determining the demand loads for farm buildings. These methods use amperes for the calculations, where the methods of the previous parts of *Article 220* use volt-amperes. It is important to keep in mind that amperes at 120 volts are different than amperes at 240 volts for the purpose of making a service calculation for a farm building. It is suggested that all loads be converted to a 240-volt basis. For example, assume that there are lighting circuits that draw 26 amperes at 120 volts. When making the service calculation of *Section 220-40*, consider this lighting load as 13 amperes at 240 volts. Conversion of a farm building 120-volt demand load to a 240-volt basis is illustrated in Figure 4.2.

A group of farm buildings is generally supplied power from a central electrical distribution point. This may be a service at one building, or it may be a service at a central location. A common practice for a farm is to provide a meter pole, as shown in Figure 4.3, which shall have a minimum ampere rating determined by using *Section 220-41*. Overhead service drops or underground laterals run from the meter pole to the individual buildings.

The neutral of a service entrance or of a feeder shall be sized with an ampere rating not smaller than the maximum unbalance load of the ungrounded conductors. *Section 220-22* provides the rules for determining the minimum permitted maximum unbalanced load for a particular feeder or service entrance. Examples of how to make this determination are covered later in this unit. Examples in *Appendix D* of the Code show how to make neutral calculations. There is a minimum permitted size for the service entrance neutral conductor, which is given in *Section 230-42(c)*. In the case of a dwelling that is supplied at 120/240 volts, single-phase with 3 wires, the service entrance conductors or feeder conductors are permitted to be sized in accordance with *Table 310-15(b)(6)*.

Article 220, Part C describes alternative methods of making the service and feeder calculations instead of the method of *Article 220, Part B*. *Section 220-30* is an optional

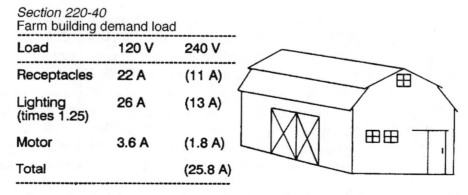

Section 220-40
Farm building demand load

Load	120 V	240 V
Receptacles	22 A	(11 A)
Lighting (times 1.25)	26 A	(13 A)
Motor	3.6 A	(1.8 A)
Total		(25.8 A)

Figure 4.2   Farm building feeder demand loads are determined as current in amperes on a 240-volt basis.

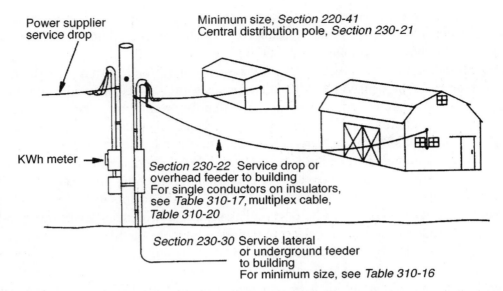

Figure 4.3   A center distribution pole provides a location for metering and a distribution point for conductors supplying power to farm buildings.

calculation method for a single-family dwelling. *Section 220-31* deals with the special case of an addition to an existing dwelling. This method can be used to determine if an addition to an existing dwelling will require an increase in the size of the service entrance. *Section 220-32* is the optional method for determining the demand load for a multifamily dwelling. *Section 220-35* is an optional calculation used to determine the demand load of a feeder or a service entrance for an addition to any building type. *Section 220-36* is an optional calculation for a new restaurant.

An optional method is included in *Part C* for determining the demand load of a school. Examples of a service or feeder calculation for a school and other buildings are not included in this text or in the Code because application of these methods is usually limited to work performed by engineers.

Article 230 covers the requirements for the installation of service entrances. Figure 4.4 shows a typical service entrance to a building with the components labeled to assist in understanding the terminology in the Code. Clearances of overhead service conductors are covered in *Section 230-24*. These clearance requirements are the same as the ones in *Section 225-19* for aerial outside feeder and branch circuit conductors. The ratings of service entrance conductors are covered in *Section 230-42*. The minimum rating of service disconnecting means is in *Section 230-79*.

The rating of the service entrance and the service entrance conductors is determined by calculating the expected load by one of the methods described in *Article 220*. The minimum size of service entrance conductors is determined by one of the ampacity tables in *Article 310*. In most cases, the service entrance conductors are sized using *Table 310-16*. The ampere rating of the service entrance conductors shall be not less than the calculated load. In some cases, it is permitted for the service overcurrent device rating to be greater than the rating of the conductors. One example occurs when applying the provisions of *Section 240-3(b)*. Another example is in the case of a single-family dwelling served by a 120/240-volt, 3-wire electrical system. In this case, it is permitted to size the service entrance conductors using *Table 310-15(b)(6)*.

The requirements for establishing a grounding electrode to earth and for bonding the electrical enclosures of service equipment are contained in *Article 250*. If any of the

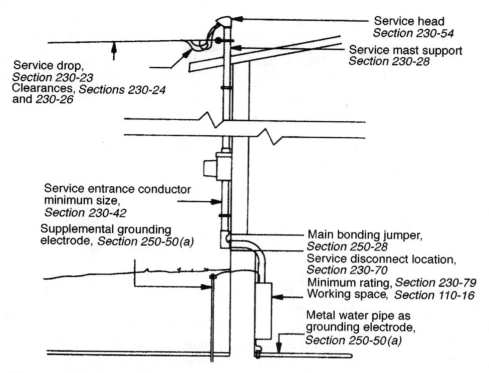

Figure 4.4 **Information needed for sizing and installing components of a service entrance is found in** *Article 230* **and** *Article 250.*

grounding electrodes of *Section 250-50* are present at the building, then they shall be used as a grounding electrode for the service. If none of the grounding electrodes of *Section 250-50* are available, then an electrode shall be established as described in *Section 250-52.* A discussion of grounding and bonding of service equipment is contained in Unit 5.

Article 600 deals with electric signs and outline lighting. This article is discussed with services because, when a service size is determined for commercial or industrial property, a load for an electric sign shall be included. The load included shall not be less than 1,200 voltamperes or the actual load if it is larger, according to *Section 600-5(b)(3).* Circuit requirements and sign and outline lighting installation requirements are covered in *Article 600.*

Appendix D provides detailed examples to show the intent of the Code in determining the size of services and feeders and how to determine the minimum number of branch circuits required for some installations. The notes at the beginning of *Appendix D* state that the nominal voltages of 120, 208, and 240 shall be used to make calculations. They also state that when making calculations, fractions of amperes may be rounded down when the fraction is less than 0.5 amperes and rounded up when the fraction is 0.5 amperes and larger. This is also stated in *Section 220-2(b).* The notes also state that when a calculation for ranges results in a fraction of a kilowatt or kilovolt-ampere, it is permitted to round down if the fraction is less than 0.5. Table 4.1 is a summary of the type of calculations found in the examples.

Several examples of service and feeder calculations are included in *Appendix D* of the Code. These examples are helpful in understanding the methods of *Article 220.* There are two variations of *Example D1,* which is of a single-family dwelling using the method of *Article 220, Part B. Example D2* is the optional calculation method of *Section 220-30* for a single-family dwelling. There are three different variations of this example.

Example D3 is the service calculation for a commercial store building. This is an important example because it illustrates the use of the 1.25 factor for continuous loads. It also

Table 4.1. Summary of the type of calculations found in the *Examples* of *Appendix D*.

Example number	Number of branch circuits	Service calculations Section 220, Part B method	Optional method	Neutral calculation	Range demand load
D1(a)	Yes	Yes	—	Yes	Yes
D1(b)	—	—	—	Yes	—
D2(a)	—	—	Yes	Yes	Yes
D2(b)	—	—	Yes	Yes	Yes
D2(c)	—	—	Yes	—	—
D3	Yes	Yes	—	—	—
D4(a)	Yes	Yes	—	Yes	Yes
D4(b)	Yes	—	Yes	Yes	Yes
D5(a)	Yes	Yes	—	Yes	Yes
D5(b)	Yes	Yes	—	Yes	—
D6	—	—	—	—	Yes

illustrates which loads are taken as continuous loads. Receptacle outlets in a commercial building are not always considered continuous loads.

Example D3 can be confusing. The service entrance main overcurrent device is required to have a rating not less than the noncontinuous load plus 1.25 times the continuous load. This turns out to be 135 amperes for the example. Looking at *Section 240-6* for the next standard overcurrent device rating higher than 135 amperes results in a minimum service entrance rating of 150 amperes. Keep in mind that the Code specifies the minimum permitted, and not necessarily the recommended size for a particular application. The service entrance conductors are sized based on the same current. The allowable ampacity values listed in the tables of *Article 310* are continuous currents. Therefore, the current used to size the conductors is the sum of the noncontinuous loads and 125 percent of the continuous loads. For *Example D3*, the current used to size the conductors is 135 amperes. The 75°C column of *Table 310-16* is used to select the minimum permitted service entrance conductor size, which is AWG number 1/0 rated at 150 amperes.

Example D4 is for a multifamily dwelling served with a single-phase electrical system. The service is sized for the entire building, and a calculation of the feeder size for the individual living units is included. This calculation is performed using the method of *Article 220, Part B* and the optional method of *Section 220-32*. *Example D5* is a service calculation for a multifamily dwelling served with a 208/120-volt, 3-phase electrical system. Look in the major Code change section for a discussion of how to include electric clothes dryers in the service calculation. The example is performed using the method of *Article 220, Part B* and the optional calculation of *Section 220-32*. *Example D6* illustrates how to determine the demand load for multiple electric ranges in a dwelling for the purpose of determining the size of service entrance conductors or feeder conductors. Methods are described for determination of the load on the service neutral in several of the examples. There are also several calculations to determine the minimum number of lighting and receptacle branch circuits.

SAMPLE CALCULATIONS

Examples of calculations of service entrances are included in this section to supplement the examples of *Appendix D*. The examples in the Code should be studied with these service calculations.

Example of a Commercial Service Calculation

A store has an area of 3,850 square feet. There are 70 receptacle outlets and 1,400 watts of outside lighting, which may operate during store hours. There is also an outside electric sign. The store also contains three refrigerated coolers with 240-volt, 1/2-horsepower single-phase motors and a walk-in cooler with a 3-horsepower, 240-volt single-phase motor. The central furnace has a 3/4-horsepower, 240-volt single-phase motor on the blower.

1. Determine the single-phase 120/240-volt service demand load for the store to be used to size the ungrounded service entrance conductors.

Before proceeding with the calculation of the demand load for this store building, read *Section 215-2(a)* and *Section 230-42(a)*. The overcurrent device rating protecting the service entrance conductors shall have an ampacity not less than 100 percent of the noncontinuous load, plus 125 percent of the continuous load. It is now required that the feeder conductors be sized to carry 125 percent of the continuous load. There is a similar requirement for overcurrent devices in panelboards in *Section 384-16(d)*. The demand load used for determining the minimum size ungrounded conductors for the service is 125 percent of the continuous load, plus 100 percent of noncontinuous loads.

Noncontinuous loads:
    Receptacles (*Sections 220-3(b)(9)* and *220-13*)
        70 × 180 VA = 12,600 VA

First 10 kVA at 100%	10,000 VA	
Remainder over 10 kVA at 50%	1,300 VA	
Total noncontinuous load		11,300 VA

Continuous loads
    General lighting load (*Section 220-3(a)*) (assume continuous)

3,850 sq. ft × 3 VA / sq. ft. =	11,550 VA	
Outside sign circuit (*Sections 220-3(b)(6)* and *600-5(b)(3)*)		
Not specified, so use minimum	1,200 VA	
Outside lighting (*Section 220-3(b)(1)*) (assume continuous)		
1,400 VA =	1,400 VA	
Total continuous load		14,150 VA

Motor loads (*Sections 220-3(b)(3)* and *430-24*)

3 hp	240 V × 17 A × 1.25	=	5,100 VA	
1/2 hp	240 V × 4.9 A × 3 motors	=	3,528 VA	
3/4 hp	240 V × 6.9 A	=	1,656 VA	
Total motor load				10,284 VA
Total				35,734 VA

$$\frac{35,734 \text{ VA}}{240 \text{ V}} = 149 \text{ A}$$

2. Determine the minimum rating permitted for the overcurrent device protecting the service entrance conductors, assuming one circuit-breaker or set of fuses is protecting the ungrounded conductors.

According to *Section 215-2(a)* and *Section 384-16(d)*, the minimum permitted rating of the single overcurrent device protecting the service entrance conductors shall be not less than the noncontinuous load plus 125 percent of the continuous load.

Noncontinuous load	11,300 VA
Continuous load, 14,150 VA $\times$ 1.25 =	17,688 VA
Motor load	10,284 VA
Total	39,272 VA

$$\frac{39,272 \text{ VA}}{240 \text{ V}} = 164 \text{ A}$$

The next higher standard rating of overcurrent device listed in *Section 240-6* is 175 amperes, but from a practical standpoint, the size of main overcurrent device in a panelboard used as service equipment for this building would be 200 amperes.

3. Determine the minimum permitted size of ungrounded conductors for the service entrance to this store building, assuming that the conductors are copper with 75°C insulation and terminations and a single service entrance main overcurrent device rated at 200 amperes.

The demand load for determining the minimum size service entrance conductors is the same used for determining the rating of the service, *Section 215-2(a)*. The minimum size copper conductor, with 75°C insulation and terminations permitted for the calculated demand load of 164 amperes, is AWG number 2/0 as determined using *Table 310-16* and *Section 240-3(b)*. But the maximum overcurrent device permitted for this conductor is 175 amperes. Therefore, the minimum permitted size of service entrance conductors when the main overcurrent device is 200 amperes is AWG number 3/0 copper with 75°C insulation.

If according to *Section 230-90, Exception 3* there were two to six overcurrent devices rather than only one main overcurrent device, then the service entrance conductor would only have to be sized to carry the load of 164 amperes. In that case, the minimum permitted copper service entrance conductor size with 75°C insulation and terminations would then be AWG number 2/0.

4. Determine the minimum permitted size of neutral service conductor for the store, assuming that the conductor is copper with type THWN insulation.

The minimum size of neutral conductor permitted for this service is determined by the rules of *Section 220-22*. The maximum unbalanced load is 50 percent of all line-to-neutral load in the case of a 3-wire single-phase and 34 percent of line-to-neutral load in the case of a 4-wire wye 3-phase system.

General lighting (120 V)	11,550 VA $\times$ 1.25 =	14,438 VA
(The last sentence of *Section 220-22* applies only to 3-phase, 4-wire wye)		
Receptacles (120 V)		
First 10 kVA at 100%		10,000 VA
Remaining 2.6 kVA at 50%		1,300 VA
Outside sign circuit (120 V)	1,200 VA $\times$ 1.25 =	1,500 VA
Outside lighting (120 V)	1,400 VA $\times$ 1.25 =	1,750 VA
Total		28,988 VA

$$\frac{28,988 \text{ VA}}{240 \text{ V}} = 121 \text{ A}$$

The minimum size neutral conductor permitted is AWG number 1, with 75°C insulation and terminations. It may be necessary to check to see if this is smaller than the minimum permitted neutral conductor size, according to *Section 230-42(c)*.

5. Establish a grounding electrode and determine the minimum copper grounding electrode conductor size permitted. See Unit 5 for more information.

Select an available electrode from *Section 250-50*. If a metal underground water pipe is used, then supplement it with one additional electrode (*Section 250-50(a)*). If electrodes listed in *Section 250-50* are not available, then choose one of the electrodes listed in *Section 250-52*. The minimum size grounding electrode conductor is found in *Table 250-66*. If the only electrode available is a rod, pipe, or plate, then the minimum size required is AWG number 6 copper. In this example of a 200-ampere service, the minimum size bare copper grounding electrode conductor is AWG number 4.

6. Determine the minimum number of 20-ampere, 120-volt general lighting circuits for the inside of the store. Continuous load is considered because the rating of the circuit is the size of overcurrent protection. The overcurrent device is not permitted to be loaded continuously at more than 80 percent of the overcurrent device rating. The minimum number of general illumination branch circuits is determined based on the actual lighting load even if it is less than the calculated load based on *Section 220-3(a)*. The actual general illumination load was not stated in the problem, so the calculated value will be used to determine the minimum number of lighting circuits.

$11{,}550 \times 1.25 = 14{,}438$ VA

$$\frac{14{,}438 \text{ VA}}{120 \text{ V}} = 120 \text{ A}$$

$$\frac{120 \text{ A}}{20 \text{ A / circuit}} = 6 \text{ circuits}$$

7. Determine the minimum number of 20-ampere, 120-volt receptacle circuits for the store. Some inspectors may judge a receptacle circuit to be a continuous load; however, this issue is not clear in the Code. It depends on the use of the circuit. For this example, we will consider the receptacle load to be a noncontinuous load. A minimum of five circuits is required.

$$\frac{11{,}300 \text{ VA}}{120 \text{ V}} = 94 \text{ amperes}$$

$$\frac{94 \text{ A}}{20 \text{ A / circuit}} = 4.7 \text{ circuits}$$

Example of a Single-Family Dwelling Demand Load

A single-family dwelling has a living area of 1,800 square feet, and the dwelling is to contain the following appliances at the time of construction:

Appliance	Current	Voltage	Power
Electric range		240 V	14 kW
Microwave oven (built-in)	12 A	120 V	
Electric water heater		240 V	6 kW
Dishwasher		120 V	1.8 kW
Clothes dryer		240 V	5 kW
Water pump	8 A	240 V	
Food waste disposer	7.2 A	120 V	
Baseboard electric heat (8 total units)		240 V	15 kW
Air conditioner	3 at 8 A	240 V	

1. Determine the service demand load using the method of *Article 220, Part B.*

General lighting load (*Section 220-3(a)*)
      1,800 sq. ft. $\times$ 3 VA/sq. ft. =   5,400 VA

Small-appliance circuits (*Section 220-16(a)*)
      2 $\times$ 1,500 VA =   3,000 VA

Laundry circuit (*Section 220-16(b)*)
      1 $\times$ 1,500 VA =   1,500 VA

Subtotal   (*Sections 220-11, 220-16*)   9,900 VA

      First 3,000 VA at 100%   3,000 VA

      Remainder at 35%   2,415 VA

Electric range (*Table 220-19*)   8,800 VA

Electric space heating (*Section 220-15*)   15,000 VA

Air conditioning (*Section 220-21*)
      3 $\times$ 8 A = 24 A
      24 A $\times$ 240 V = 5,760 VA   0 VA

Clothes dryer (*Section 220-18*)
      5 kW at 100%   5,000 VA

Other appliances (*Section 220-17*)
  Microwave oven (built-in)
      12 A $\times$ 120 V =   1,440 VA

  Electric water heater =   6,000 VA

  Dishwasher   1,800 VA

  Water pump (*Section 430-24*)
      8 A $\times$ 240 V $\times$ 1.25 =   2,400 VA

  Food waste disposer
      7.2 A $\times$ 120 V =   864 VA

Subtotal   12,504 VA

      12,504 VA $\times$ 0.75 =   9,378 VA

Total demand load   43,593 VA

Service load

$$\frac{43,593 \text{ VA}}{240 \text{ V}} = 182 \text{ amperes}$$

2. Determine the minimum size neutral conductor for the service entrance. The rule for sizing the neutral is found in *Section 220-22.*

General lighting, small-appliance, and laundry load   5,415 VA

Electric range   8,800 VA $\times$ 0.7 =   6,160 VA

Electric clothes dryer   5,000 VA $\times$ 0.7 =   3,500 VA

Other electric appliances
      Microwave oven (built-in)   1,440 VA
      Dishwasher   1,800 VA
      Food waste disposer
        7.2 A $\times$ 120 V =   864 VA

                                       4,104 VA

Total 120-volt load (unbalanced load)   19,179 VA

Unbalanced load

$$\frac{19,179 \text{ VA}}{240 \text{ V}} = 80 \text{ amperes}$$

Optional Dwelling Service Calculation

The optional method of determining demand load for a single-family dwelling usually results in a smaller value than the previous method in *Section 220, Part B.* The minimum size service wires, however, is 100 amperes.

Electric space heating $15,000 \times 0.4 =$		6,000 VA
(Each room separately controlled)		
Air conditioning (*Section 220-30(c)*)		OMIT
General lighting load		
1,800 sq. ft. $\times$ 3 VA/sq. ft. =	5,400 VA	
Small-appliance circuits		
2 $\times$ 1,500 VA =	3,000 VA	
Laundry circuit		
1 $\times$ 1,500 VA =	1,500 VA	
Electric range	14,000 VA	
Clothes dryer	5,000 VA	
Microwave oven (built-in)	1,440 VA	
Electric water heater	6,000 VA	
Dishwasher	1,800 VA	
Water pump	2,400 VA	
Garbage disposer	864 VA	
Subtotal	41,404 VA	

Apply the demand factors of *Section 220-30:*		
First 10 kVA of all other load at 100%		10,000 VA
Remainder of other load at 40%		
31,404 VA $\times$ 0.4 =		12,562 VA
Total load		28,562 VA
Service load		

$$\frac{28,562 \text{ VA}}{240 \text{ V}} = 119 \text{ amperes}$$

Note: Earlier, the demand load was determined to be 182 amperes. The minimum neutral size is determined using *Section 220-22,* as shown earlier.

Farm Building Demand Load

A hog farrowing barn contains outlets for twenty heat lamps at 250 watts each, six 2-lamp, 40-watt fluorescent strips with a load of 1.6 amperes at 120 volts, and three electric fans operating at 240 volts: 1/6 horsepower, 2.2 amperes; 1/4 horsepower, 2.9 amperes; and 1/2 horsepower, 4.9 amperes. In addition, there are eight general-purpose 120-volt receptacle outlets.

1. Determine the minimum service demand load for the building. All loads are considered to operate without diversity except the general-purpose receptacle outlets. The heat lamps are considered a continuous load. The 120-volt heat lamp, receptacles, and lighting loads are divided by 240 volts rather than 120 volts, because all amperes for the service calculation must be on a 240-volt basis.

Loads operating without diversity

Heat lamps

$$\frac{20 \times 250\ \text{W} \times 1.25}{240\ \text{V}} =$$ 26.0 A

Lights

$$\frac{1.6\ \text{A} \times 120\ \text{V} \times 6 \times 1.25}{240\ \text{V}} =$$ 6.0 A

Fans

4.9 A × 1.25 =   6.1 A

2.9 A =   2.9 A

2.2 A =   2.2 A

43.2 A

Loads operating with diversity:

Receptacles

$$\frac{8 \times 180\ \text{VA}}{240\ \text{V}} =$$ 6.0 A

Loads operating *without* diversity 100%   43.2 A

Loads operating *with* diversity, but not   6.0 A

less than first 60 amperes of all loads at 100%

Total load   49.2 A

MAJOR CHANGES TO THE 1999 CODE

These are the changes to the 1999 *National Electrical Code*® that correspond to the Code sections studied in this unit. The following analysis explains the significance of only the changes from the 1996 to the 1999 Code, and this analysis is not intended to be used in place of the Code. Refer to the actual section of the 1999 Code for the exact wording and meaning of each section discussed. Changes are indicated in the Code with a vertical line in the margin. If material has been deleted or moved to another part of the Code, the location of the deletion is indicated with a dark dot in the margin.

Article 215   **Feeders**

There was some rearranging of the material in this article, and the minimum conductor size requirement of *Section 220-10(b)* was moved to *Section 215-2(a)*. Also, the minimum rating of overcurrent device for a feeder previously specified in *Section 220-10(b)* was moved to *Section 215-3*. No actual changes of significance occurred in this article.

215-1: The scope was modified by adding the words **overcurrent protection** to the list of feeder requirements included as a part of this article. This is not new. Overcurrent protection was omitted from the list in previous editions of the Code.

Article 220, Parts B, C, and D   **Feeder and Service Calculations**

220-12(b): This is a new paragraph that now makes it clear that when track lighting is added to a show window, it is to be included in any load calculation at 150 volt-amperes for each 2 feet of track and that this is in addition to the 200 volt-amperes per linear foot of show window. This is not a change of intent, but it was not clear that track lighting was to be included as an additional load.

220-16(a) Exception: When the kitchen refrigerator is installed on a separate 15-ampere branch circuit in a dwelling, it is not required to make any allowance for this load in the dwelling demand load calculation. It is already included as a part of the 1,500 volt-amperes for each of the small-appliance branch circuits.

220-18: A new sentence was added specifying how the electric dryer load is to be calculated when two or more single-phase clothes dryers are supplied from a 3-phase, 4-wire electrical system, which is frequently the case for multifamily dwellings. The method was never specified in the Code, but an example dealing with electric ranges was worked out in *Example 5(a)* in *Chapter 9* of the previous edition of the Code. Here is how the method works. Assume the electric clothes dryers are divided up as equally as possible between the phases. Assuming an apartment building has eight living units with an electric clothes dryer, then the maximum number of clothes dryers connected between any two phases would be three. That number is to be doubled (2 × 3 dryers = 6 dryers). Now look up the demand factor for six clothes dryers in *Table 220-18,* which is 0.70. Assuming the nameplate rating of the clothes dryers does not exceed 5 kilowatts, then the demand load for the electric clothes dryers for this multifamily dwelling is 5 kilovolt-amperes times six dryers times 0.70 (5 kVA × 6 dryers × 0.70 = 21 kVA). This is illustrated in Figure 4.5.

The remainder of the method for determining the total electric clothes dryer load required to be included as a part of the main 3-phase service demand load calculation is not described in the section. It is necessary to study *Example D5(a)* in *Appendix D* to determine the remainder of the method. In the previous example, the maximum demand load to be included for any two phases was 21 kilovolt-amperes. But that was two times the maximum load on any two phases, so that number is divided by two (21 kVA × 2 = 10.5 kVA), then multiplied by three to put it on a 3-phase basis (10.5 kVA × 3 = 31.5 kVA). The 31.5 kilovolt-amperes is the demand load on a 3-phase basis to be included in the service demand load calculation. A fine print note was added to this section referring to the calculation in *Example D5(a),* but this is an electric range calculation and it is not exactly the same for clothes dryers. The electric clothes dryer demand load calculation is to be performed as follows:

1. Determine the maximum number of clothes dryers connected between any two phases.

Section 220-18

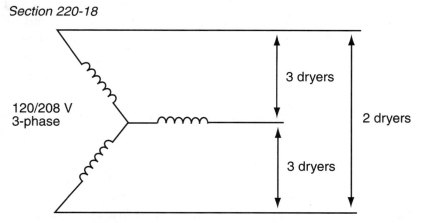

Figure 4.5   When multiple single-phase electric clothes dryers are supplied from a 3-phase, 120/208-volt, 4-wire service, the demand factor from *Table 220-18* is determined by multiplying the maximum number of dryers connected between any two phases by two.

2. Multiply the number from step 1 by two.
3. Look up the demand factor from *Table 220-18* for the number determined in step 2.
4. Multiply the demand factor from step 3 by 5 kilovolt-amperes, or the nameplate rating if higher, times the number determined in step 2.
5. Divide the number of kilovolt-amperes in step 4 by two.
6. Multiply the number of kilovolt-amperes from step 5 by three, and use this number in the service calculation.

220-21: For loads that are unlikely to be operating at the same time, this section permits the omission of the smaller load while only including the largest for load calculation. The previous edition of the Code required the loads to be **dissimilar** such as heating and air conditioning that did not operate at the same time. Now the loads are not required to be dissimilar. For example, two cooling compressors in a commercial building may be interlocked so they cannot operate at the same time. In this case, the larger of the two compressor loads will be included in the calculation.

220-30: This is an optional method for determining feeder and service loads for a dwelling unit. The section was completely revised; however, the only change applies when there is a heat pump in a dwelling with supplemental electric heating where the compressor and electric heaters are permitted to operate at the same time. The previous edition of the Code required the heat pump load to be added to the electric heat load and then multiplied by 0.65. Not both the heat pump load and the electric heat load are to be included at 100 percent. This is shown in *Example D2(c)*.

220-35(1) Exception: This section specifies the method of determining the demand load for an existing installation where there will be an additional load. It is permitted to determine the maximum demand load for the existing load in amperes on the most heavily loaded conductor over a period of not less than 30 days. The previous edition of the Code did not define demand load. Maximum demand is now defined as the average ampere load recorded over a 15-minute period. It is not the maximum amperes that would occur for a short time interval such as the in-rush current of a motor.

Article 230   **Services**

230-40 Exception 1: If there are several occupancies in a building, it is permitted to run one service drop or lateral to each occupancy. But what if one of the occupancies had a need for two different types of power, requiring a second service drop or lateral to that occupancy? The previous edition of the Code did not permit a second service drop or lateral to a single occupancy. Now more than one lateral or service drop is permitted to be run to a single occupancy in a multi-occupancy building if they are supplying different power or power for a different purpose that requires a separate service drop or lateral.

230-40 Exception 4: This is a new exception that permits an additional set of service conductors to be installed for the purpose of providing power to common area branch circuits for two-family or multifamily dwellings as illustrated in Figure 4.6.

230-42(a): The previous edition of the Code required that service entrance conductors be of sufficient size to carry the load as determined in *Article 220*. That is still true; however, now it states in this section as well as in *Article 220* that the conductors shall have an ampacity of not less than the noncontinuous load plus 1.25 times the continuous load. This is not a new rule for service conductor sizing because this language was in *Section 220-10(b)* of the previous edition of the Code.

Section 230-40 Exception 4

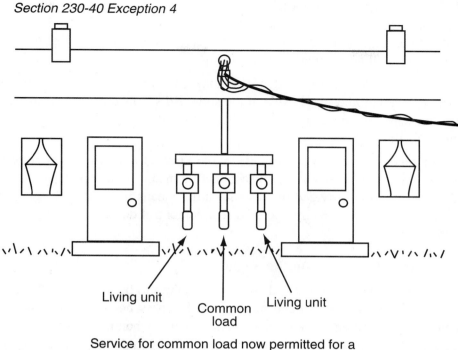

Living unit

Common load

Living unit

Service for common load now permitted for a two-family dwelling

Figure 4.6 **A third service is permitted for a multifamily dwelling when the third service is for the purpose of supplying a common load.**

230-46: Service entrance conductors may now be spliced or tapped by the use of clamped or bolted connections. Unless a listed underground splice kit is used for direct buried installation, enclosures are required for these splicing connections. The previous edition of the Code permitted these types of splices, but they were required to be placed in metering equipment enclosures. Another significant change in this section is that the previous edition of the Code required special permission of the authority having jurisdiction to extend underground service entrance conductors. Now this is no longer the case.

230-50: Protection requirements for cables used in the installation of a service entrance are outlined in this section. To make it clear that the protection requirements for cables apply to all types of service cables, the phrase **service entrance cable** was changed to **service cable.**

230-51(a): The phrase **service entrance cable** was changed to **service cable** to make it clear that if type MI, MC, or SE is used as the service cable, it is required to be supported within 12 inches of every weather head, gooseneck, or termination point and at intervals not exceeding 30 inches. Using the language from the previous edition of the Code, it could be interpreted that these support requirements applied only to type SE cable.

230-54(b): Now only type SE cable is permitted to be formed into a gooseneck in place of a raintight service head. If other types of service cables are installed, such as type MI or MC, a weather head is now required to be used.

230-79(c): All 3-wire, one-family dwellings are to have a service disconnecting means with a rating of at least 100 amperes. In the previous edition of the Code, this 100-ampere requirement applied only when the computed load was greater than 10 kilovolt-amperes or when there were six or more 2-wire branch circuits. Now there is no case where a single-family dwelling is permitted to have a service rated less than 100 amperes.

230-92: The word **occupant** was added to this section to make it clear to whom the service overcurrent device was not accessible. If the service overcurrent device is not readily accessible to an **occupant,** then branch circuit devices must be installed on the load side at a readily accessible location to the occupant.

230-204(a): An **air-break** isolation switch is no longer the only type of device that is permitted to be installed on the supply side of the service disconnecting means as visible break contacts for services operating at more than 600 volts when nonvisible means of overcurrent protection are employed.

230-205(c): For services exceeding 600 volts that are part of a multibuilding industrial complex under single management, the disconnecting means is now permitted to be located in a separate building provided the disconnecting means is capable of being opened by remote control. The electrically operated remote-control device is required to be readily accessible.

Article 600   **Electric Signs and Outline Lighting**

600-4(b): Incandescent lamp illuminated signs are required to be marked with the input voltage and current rating. A wattage marking with letters not less than 1/4 inch high is required to be located on the sign in a location visible when relamping.

600-5(b): The previous edition of the Code limited the rating of a branch circuit to 20 amperes for circuits that supplied incandescent, fluorescent, and high-intensity discharge lighting systems for electric signs and outline lighting. The 20-ampere maximum restriction no longer applies to high-intensity discharge lighting systems for signs and outline lighting.

600-6: This section deals with disconnecting means for electric signs. There is a new requirement for signs and outline lighting systems installed within fountains. For these installations, a disconnect must be located within sight of the sign and at least a horizontal distance of not less than 5 feet from the inside of the fountain.

600-7: Listed liquidtight flexible metal conduit in lengths not to exceed 100 feet is permitted to enclose the secondary conductors. Also, small metal parts with a dimension of not greater than 2 inches and not likely to become energized are not required to be grounded if spaced not more than 3/4 inch from electric discharge tubing and the system operates at not over 100 hertz. If the system operates at more than 100 hertz, the metal parts must be spaced 1 3/4 inches from electric discharge tubing.

600-23(f): This new paragraph requires marking of transformers and electronic power supplies if secondary fault protection is provided.

600-32(a): This section specifies the wiring methods permitted for the secondary circuit operating at more than 1,000 volts from a transformer or an electronic power supply. Electrical nonmetallic tubing, liquidtight flexible nonmetallic conduit, and rigid nonmetallic conduit are no longer permitted to be used for such installations unless they contain only one conductor and, depending upon operating frequency, are installed with a minimum spacing from grounded metal parts. When exposed in close proximity to more than 1,000 volts, the nonmetallic materials tend to break down.

Appendix D   **Examples**

There were no changes in the examples except *Example D2(c)*. See explanation of *Section 220-30* for the change. The examples were moved from *Chapter 9* in the previous edition of the Code to *Appendix D* at the end of the Code.

WORKSHEET NO. 4
SERVICES AND
FEEDER CALCULATIONS

These questions are considered important to understanding the application of the *National Electrical Code®* to electrical wiring, and they are questions frequently asked by electricians and electrical inspectors. People working in the electrical trade must continue to study the Code to improve their understanding and ability to apply the Code properly.

DIRECTIONS: Answer the questions and provide the Code reference or references where the necessary information leading to the answer is found. Space is provided for brief notes and calculations. An electronic calculator will be helpful in making calculations. You will keep this worksheet; therefore, you must put the answer and Code reference on the answer sheet as well.

1. Service entrance cable, type SE, used for a service entrance to a building, as shown in Figure 4.7, shall be supported by straps or similar means within 12 inches of the service head or gooseneck and at intervals not to exceed _____ inches.

   Code reference _____

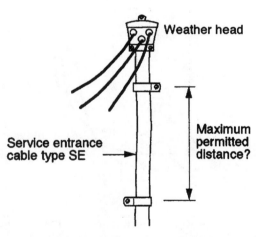

Weather head

**Service entrance
cable type SE**

**Maximum
permitted
distance?**

Figure 4.7   **What is the maximum spacing permitted for support of service entrance cable on the outside of a building?**

2. If you have an air-conditioning load and a space heating load in a building, do they both have to be included in the service load calculation? (Yes or No)

   Code reference _____

3. A 3-phase, 208/120-volt wye service entrance for a commercial building is rated at 1,200 amperes. Is this service entrance required to be provided with ground-fault protection of the equipment? (Yes or No)

   Code reference _____

4. A small retail store has an area of 2,100 square feet and contains 30 general-purpose receptacle outlets, each on a separate strap and not considered to be supplying continuous loads. Outside lighting, not for a sign, at the front of the store is 800 watts. There is a furnace with a 1/3-horsepower electric motor operating at 120 volts. This is a 120/240-volt, 3-wire, single-phase service. The store does not have a show window. Determine the minimum service feeder demand load in amperes for the store, which is at ground level for the purpose of determining the rating of a single service overcurrent device.

   Code reference _____

5. Is the service head required to extend above the point where the service drop attaches to a building, as shown in Figure 4.8? Assume no obstructions are above the point of attachment. (Yes or No)

   Code reference _____

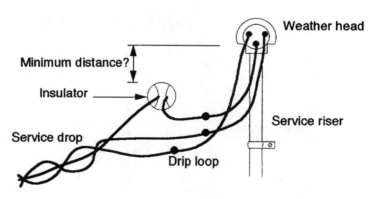

Figure 4.8   A service drop is terminated to the side of a building adjacent to a service mast.

6. A single-family dwelling service by a 120/240-volt, 3-wire service entrance with a single main overcurrent device rated at 150 amperes is installed using aluminum type SE cable that has conductors with an insulation rating of 75°C. Determine the minimum size aluminum conductor permitted for the service entrance.

   Code reference _____

7. A multifamily dwelling contains the following electric ranges: 8 ranges at 10 kilowatts, 12 ranges at 12 kilowatts, and 10 ranges at 14 kilowatts. Determine the minimum range demand load used for sizing the service entrance for this multifamily dwelling.

   Code reference _____

8. A multifamily dwelling consists of 12 single-family living units, each with a 1.2 kilovolt-ampere dishwasher. Determine the minimum permitted demand load for the 12 dishwashers to be used in the service entrance calculation, using the method of *Article 220, Part B*.

   Code reference _____

9. A dwelling service mast extends up through the roof overhang with the service drop connected to the 2-inch rigid conduit mast, as shown in Figure 4.9. The roof slope is greater than 4 inches of rise in 12 inches of run, and the feeder operates with 240 volts maximum between conductors. The minimum permitted distance between the insulator and the roof as shown in Figure 4.9 is _____ inches.

   Code reference _____

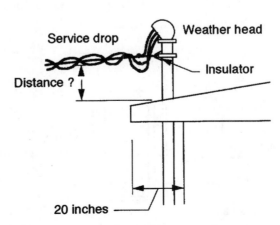

Figure 4.9   A mast riser for a service entrance extends up through a roof overhang.

10. A single-family dwelling unit in an apartment building has a total living area of 958 square feet. Laundry facilities are provided elsewhere in the building, available to all residents. The living unit contains a 12-kilowatt electric range, 3.5-kilowatt electric water heater at 240 volts, two air conditioners, each of which draws 8 amperes at 240 volts, a furnace that draws 6 amperes at 120 volts, and a 1.2-kilowatt dishwasher at 120 volts. Determine the minimum demand load in amperes of the 120/240-volt subfeeder to each living unit, using the method of *Article 220, Part B.*

   Code reference _____

11. Determine the minimum permitted demand ampere rating (unbalanced load) of the neutral conductor for the subfeeder to the living unit in the apartment building described in Problem 10.

   Code reference _____

12. Determine the minimum number of 15-ampere general-illumination branch circuits required for a 660-square-foot addition to a dwelling. The only electrical loads in the addition are lights and receptacles.

   Code reference _____

13. A service drop terminates at an insulator attached to the side of a building. The insulator is attached beside a window, as shown in Figure 4.10. The minimum permitted distance between the insulator and edge of the window that can be opened is _____ feet.

   Code reference _____

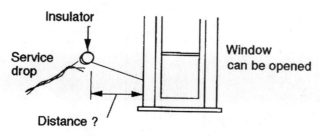

Figure 4.10   A service drop is terminated to the side of a building adjacent to a window that can be opened.

14. A single-family dwelling has a total living area of 1,850 square feet. Appliances are a 12-kilowatt electric range, a 4.5-kikowatt electric water heater, a 1.2-kilovolt-ampere dishwasher (120 volts), central air conditioning consisting of 17 amperes at 240 volts, a furnace of 6 amperes at 120 volts, and a 5-kilowatt clothes dryer. Determine the minimum demand load in amperes for the 120/240-volt service entrance of the dwelling using the optional method.

    Code reference _____

15. The service entrance in an older dwelling with a 120/240-volt, 100-ampere service is to be upgraded for resale. The living area is 2,200 square feet, and the loads are a 4-kilowatt water heater, a 5-kilowatt clothes dryer, a 1.2-kilovolt-ampere dishwasher, a 12-kilowatt electric range, and a furnace that draws 6 amperes at 120 volts. Determine the minimum permitted service entrance demand load using the optional method. The dwelling will have two small appliance circuits.

    Code reference _____

16. A farm building electrical load consists of mostly 240-volt, single-phase electric motors with only two 20-ampere, 120-volt general circuits. It is unlikely that additional 120-volt load will be added. If the service is 150 amperes using AWG number 1/0, type THWN copper ungrounded service conductors, what is the minimum size copper neutral service entrance conductor permitted for this service?

    Code reference _____

17. A farm building load consists of 92 amperes of load expected to operate without diversity and 76 amperes of load that operates with diversity. These load amperes are on a 240-volt basis. Determine the minimum ampere demand load used to size the service equipment for the building.

    Code reference _____

18. A farm consists of the following buildings and building demand loads. Determine the calculated minimum ampere rating of the main farm service.

    dwelling   74 amperes
    barn   66 amperes
    shop   36 amperes
    cattle   24 amperes
    storage   20 amperes

    Code reference _____

19. A commercial building has a demand load of 130 amperes; however, the electrician will install a 200-ampere service. Determine the minimum size type THWN aluminum service entrance conductors permitted if the wires are in conduit.

    Code reference _____

20. A large electric motor installation is served with a 480-volt corner grounded delta 3-phase system as shown in Figure 4.11. Running overcurrent protection for the motor is provided by thermal devices in the motor starter. Is a fuse, which does not serve as overload protection, permitted to be installed in the grounded phase conductor of the service equipment serving the motor? (Yes or No)

    Code reference _____

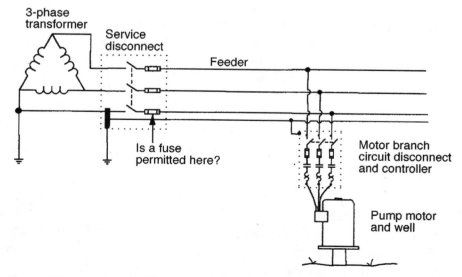

Figure 4.11 A fusible disconnect switch is installed for a single large motor containing fuses sized for short-circuit and ground-fault protection that have a rating too high to serve as running overcurrent protection.

21. A multifamily dwelling consists of eight individual dwelling units. Each dwelling unit is 800 square feet and contains a 10-kilowatt range, 2.5-kilowatt water heater, 1.2-kilowatt dishwasher, 2.0-kilowatt electric space heat, and an 8-ampere, 240-volt air conditioner. The house load consists of 2,600 watts of general lighting, two 5-kilowatt clothes dryers, and two laundry circuits. Determine the minimum demand load in amperes used to determine the minimum permitted rating for the 120/240-volt single-phase service.

    Code reference _____

22. The minimum rating permitted for a 3-wire service entrance for a dwelling with six or more branch circuits is _____ amperes.

    Code reference _____

23. If a commercial storage building only has a limited demand load, it is permitted to be served by a 60-ampere service. (True or False)

    Code reference _____

24. A service entrance to a building is permitted to consist of a panelboard with a single main overcurrent device and three fusible disconnect switches grouped together, as shown in Figure 4.12. The panelboard and disconnect switches are marked as suitable for use as service equipment, and all four enclosures are tapped from the service entrance conductors. (True or False)

    Code reference _____

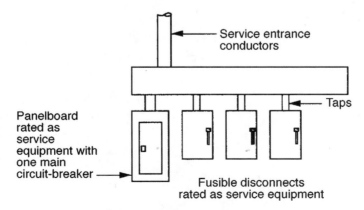

Figure 4.12 A service entrance consists of a panelboard and three fusible disconnect switches tapped from the service entrance conductors.

25. An electric sign is not protected from physical damage, and it is located above an area accessible to vehicles. The minimum clearance from the bottom of the sign to the paved vehicle drive surface shall not be less than _____ feet.

Code reference _____

ANSWER SHEET

Name _____

No. 4   SERVICES AND FEEDER CALCULATIONS

Answer	Code reference
1. _____	_____
2. _____	_____
3. _____	_____
4. _____	_____
5. _____	_____
6. _____	_____
7. _____	_____
8. _____	_____
9. _____	_____
10. _____	_____
11. _____	_____
12. _____	_____
13. _____	_____
14. _____	_____
15. _____	_____
16. _____	_____
17. _____	_____
18. _____	_____
19. _____	_____
20. _____	_____
21. _____	_____
22. _____	_____
23. _____	_____
24. _____	_____
25. _____	_____

UNIT 5

Grounding
and Bonding

OBJECTIVES

Upon completion of this unit, the student will be able to:

- explain the purpose of equipment grounding.
- explain the purpose of electrical system grounding, and define bonding.
- diagram a single-phase, dual-voltage, 3-wire electrical system and label the voltages between the wires.
- diagram a 3-phase wye and a 3-phase delta electrical system and label the voltages between the wires.
- state when a single-phase and a 3-phase electrical system are required to be grounded if the voltages between the wires are known.
- show which wire of an electrical system is required to be grounded if the electrical system is one required to be grounded.
- name at least five materials considered by the Code as acceptable as a means of grounding electrical equipment not in a location with specific requirements.
- determine the minimum size equipment grounding conductor permitted for a branch circuit and feeder if the rating of overcurrent protection is known.
- state the minimum requirement for a grounding electrode for a service entrance to a building.
- determine the minimum size of grounding electrode conductor permitted for a particular service entrance.
- specify the type required and the minimum size permitted of bonding conductor for a swimming pool.
- answer wiring installation questions relating to *Articles 250, 280, 680,* and *Appendix E.*
- state at least five significant changes that occurred from the 1996 to the 1999 Code for *Articles 250, 280, 680,* or *Appendix E.*

CODE DISCUSSION

The emphasis of this unit is to discuss the purpose of equipment grounding and of system grounding. Equipment grounding is easily understood if the electrical installer has a clear understanding of the purpose of equipment grounding, an understanding of the circuit involved, and an understanding of the requirements of a good equipment grounding conductor. This unit deals with grounding and bonding of electrical systems in general. There are several locations or types of materials with special grounding requirements. These cases will be discussed in later units. For example, requirements for cable tray to be used as an equipment grounding conductor will be covered in Unit 12, grounding of electrical equipment in agricultural buildings will be covered in Unit 14, hazardous locations in Unit 9, and health care facilities in Unit 10. Only the specific grounding and bonding requirements different from *Article 250* are covered in other articles of the Code.

Article 250 deals with the grounding of an electrical system and the grounding of equipment. The requirements of this article apply to all wiring installations, unless specifically different or additional grounding requirements are covered elsewhere in the Code. *Section 250-4* references the other locations in the Code where there are different or additional grounding requirements for specific materials or locations. System grounding and equipment grounding are illustrated in Figure 5.1.

Parts B and *C* of *Article 250* cover requirements for the grounding of an electrical system. *Section 250-20* specifies when an electrical system is to be grounded, and *Section 250-26* specifies which wire of the electrical system is to be grounded. *Section 250-32* covers the situation where one building is supplied power from another building. This section should be studied carefully. In the case of agricultural buildings, some additional grounding requirements are covered in *Article 547*.

Parts E and *F* cover which enclosures and equipment shall be grounded and how the grounding is accomplished. In general, it is not permitted to connect the grounded circuit conductor to the equipment grounding conductor except at the service entrance. There are a few exceptions to this rule, and they are covered in *Sections 250-140* and *250-142*.

Bonding is the electrical connection of metal equipment so it acts as one solid piece of equipment from the standpoint of electrical current flow. Bonding is done when there is question about the effectiveness of the connection between pieces of noncurrent-carrying metal equipment likely to become energized. An example is where a nonmetallic piece of conduit is used to connect two metal enclosures. A wire is run from one enclosure to the other to bond them together. Bonding of the service equipment is of particular importance. Extra precautions are taken at the service equipment that are not usually taken in other locations. The reason is that the service entrance conductors are protected from overcurrent at the load end of the

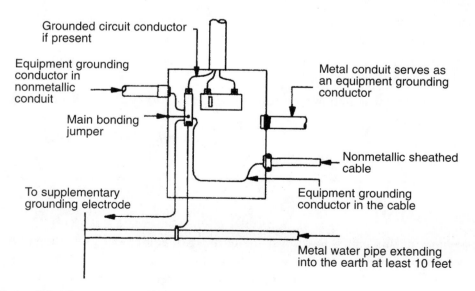

Figure 5.1   System grounding and equipment grounding are required to be considered in all electrical systems.

conductors, not at the supply end. If a fault should occur, it is particularly important that the metal equipment be capable of conducting the available fault current.

Part C covers grounding electrodes for service equipment. The grounding electrodes listed in *Section 250-50* are required to be used if present, and those listed in *Section 250-52* are used if the previous ones are not available. *Section 250-118* covers the materials permitted to be used as equipment grounding conductors. *Section 250-64* covers the installation of the grounding electrode conductor, which is the conductor from the service entrance to the grounding electrode. The minimum size permitted for this grounding electrode conductor is determined from *Table 250-66*. The minimum size equipment grounding conductor permitted to be used for a circuit or feeder is found in *Table 250-122*. It is important to understand the purpose of each of these tables so they will not be confused when selecting grounding wire sizes. Equipment grounding conductor size is determined from *Table 250-122* based on the rating of the circuit or feeder overcurrent device. System grounding electrode conductor size is determined from *Table 250-66* based on the size of the ungrounded service conductors.

Article 280 covers specifications for and the installation of a surge arrester installed on an electrical system. These requirements do not apply to a surge arrester associated with a primary electrical distribution system of an electrical power supplier. The *National Electrical Safety Code®* covers these installations.

A surge arrester is installed between the ungrounded conductors and the grounded conductor or earth. Figure 5.2 shows a metal oxide surge arrester installed at the service entrance drip loop of a building. A typical application for a surge arrester is to allow an elevated voltage on an ungrounded conductor due to lightning to discharge directly to earth through the arrester. Typically, arresters conduct current when a design breakdown voltage is exceeded. Below the breakdown voltage, the arrester acts like an open circuit.

Article 680 covers the wiring of equipment associated with swimming pools, fountains, and similar equipment. Requirements are placed on the wiring installed in the area of pools, fountains, and similar installations. The types of installations covered are swimming, wading, therapeutic and decorative pools; fountains; hot tubs; spas; and hydromassage bathtubs. The terms **pool** and **fountain** are defined in *Section 680-4*. Definitions important to the installation of equipment in these areas are covered in *Section 680-4*.

Differences in voltage between two places around the pool, fountain, or similar installation may be of a sufficient level to create a hazard. To help prevent hazardous conditions from developing, metal parts of the swimming pool are required by *Section 680-22* to be bonded

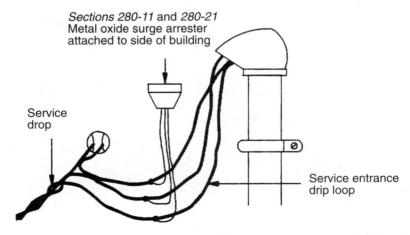

Figure 5.2 A surge arrester installed at a service entrance drip loop for lightning protection.

together with a solid copper wire not smaller than AWG number 8. Metal raceway, metal-sheathed cable, metal piping, and other metal parts fixed in place within 5 feet horizontally of the inside edge of the pool shall also be bonded together. All metal parts shall be bonded to a common bonding grid. This is illustrated in Figure 5.3. The specifications for a common bonding grid are not specifically stated in the Code except in *Section 680-22(b)*, which states that structural reinforcing steel of a concrete pool or the wall of a bolted or welded metal pool is permitted to serve as the common bonding grid. A solid copper wire with a minimum size of AWG number 8 and either insulated, covered, or bare is permitted to serve as the common bonding grid. No specifications are given concerning the installation of this solid copper wire to form the common bonding grid.

At least one 125-volt receptacle outlet, not for the water-circulating pump, is required to be installed not more than 20 feet and not closer than 10 feet from the inside edge of a permanent swimming pool as stated in *Section 680-6(a)*. A fine print note in this section covers the situation in which a receptacle outlet beyond a barrier or wall is within 20 feet of the edge of a permanent swimming pool. All 125-volt receptacle outlets within 20 feet of the inside edge of the pool shall be protected by a ground-fault circuit-interrupter. These same rules apply in the case of a spa or hot tub installed on the outside of a building, as stated in *Section 680-40* and illustrated in Figure 5.4. In the case of a spa or hot tub installed on the inside of a building, at least one 125-volt rated receptacle outlet, which is required to be ground-fault circuit-interrupter protected, is required to be installed at a distance of not more than 10 feet from the inside edge of the spa or hot tub, and not closer than 5 feet.

GROUNDING AND BONDING FUNDAMENTALS

Grounding and bonding is one of the most important parts of an electrical wiring system, and at the same time it is one subject of the Code that seems to be confusing. It is essential to understand the purpose of grounding and bonding, and then it will be easier to make sure the installation has met the requirements of the Code. Grounding and bonding is the key element to the backup safety system for an electrical circuit or electrical equipment. If the grounding system does not function properly, then other safety devices such as circuit-breakers and fuses may not function when a problem develops. Grounding and bonding must be installed with as

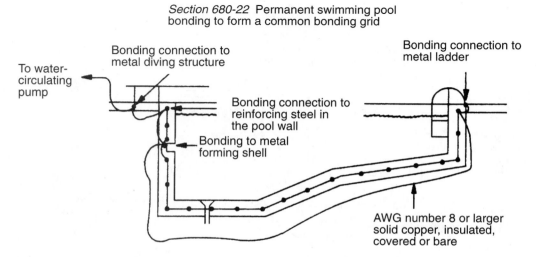

Figure 5.3   Metal parts of a permanent swimming pool are required to be bonded together to reduce the likelihood of voltage gradients in the pool area.

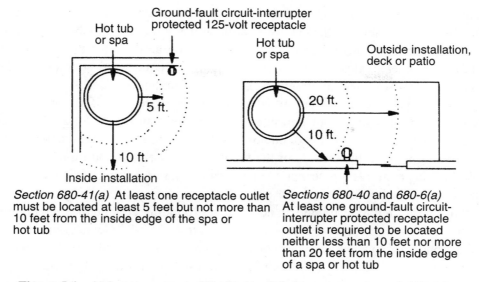

Hot tub or spa

Ground-fault circuit-interrupter protected 125-volt receptacle

Hot tub or spa

Outside installation, deck or patio

5 ft.

20 ft.

10 ft.

10 ft.

Inside installation

Section 680-41(a) At least one receptacle outlet must be located at least 5 feet but not more than 10 feet from the inside edge of the spa or hot tub

Sections 680-40 and *680-6(a)* At least one ground-fault circuit-interrupter protected receptacle outlet is required to be located neither less than 10 feet nor more than 20 feet from the inside edge of a spa or hot tub

Figure 5.4   At least one ground-fault circuit-interrupter protected 125-volt receptacle outlet for an inside installation is required to be installed at least 5 feet but not more than 10 feet from the inside edge of an inside spa or hot tub.

much care as any other part of the electrical system. One Code section in particular discusses important information necessary to understand system and equipment grounding and bonding. Read *Section 250-2* to get an explanation of the purpose of grounding.

Equipment Grounding

The equipment grounding conductor is permitted to be a metal raceway or a metal box, cabinet, or frame of equipment. These and other permitted means of providing an equipment grounding conductor are covered in *Section 250-118*. The equipment grounding conductor is permitted to be copper or aluminum wire. There is a restriction in the case of aluminum wire used for grounding. Aluminum as an equipment grounding wire is not permitted to be installed where it is in direct contact with masonry or the earth, or where subject to corrosive conditions. When aluminum wire is exposed and used outside for grounding, it is not permitted to be installed within 18 inches of the earth. These restrictions on aluminum and copper-clad aluminum conductors are found in *Section 250-120* and *Section 250-64*.

Equipment grounding wires are permitted to be solid or stranded, insulated or bare. If the wire is insulated, it is required to have insulation that is green or green with yellow stripes. These requirements are found in *Section 250-119,* and for flexible cords, *Section 400-23*. If the equipment grounding wire is AWG number 4 or larger, it is permitted to be identified as an equipment grounding wire at the time of installation at every location where it is accessible. Acceptable means of identifying an equipment grounding conductor of size AWG number 4 or larger that is not bare, green, or green with yellow stripes are: (1) to strip the insulation from the entire exposed wire, leaving the wire bare, or (2) to cover the entire exposed wire with green tape or green paint.

All fittings, conduit, splices, and any other connections in the equipment grounding conductor shall be made up tight with proper tools. Flexible metal conduit is not considered to be as effective an equipment grounding conductor as rigid metal conduit, intermediate metal conduit (IMC), or electrical metallic tubing (EMT). Therefore, limitations are placed on the use of flexible metal conduit and liquidtight flexible metal conduit for use as an equipment grounding conductor. These restrictions are given in *Section 250-118*. If these requirements are not satisfied, then an equipment grounding wire shall be provided to bond from the enclosure at one end of the flexible metal conduit to the enclosure at the other end of the flexible metal conduit.

There are several methods of installing and terminating an equipment bonding jumper from one end to the other of flexible conduit. Installation of equipment bonding jumpers is covered in *Section 250-102*. The equipment bonding or grounding jumper is permitted to be run on the inside of a raceway, or it is permitted to be run on the outside. If run on the outside, the equipment bonding jumper is not permitted to be more than 6 feet long. The bonding jumper shall be routed with the raceway or enclosure. Special fittings are manufactured for terminating the bonding wire on the outside of the flexible metal conduit and liquidtight flexible metal conduit. Various methods are acceptable for terminating an equipment bonding jumper or an equipment grounding wire at an enclosure.

The methods permitted to terminate an equipment grounding conductor or an equipment bonding jumper to an enclosure are covered in *Section 250-8*. Connecting an equipment grounding conductor to a metal box or enclosure is required to be done with a device listed for the purpose or by means of a grounding screw used for no other purpose. When an equipment grounding conductor is terminated at a lug in an enclosure, the screw or bolt used to connect the lug to the enclosure shall be used for no other purpose. Connections or fittings that depend on solder are not permitted. This does not prevent the use of solder at a connection where the connection is made mechanically secure before soldering. *Section 250-148* requires that the removal of a device at a box or enclosure shall not interrupt the continuity of the equipment grounding conductor.

Sizing Equipment Grounding Conductors

An equipment grounding wire size is based on the size of the overcurrent device protecting the circuit. *Table 250-122* in the Code lists the minimum wire size permitted for various overcurrent device ratings. In the case of electrical cable, the manufacturer has installed the correct size of equipment grounding wire based on the maximum size overcurrent device rating permitted to be used for that cable. When raceway is permitted as an equipment grounding conductor, cross-sectional area of the metal is adequate for the largest size wires permitted in the raceway. Usually, the only time when an equipment grounding wire must be sized is in the case of nonmetallic conduit or for bonding jumpers. The following examples illustrate how to determine the minimum permitted size of equipment grounding conductor for a branch circuit or a feeder.

Example 5.1   Determine the minimum size copper equipment grounding conductor for a 30-ampere circuit consisting of type THWN copper wires size AWG number 10 installed in rigid nonmetallic conduit.

Answer: From *Table 250-122,* the minimum size copper equipment grounding conductor permitted is AWG number 10.

Example 5.2   A 7-foot length of liquidtight flexible metal conduit extends from a fusible disconnect switch to an electric furnace. The wires inside the flexible conduit are type THHN copper size AWG number 4 with 70-ampere overcurrent protection. Determine the minimum size copper equipment grounding conductor to bond around the liquidtight flexible metal conduit. There is an equipment grounding lug in the furnace for terminating the equipment grounding conductor.

Answer: A 70-ampere overcurrent device is not listed in *Table 250-122*; therefore, the next larger overcurrent device rating shall be used. For this example, use the equipment grounding wire size required for a 100-ampere overcurrent device. The minimum size copper equipment grounding wire size for this 70-ampere circuit is AWG number 8.

Parallel Equipment Grounding Conductors

In *Section 310-4,* conductors are permitted to be run in parallel. In general, the minimum conductor size permitted to be installed in parallel is AWG number 1/0. This minimum size applies to ungrounded conductors and to grounded conductors. There is no mention of equipment grounding conductors, bonding conductors, or grounding electrode conductors. *Section 250-122(f)* covers the situation where conductors are run in parallel. When the conductors are run in parallel sets within more than one raceway where an equipment grounding conductor is required, the equipment grounding conductors are required to be run in parallel. Each of the separate equipment grounding conductors required is determined using *Table 250-122* and based on the size of the circuit overcurrent device rating. Example 5.3 illustrates how the size of parallel equipment grounding conductors is determined.

Example 5.3 A feeder protected with 500-ampere time-delay fuses is run with two parallel sets of conductors in separate rigid nonmetallic conduits. Determine the minimum size of copper equipment grounding conductor required in each conduit run of the feeder.

Answer: A separate equipment grounding conductor is required to be run with each set of conductors. Each of the equipment grounding conductors is required to be sized according to *Table 250-122.* The minimum size copper equipment grounding conductor required for a 500-ampere overcurrent device is AWG number 2.

Electrical Shock

Electrical shock is caused by the flow of electrical current through the body or a part of the body. Current passing through the body can cause burns, muscle reaction, and injury to body organs. If the current travels through the head or central part of the body, severe injury or even death may occur. Dry skin has a high resistance to the flow of electricity. With under 50 volts, enough current generally will not flow through the body to be harmful to a human. This is not necessarily true for a human in wet conditions or for an animal standing in a moist environment. The moist hoof or foot of a farm animal offers low resistance to the flow of electrical current. The main factors involved in electrical shock to an animal or human are:

- voltage
- duration of the electrical shock
- condition of skin or body surface
- surface area contact with the source
- path the current takes

Voltage and condition of the skin usually determine the amount of current that flows. A normal person usually will just begin to feel the sensation of shock if 1/1000 ampere or 1 milliampere of 60 hertz alternating current flows through the body. Shock may become painful to humans with a continuous flow of 8 milliamperes or more. Muscular contraction (cannot let go) may occur in humans with a continuous current flow of 15 milliamperes or more. Greater flows of current through the body are usually very serious. An effective equipment grounding system is important for the protection of people and animals with the damp and wet conditions that exist in many residential, commercial, and farm locations.

Electrical System Grounding

A building wiring system usually must have one conductor grounded to the earth. This grounding helps limit voltages caused by lightning surges. Grounding prevents high voltages from occurring on the wiring if a primary line should accidentally contact a secondary wire. Grounding limits the maximum voltage to ground from the hot wires. For most residential, commercial, and farm wiring systems, the maximum voltage to ground is not more than 125 volts. Several decisions must be made in the grounding of an electrical system to the earth. Three major decisions are which wire is permitted to be grounded, the type of grounding electrode, and determining the minimum permitted size of grounding electrode conductor.

Which Conductor of the Electrical System to Ground

The service entrance wire required to be grounded is described in Code *Section 250-20(b)* for an alternating current system. The grounded circuit conductor shall have white or gray insulation. Identification of the grounded circuit conductor is covered in *Section 200-6*. If the wire is larger than AWG number 6, the wire is permitted to be labeled with white or gray tape or paint where the wire is visible. The following electrical systems are required to be grounded:

- Single-phase, 2-wire, 120 volts
- Single-phase, 3-wire, 120/240 volts
- Three-phase, 4-wire, 120/208 volts, wye
- Three-phase, 4-wire, 120/240 volts, delta
- Three-phase, 4-wire, 277/480 volts, wye, when supplying phase to neutral loads
- Three-phase, 4-wire, 347/600 volts, wye, when supplying phase to neutral loads

Grounding Electrode

The grounding electrode is the means by which electrical contact is made with the earth. The type of electrode permitted to be used for grounding an electrical system is given in *Sections 250-50* and *250-52*. The following grounding electrodes are covered in *Section 250-50* and shall be used if available.

- Metal underground water pipe in direct contact with the earth for at least 10 feet. A metal underground water pipe shall be supplemented by at least one additional electrode
- The metal frame of the building, where the metal frame is effectively grounded
- A bare copper wire at least 20 feet long and not smaller than AWG number 4 encased in concrete in contact with the earth, with at least 2 inches of concrete around all sides of the wire
- A bare copper wire circling the building at a depth of at least 2 1/2 feet, size AWG number 2 or larger, and at least 20 feet long

If more than one of the grounding electrodes described in *Section 250-50* are available at the building, they all shall be electrically connected together with a grounding wire to form a grounding electrode system. If none of the grounding electrodes are available, then one grounding electrode described in *Section 250-52* shall be used. The grounding electrodes permitted to be used, as described in *Section 250-52,* are:

- A rod or pipe electrode driven to a depth of at least 8 feet into the soil. If rock bottom is encountered, it shall be permitted to drive the pipe or rod at an angle of not less than 45 degrees from the horizontal. Or it may be laid in a trench at least 2 1/2 feet deep. The acceptable rods and pipes shall have the following minimum diameters: (1) 3/4-inch galvanized steel pipe, (2) 5/8-inch iron or steel rod, and (3) 1/2-inch copper-coated rod

- A plate electrode shall be used in areas where soil conditions prevent the use of a pipe or rod electrode. The plate shall make contact with at least 2 square feet of soil
- Other metal underground structures or equipment

Section 250-104(b) requires that the above-ground portion of a metal gas piping system upstream from the equipment shutoff valve be bonded to the grounding electrode system of the electrical system. *Section 250-52(a)* does not permit a metal underground gas piping system to serve as a grounding electrode for the electrical system. This required bonding may make the metal gas piping system a grounding electrode, but it is not permitted to be counted as one of the required grounding electrodes.

Rod, pipe, and plate electrodes sometimes do not make good low-resistance contact with the earth. They must be placed in areas not subject to damage but where the soil is most likely to be moist. They should not be installed in areas protected from the weather, such as under-roof overhangs. The Code specifies in *Section 250-56* that a single rod, pipe, or plate electrode shall have a resistance to ground not exceeding 25 ohms. If the resistance to earth is greater than 25 ohms, then one additional rod, pipe, plate, or other electrode shall be installed. These electrodes shall be at least 6 feet apart.

Lightning protection system grounding electrodes are not permitted to be used as one of the required electrodes for the electrical system. However, electrodes for the electrical and lightning systems shall be electrically connected together, as specified in *Section 250-60*.

Grounding and Bonding for Service with Parallel Conductors

Several components of the grounding and bonding system must be sized when a service entrance consists of two or more parallel sets of service entrance conductors. The grounding and bonding conductors may all be different sizes as illustrated in Figure 5.5. The minimum permitted size of grounding electrode conductor is found in *Section 250-66*. When there are

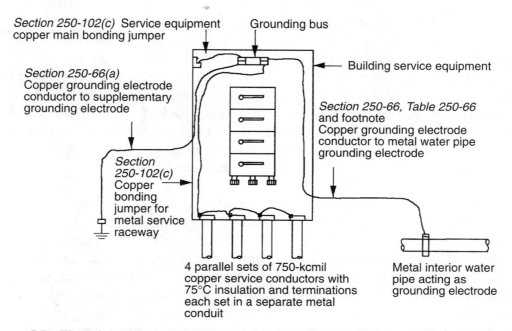

Figure 5.5   The minimum permitted size of the grounding electrode conductor and the main bonding jumper are determined based on the sum of the cross-sectional areas of the conductors of any one phase.

two or more parallel sets of service entrance conductors, the minimum size of grounding electrode conductor is determined based on the sum of the cross-sectional areas of the ungrounded conductors for any one particular leg or phase. This information is found in *Note 1 to Table 250-66*. With some 3-phase electrical systems, such as a 4-wire delta system, the phase conductors may not all be the same size. In this case, the phase that yields the largest sum of the cross-sectional areas of the conductors is used for the grounding electrode conductor determination. Note from *Table 250-66* that the largest size of grounding electrode conductor required to be installed for the service of Figure 5.5 is AWG number 3/0 copper or 250-kcmil aluminum. Some grounding electrodes, such as a ground rod, have a limited ability to dissipate ground fault current. Therefore, *Section 250-66(a)* permits the grounding electrode conductor to be smaller in some cases than would normally be required by *Table 250-66*.

Bonding of metal enclosures and raceways is required for service equipment. What must be bonded is covered in *Section 250-92*, the method of bonding is covered in *Section 250-94*, and the minimum permitted size of bonding jumper is specified in *Section 250-102*. A bonding means may be provided with the service equipment, or a copper or aluminum wire may be used as the bonding means. The main bonding jumper connects the service equipment enclosure to the grounding electrode conductor and to the grounded service conductor if the supply system has a grounded conductor. In the case of Figure 5.5, the main bonding jumper connects the grounding bus to the service equipment enclosure. The minimum permitted size of this main bonding jumper is determined from *Section 250-102(c)*. This section is somewhat confusing in the case of parallel service entrance conductors. First, the main bonding jumper is not permitted to be sized smaller than the size specified in *Table 250-66* for the service entrance conductors where the cross-sectional area to use is the sum of the cross-sectional areas of all conductors of one phase. For example, if there are two sets of 500-kcmil conductors in parallel for the service entrance, the cross-sectional area to be used is 1,000 kcmil. Secondly, when the cross-sectional area of the service entrance conductors for one phase is determined to be larger than 1,100-kcmil copper or 1,750-kcmil aluminum, then the minimum permitted size of conductor for the main bonding jumper is 12.5 percent (0.125 times the cross-sectional area) of the cross-sectional area of the conductors. The main bonding jumper in this latter case will be larger than the grounding electrode conductor.

If the service entrance conductors are run in metal raceway, then the metal raceway must be bonded to the service equipment enclosure or to the service grounding bus. *Section 250-102(c)* explains how to determine the minimum size of service raceway bonding jumper. The minimum size is based on the size of the largest conductors in the raceway. If more than one set of parallel conductors are in a single raceway for each phase, then the cross-sectional area is taken as the sum of the cross-sectional areas of the individual conductors for that phase. Example 5.4 illustrates sizing of grounding electrode conductors and bonding jumpers for a service entrance with parallel sets of service entrance conductors.

Example 5.4   The conductors for a service entrance consist of four parallel sets of 750-kcmil copper conductors with 75°C insulation and terminations. Each set of conductors is run in a separate rigid metal conduit as shown in Figure 5.5. All bonding and grounding electrode conductors are copper. The metal water pipe is used as a grounding electrode, and it is supplemented with a driven ground rod. A separate copper grounding electrode conductor is run from the grounding bus in the service equipment to each grounding electrode. The grounding bus is bonded to the service equipment enclosure with a copper main bonding jumper. A single copper conductor is run from the service equipment grounding bus to a bonding bushing on each service entrance conduit. Determine the minimum permitted size of the following:

a. the copper grounding electrode conductor to the water pipe
b. the copper grounding electrode conductor to the ground rod
c. the copper main bonding jumper from the grounding bus to the service equipment enclosure
d. the copper bonding jumper from the grounding bus to each service conductor conduit

Answer: a. The minimum permitted size of the copper grounding electrode conductor to the water pipe is determined from *Section 250-66, Table 250-66,* and the footnote to the table. First, determine the total cross-sectional area of the conductors of one phase by adding the cross-sectional areas of all conductors of that phase to get 3,000 kcmil. This is larger than 1,100 kcmil; therefore, the minimum permitted size is AWG number 3/0.

$$750 \text{ kcmil} \times 4 = 3{,}000 \text{ kcmil}$$

b. *Section 250-66* does not require the grounding electrode conductor to a made electrode such as a ground rod to be larger than AWG number 6.

c. *Section 250-102(c)* requires the main bonding jumper to not be smaller than specified in *Table 250-66* or not less than 0.125 times the cross-sectional area of the conductors of one phase, whichever is larger. In this case, it is 0.125 times 3,000 kcmil that gives 375 kcmil. If one of the standard wire sizes from the Code is used, the minimum wire size can be determined from *Table 310-16* or *Table 8* of *Chapter 9.* The minimum standard wire size would be 400 kcmil.

$$3{,}000 \text{ kcmil} \times 0.125 = 375 \text{ kcmil}$$

d. The minimum permitted size of bonding jumper for the service entrance conduit is based on the phase conductor cross-sectional area contained in the conduit according to *Section 250-102(c).* In this case, the phase conductor size is 750 kcmil, and the minimum permitted raceway bonding jumper size from *Table 250-66* is AWG number 2/0 copper.

Electrical System Types and Voltages

Selecting the type of electrical supply system for a commercial, farm, or industrial application involves careful consideration of several factors, some of which are: (1) the size of the electrical service required, (2) the voltages required, (3) the anticipated kilowatt-hour usage, (4) the electrical energy rates available, and (5) the size and number of electrical motors. An electric power supplier customer service representative is usually available to help evaluate these points when making a choice for the type of electrical system for a particular application.

Single-Phase Electrical Systems

Some buildings are served by a 2-wire, 120-volt electrical system, as shown in Figure 5.6. These systems are permitted as long as the building needs are limited to two circuits. This type of system is permitted to be installed in an outbuilding on a property where there is little electrical usage in the building. A 2-wire, 120-volt electrical system can be derived from a separate transformer, or it can be derived from a 120/240-volt, 3-wire, single-phase system or from a 120/208-volt, 4-wire, 3-phase electrical system.

The most common electrical system for dwellings is the 120/240-volt, 3-wire, single-phase system shown in Figure 5.7. The wire originating at the transformer center tap is grounded. The voltage from the grounded neutral wire to each ungrounded wire is 120 volts. A load powered at 240 volts, such as a motor, is supplied by the two ungrounded wires. The power

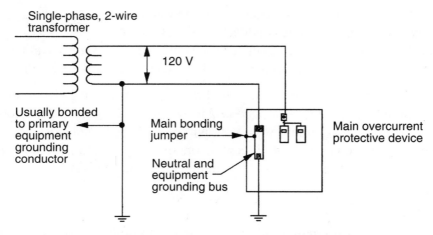

Figure 5.6   A 120-volt, 2-wire, single-phase electrical system.

supplier grounds the common conductor at the transformer, and the common conductor is required by *Section 250-26* to be grounded at the service disconnecting means.

It is important to balance the 120-volt load so that half is connected between the neutral and each ungrounded wire. If the 120-volt loads are perfectly balanced, the current flowing on the neutral will be zero. Loads operating at 240 volts do not use the neutral; therefore, they do not place current on the neutral.

Three-Phase Electrical Systems

Three-phase electrical systems are of different types and voltages to fit different needs. There are three common types of delta electrical systems. One type of delta 3-phase electrical system is the 3-wire ungrounded electrical system. *Section 250-20* permits the 240-volt and the 480-volt 3-phase delta 3-wire systems to be operated without one of the conductors grounded. A 480-volt ungrounded 3-wire delta electrical system is illustrated in Figure 5.8.

It is important to remember that equipment is required to be grounded as specified in *Sections 250-110* and *250-112*. *Section 250-130(b)* states that the equipment grounding conductors shall be bonded to the grounding electrode conductor of the ungrounded system.

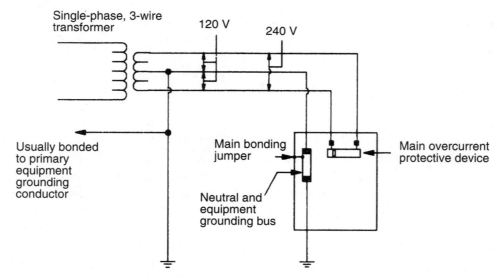

Figure 5.7   A 120/240-volt, single-phase, 3-wire electrical system.

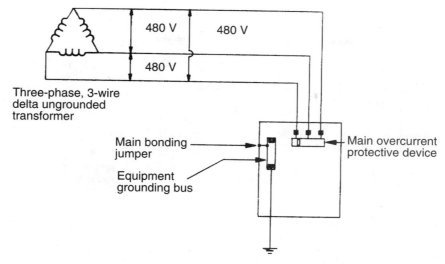

Figure 5.8 **Three-phase delta 3-wire ungrounded 480-volt electrical system.**

This can be confusing. The system is considered ungrounded because one circuit conductor is not grounded. But a grounding electrode is required for the service.

A 3-phase, 3-wire electrical system operating either at 240 volts between phases or 480 volts between phases is permitted to be operated with one of the phase conductors grounded. This is frequently called a corner grounded delta 3-phase electrical system. A 3-wire, corner grounded, 3-phase electrical system is shown in Figure 5.9. *Section 240-22* states that an overcurrent device is not permitted to be installed in a grounded conductor unless all ungrounded conductors are opened when the overcurrent device operates. An overcurrent device is permitted to be installed in the grounded phase conductor when it serves as overload protection for an electric motor, as permitted in *Section 430-36.*

The 4-wire, 120/240-volt delta 3-phase electrical system provides single-phase and 3-phase service at 240 volts and single-phase at 120 volts. This system is shown in Figure 5.10. There is a single-phase neutral conductor with this system. The voltage from two of the

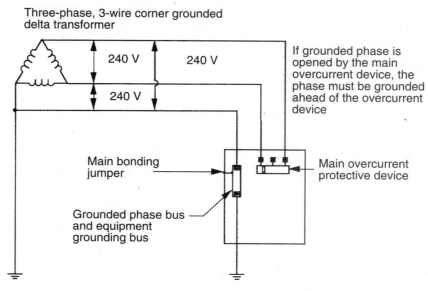

Figure 5.9 **A three-phase, 3-wire delta electrical system is permitted to have one of the phase conductors grounded.**

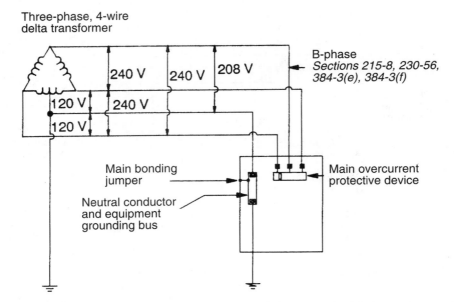

Figure 5.10   A 3-phase, 4-wire delta, 120/240-volt electrical system.

ungrounded phase conductors to the neutral is 120 volts. Voltage from one of the phase conductors to the neutral is 208 volts. This phase conductor is called the phase with the higher voltage to ground in *Sections 215-8, 230-56,* and *384-3(e).* In *Section 384-3(f),* the phase conductor with the higher voltage to ground is required to be the B-phase, which is required to be the center conductor in a disconnect or panelboard. There is an exception in the case of some metering equipment. The phase conductor with the higher voltage to ground is not intended to be combined with the neutral because the voltage is 208 instead of 120. Single-phase and 3-phase loads are permitted to be served from the same panel, but this must be done with care to keep the loads balanced on the phase conductors and the neutral conductor.

Any 3-phase delta electrical system can be operated with only two transformers. This is called an open delta system. If the electrical system of Figure 5.10 was operated as an open delta electrical system, the transformer between the B-phase and the A-phase or the transformer between the B-phase and the C-phase would be missing. The voltages delivered by this open delta electrical system should be the same as for the full delta 3-phase system. There may be a slight voltage variation between phase conductors.

Three-phase power may be supplied using a 4-wire wye electrical system. Figure 5.11 shows a 120/208-volt, 3-phase wye electrical system. A 277/480-volt, 4-wire wye, 3-phase system is also available. The voltage from any one of the ungrounded phase conductors to the neutral is the same with this system.

It is possible to install the neutral-to-phase circuits to balance the neutral-to-phase loads on the three transformers. This is frequently done using multiwire branch circuits with one neutral common to three circuit conductors. Motors normally designed for 240-volt operation must not be connected to a 120/208-volt wye system unless marked on the nameplate as suitable for operation at 208 volts. Motors rated at 200 to 240 volts are available. With this system, equipment must be specified with 208-volt motors. The 3-phase, 4-wire, 277/480-volt wye electrical system is used commonly for industrial applications. Motors are powered using 3-phase, 480-volt circuits, and single-phase electric discharge lighting circuits are operated at 277 volts.

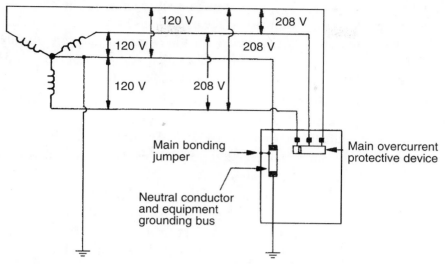

Figure 5.11   A 3-phase, 4-wire wye, 208/120-volt electrical system.

High-Impedance Grounding

High-impedance (high-resistance) grounding is permitted for some industrial applications at 480 volts and above and where qualified maintenance staff are always on duty to immediately investigate ground-faults. This type of electrical system is carefully engineered for each application. It provides an alternative to the ungrounded 480- and 600-volt industrial electrical system. The typical application is for industrial processes where orderly shutdown is necessary. If a ground-fault occurs somewhere on the system, it may not be desirable to allow that circuit or feeder to immediately shut down. On an ungrounded or high-resistance grounded system, a signal indicates that a ground-fault has occurred on a particular phase, and the maintenance personnel immediately find the fault and assess the seriousness of the condition. To make repairs, usually a controlled shutdown can be accomplished. Sometimes repairs can be made without shutting down the system. The resistor installed in series with the equipment grounding conductor limits the value of fault current to a level that an overcurrent device will not open, but the fault condition will be indicated.

With this type of system, the neutral conductors and the equipment grounding conductors are separated, and a resistor is connected between the neutral bus and the equipment grounding bus. Specifications for installing a high-impedance grounded electrical system are found in *Section 250-36.*

MAJOR CHANGES TO THE 1999 CODE

These are the changes to the 1999 *National Electrical Code®* that correspond to the Code sections studied in this unit. The following analysis explains the significance of only the changes from the 1996 to the 1999 Code, and this analysis is not intended to be used in place of the Code. Refer to the actual section of the 1999 Code for the exact wording and meaning of each section discussed. Changes are indicated in the Code with a vertical line in the margin. If material has been deleted or moved to another part of the Code, the location of the deletion is indicated with a dark dot in the margin.

Article 250   **Grounding**

The entire article was reorganized to group similar topics together in one place. Also there was major rewriting to eliminate as many of the exceptions as practical. Requirements about grounding in specific articles were deleted if the requirements were already covered in *Article 250*. A new *Appendix E* lists grounding topics along with the section number in the 1996 Code and the new section number in the 1999 Code.

250-2(c): This section is similar to *Section 250-80* of the previous edition of the Code that required bonding to interior metal water piping systems, structural steel, and similar metal piping systems likely to become energized. The only change in this requirement is that it now specifically states that an interior metal gas piping system shall be bonded to the supply system grounded conductor.

250-6(e): This is a new section directed at addressing the stray voltage issue on farms where a dc current presumably from a cathodic protection system is the source of the dc current. A listed dc isolating device is permitted to be installed in series with the equipment grounding conductor of a circuit or feeder to block the dc current but provides ac current coupling so that any ac fault current can pass through the device.

250-24(b)(2): The change in this section deals with determining the minimum permitted size of grounded circuit conductor in each raceway when service conductors are run in parallel with service conductors run in separate raceways. The minimum size under any circumstance is AWG number 1/0. It is now made clear that the minimum size of grounded circuit conductors in each raceway is determined from *Table 250-66* using the largest ungrounded conductor, or 12.5 percent of the largest conductor cross-sectional area in that raceway. This is illustrated in Figure 5.12. Other than requiring the grounded circuit conductor to be not smaller than AWG number 1/0, there was no method for determining the minimum size of conductor when the grounded circuit conductors were run in parallel in two or more raceways.

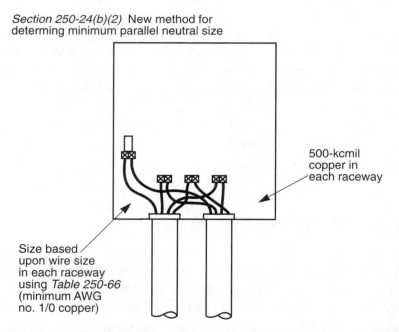

Section 250-24(b)(2) New method for determing minimum parallel neutral size

500-kcmil copper in each raceway

Size based upon wire size in each raceway using *Table 250-66* (minimum AWG no. 1/0 copper)

Figure 5.12   The neutral size in each raceway is based on the largest ungrounded conductor size in each raceway, but never smaller than AWG number 1/0. Make sure the combined ampacity of the parallel neutrals is sufficient to carry the load as calculated using *Section 220-22*.

250-24(d): This is a new requirement, but the intent is not new. The previous edition of the Code did not have a requirement of how a grounding electrode conductor was to be connected for an ungrounded ac electrical system. It now states that a grounding electrode conductor be connected to the **metal enclosure of the service conductors** at any accessible point from the load end of the service drop or lateral to the disconnecting means. Usually the most convenient means of connection is to an equipment grounding terminal or bus located inside the service disconnect enclosure.

250-30: This section specifies the grounding and bonding required for separately derived systems. There are some changes. This section is now separated into subsection (a), dealing with separately derived systems required to be **grounded,** and subsection (b), dealing with separately derived systems that are **not** grounded. In the previous edition of the Code, *Section 250-26* did not require a grounding electrode and grounding electrode conductor for separately derived systems that were not required to be grounded.

250-30(a)(1): This section requires a bonding jumper to connect a separately derived system grounded circuit conductor to the equipment grounding conductors. This was *Section 250-26(a)* in the previous edition of the Code; it required a bonding jumper and seemed to imply only one, but did not make that clear. The bonding jumper is now required to be at the same location as the grounding electrode conductor connection, as illustrated in Figure 5.13. The previous edition of the Code permitted the bonding jumper to be at the separately derived system, such as at a transformer, and the grounding electrode conductor to be at the disconnecting means supplied by the separately derived system. Now they both must be either at the transformer or at the disconnecting means.

250-30(a)(1) Exception 1: A bonding jumper is now permitted to be at more than one location as long as they are installed at the separately derived system, or at the disconnecting

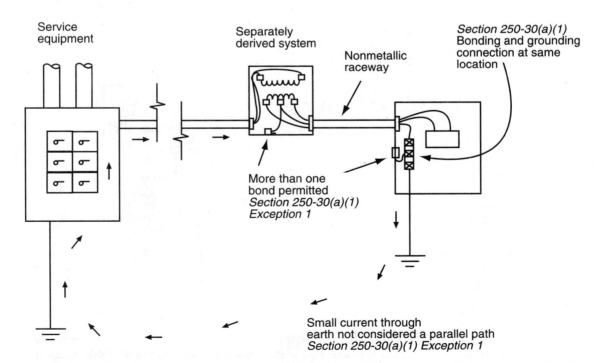

Figure 5.13   The bonding connection and the grounding electrode connection are required to be made at the same location; however, more than one bonding connection is permitted.

means, or at any point between these locations. The restriction is that a parallel path is not permitted for grounded circuit conductor current to follow except that some current flow in the earth is permitted. These points are also illustrated in Figure 5.13.

250-30(a)(3): This section states the grounding electrode requirement for a separately derived system. The change is with respect to the location for connection to a metal water piping system when it is used as the grounding electrode for a separately derived system. Only one grounding electrode is required, and it is not required to use the metal water pipe if another grounding electrode is used. But if the metal water pipe is used, connection is now required to be made within 5 feet of the point where it enters the building regardless of the distance from the separately derived system to that point, as illustrated in Figure 5.14. The previous edition of the Code seemed to imply that connection to a metal water pipe was to be at the nearest available point. This was stated in *Section 250-26(c)* and *Section 250-80(a)* of the previous edition of the Code. The separately derived system is to be grounded to the nearest grounding electrode, which could be the effectively grounded structural steel of the building or a ground rod.

250-30(a)(3)(b) Exception: This exception permits, for commercial and industrial buildings, grounding to a metal water pipe near the separately derived system as long as the water pipe is completely visible from the point of connection to the point where the water pipe enters the building.

250-30(a)(3) Exception: This is a new exception that deals with the situation where the service equipment for a building also serves as the connection to a separately derived system such as an on-site generator through a transfer switch. As long as this is listed equipment, one grounding electrode system and one grounding electrode conductor is permitted to be common for the service to the building and the separately derived system.

250-30(b): This is a new subsection that now requires a grounding electrode and grounding electrode conductor to be installed for an ungrounded separately derived electrical

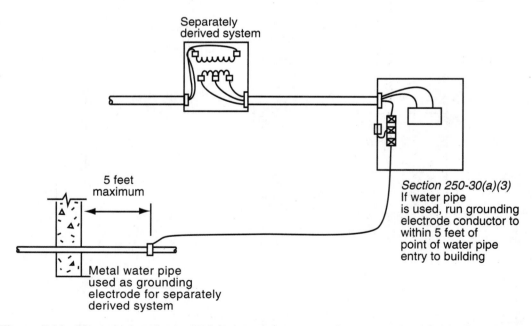

Figure 5.14 If a metal underground water piping system is used as the grounding electrode for a separately derived system, the connection to the water pipe shall be made within 5 feet of the point where it enters the building, except for commercial and industrial buildings where the water pipe is visible from the point of connection to the point of entry to the building.

system. The requirements for a grounding electrode are the same as for a grounded separately derived system. Even though the separately derived system is not grounded, grounding of equipment supplied by the system is to be as specified in other locations of the Code.

250-32(b)(2): Sentence (2) is a new requirement. If there is continuous metal, such as a metal water pipe, structural steel, or equipment making a common connection between the building supplied and the building providing the supply of power, then an equipment grounding conductor is required to be installed with the supply conductors to the building. Paragraph (1) then prohibits a connection between the grounded circuit conductor and the grounding electrode, as illustrated in Figure 5.15. The purpose is to prevent grounded conductor current (usually the neutral) from flowing through the metal connection between the two buildings rather than all of the current flowing on the conductors supplying the building.

Sentence (3) is a new requirement. If the common service supplying the buildings is provided with equipment ground-fault protection, an equipment grounding conductor is required to be run to all additional buildings and a connection between the grounded circuit conductor and the building grounding electrode is prohibited, as illustrated in Figure 5.16. This is necessary to ensure proper operation of the equipment ground-fault protection at the main service to the property.

250-50(a): A new sentence was added to make it clear that when a metal underground water pipe is supplemented with a rod, pipe, or plate electrode, the provisions of *Section*

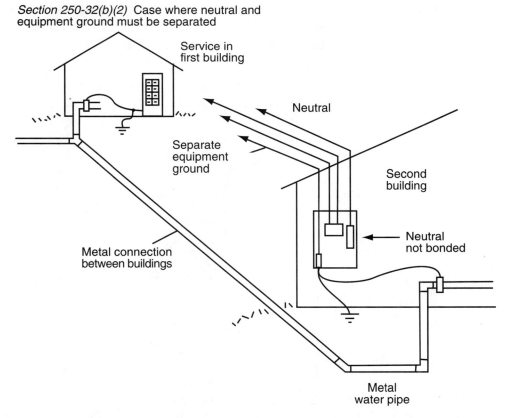

Section 250-32(b)(2) Case where neutral and equipment ground must be separated

Service in first building

Neutral

Separate equipment ground

Second building

Neutral not bonded

Metal connection between buildings

Metal water pipe

Figure 5.15   If there is a metal connection between a building supplied power from another building that would result in a parallel path to the neutral conductor, then it is required to run a separate neutral and equipment grounding conductor to the second building.

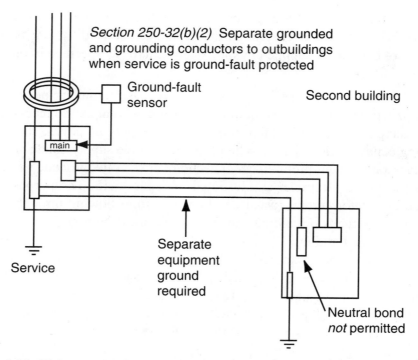

Figure 5.16 **If the service to a group of buildings is provided with equipment ground-fault protection, then the grounded circuit conductor and the equipment grounding conductor to any other building supplied from that service shall be separated.**

250-56 are required to be met. If the rod, pipe, or plate electrode does not have a resistance to earth of 25 ohms or less, it shall be supplemented with an additional electrode located not closer than 6 feet, as illustrated in Figure 5.17. This is not a change of intent, but it was not clear in the past.

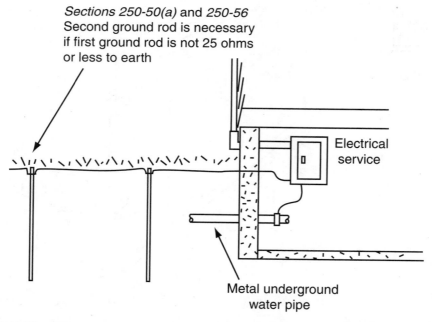

Figure 5.17 **When a grounding electrode is installed as a supplementary grounding electrode for a water pipe electrode, if the ground rod does not have a resistance of 25 ohms or less to earth, it is required to be supplemented with one additional electrode.**

250-52(d): This section describes a plate-type **made** electrode as one of several types of made electrodes that can be installed if the grounding electrodes described in *Section 250-50* are not available. The previous edition of the Code did not specify a minimum depth of burial for a plate electrode. Now the minimum depth is 2.5 feet below the surface of the earth.

Table 250-66: This table gives the minimum size of grounding electrode conductors for services and **separately derived systems.** There are no changes in this table that was *Table 250-94* in the previous edition of the Code.

250-104(a)(3): The minimum size of bonding conductor permitted from the panelboard equipment grounding terminal to the interior metal water pipe in an outbuilding supplied from another building is determined according to *Table 250-122.* This situation is related to *Section 250-32.* The bonding jumper to the water pipe is sized based on the rating of overcurrent device protecting the feeder to the building. In the previous edition of the Code, this was *Section 250-80(a),* and the bonding jumper to the water pipe was sized using *Table 250-94,* which in the new Code is *Table 250-66.*

250-104(a)(4): This section deals with the bonding of a separately derived system to the nearest available point on the interior metal water piping system in the area served by the separately derived system. This is basically the same rule that was in *Section 250-80(a)* of the previous edition of the Code. It now states that the bonding connection for the metal interior water piping system can be made at any point from the separately derived system to the first disconnecting means or overcurrent device, as illustrated in Figure 5.18. This was not specified in the previous edition of the Code. The bond to the metal water piping system is required to be made at the location where the separately derived system is connected to the grounding electrode. In the case where the separately derived system does not have a disconnecting means or overcurrent device, then the bond to the metal interior water piping system is permitted to be made only at the source of the separately derived system. It no longer states in this section that the connection is required to be accessible.

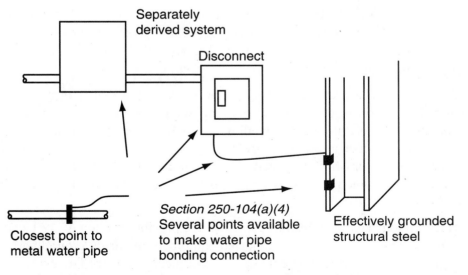

Separately derived system

Disconnect

Closest point to metal water pipe

Section 250-104(a)(4)
Several points available to make water pipe bonding connection

Effectively grounded structural steel

Figure 5.18   The bonding connection to an interior metal water piping system is made from the nearest available point on the water piping system to any one of several convenient points.

Section 250-30(a)(3) requires that if the metal water piping system is used as the grounding electrode for the separately derived system, the connection of the grounding electrode conductor is required to be made not more than 5 feet from the point where the metal water pipe enters the buildings. This section requires a connection to a metal interior water piping system at the nearest available point even if it is not being used as a grounding electrode. But if it is being used as a grounding electrode, there will most likely be two connections to the water pipe: one connection not more than 5 feet from the point where the water pipe enters the building, and the other connection at the nearest available point.

250-104(b): A bonding connection shall be made to the above-ground portion of a metal gas piping system upstream from the shutoff valve.

250-112(k): If electrical equipment is skid-mounted, this new section requires that metal skids as well as the electrical equipment enclosures mounted on the skids are required to be grounded with a bonding jumper sized the same as equipment grounding conductors in *Table 250-122*.

250-122(f): This new paragraph applies in the case where feeder conductors are run in parallel in separate cables or conductors run in separate raceways. An equipment grounding conductor is required to be run in each cable or raceway, and it is required to be sized to the overcurrent device protecting the feeder as specified in *Table 250-122*. Cables are not available with the equipment grounding conductor sized larger than required for the expected rating of the ungrounded conductors.

For example, if 500-kcmil copper ungrounded conductors are run in cable and are protected by a 400-ampere device, the equipment grounding conductor will most likely be AWG number 3. But if two of the same cables were run in parallel for an 800-ampere feeder, the minimum size equipment grounding conductor required in each cable would be AWG number 1/0. Cables are not available with 500-kcmil ungrounded conductors and an AWG number 1/0 equipment grounding conductor.

The change is that now the Code permits the equipment grounding conductor to be sized to the conductor rating in each cable or raceway if conditions of maintenance and supervision ensure that only qualified persons will service the installation, and equipment ground-fault protection is provided for the feeder and set at a level based upon the conductors in each parallel cable or raceway as illustrated in Figure 5.19. For the previous example, it would be necessary to set the ground-fault protection at 400 amperes in order to be permitted to use a cable with 500-kcmil ungrounded conductors and an AWG number 3 equipment grounding conductor.

Table 250-122 Note: The previous edition of the Code stated that it **may** be necessary to increase the size of an equipment grounding conductor in order to meet the intent of *Section 250-51*. The equipment grounding conductor is required to have sufficiently low impedance to conduct any fault current likely to occur. Now it states the equipment grounding conductor **shall be** sized larger if necessary to meet the requirements of *Section 250-2(d)*, which is the new section that contains the requirements previously contained in *Section 250-51*.

250-138(a) Exception: This was *Section 250-59* in the previous edition of the Code. This section requires that cord-and-plug-connected equipment requiring grounding be grounded by means of a fixed grounding contact on the attachment plug. The exception that permitted a self-restoring grounding contact on hand-held tools and appliances was deleted.

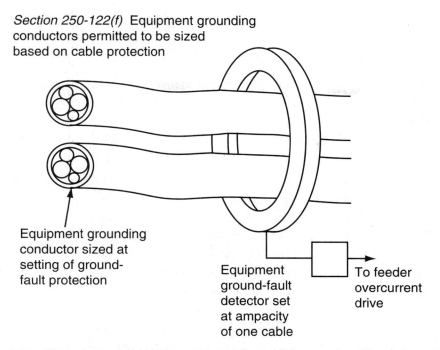

Section 250-122(f) Equipment grounding conductors permitted to be sized based on cable protection

Equipment grounding conductor sized at setting of ground-fault protection

Equipment ground-fault detector set at ampacity of one cable

To feeder overcurrent drive

Figure 5.19 **When a feeder consists of several parallel sets of cables and equipment ground-fault protection is set at the ampere rating of the cable, the equipment grounding conductor in the cable is permitted to be sized to the cable ground-fault protection rating.**

Now a fixed grounding contact is required on the attachment plug. But there is a new exception that permits a self-restoring grounding contact on plug-in type ground-fault circuit-interrupters that operate at not over 150 volts between conductors and not over 150 volts from any conductor to ground.

250-169: This is a new section dealing with the grounding of an ungrounded direct current separately derived system. The electrical system itself may not be grounded, but a grounding electrode is required to be connected using a grounding electrode wire to the enclosure of the separately derived system source or the first disconnecting means. The minimum size of the grounding electrode conductor is as specified in *Table 250-166*.

Article 280 **Surge Arresters**

There were no significant changes made to this article.

Article 680 **Swimming Pools, Fountains, and Similar Installations**

680-3(b): A new paragraph was added that requires audio equipment adjacent to pools and fountains to be installed according to the rules in *Article 640*. Underwater loudspeakers are required to meet the requirements of *Section 680-23*.

680-4: There is a new definition of **packaged therapeutic tub or hydrotherapeutic tank equipment assembly.** It is an assembly that may include pumps, air blowers, heaters, lights, controls, and sanitizer generators. Also, there is a new definition of **self contained therapeutic tubs or hydrotherapeutic tanks.** This is an assembly that contains a therapeutic tub or hydrotherapeutic tank and also may include pumps, air blowers, heaters, lights, controls, and a sanitizer generator.

680-6(a)(1): This section permits a ground-fault protected, locking-type single receptacle at a dwelling to be installed between 5 feet and 10 feet of the edge of a pool if the receptacle supplies a water pump motor. Now other ground-fault protected, locking-type single receptacles are permitted to supply other equipment associated with the water-circulating system or sanitation system if located at least 10 feet from the inside edge of the pool.

680-6(d): This is a new section that requires ground-fault circuit-interrupter protection for pump motors at locations other than dwellings if the circuit is rated 125 or 240 volts, 15 or 20 amperes. This rule applies if the motor is cord-and-plug-connected or direct wired. The section does not make reference to motors operating at 208 volts, but this situation was probably also intended to be included in the rule.

680-8: The minimum clearance from the level of water in a pool to an insulated ungrounded service drop conductor not operating at more than 750 volts to ground was raised from 18 feet to 22 feet because this change was made in the *National Electrical Safety Code®*. The minimum clearance from a diving platform or tower in any direction was increased from 16 feet to 17 feet. This clearance applies to service drop conductors or other supply conductors operating at not more than 15,000 volts. This same change in clearance was made in the *National Electrical Safety Code®*.

680-20(b)(1): An insulated copper AWG number 8 equipment grounding wire is not required to be run to the forming shell of a wet-niche lighting fixture if the lighting system is a **listed** low-voltage lighting system not requiring grounding. There are listed 12-volt lighting fixtures made for this purpose that have a transformer located up to 25 feet away. An equipment grounding conductor is run to the transformer, but one is not required to be run to the lighting fixture.

680-22(a)(1): A new last sentence was added. This section requires reinforcing bars in concrete of a pool to be bonded. The new sentence does not require bonding of reinforcing bars that at the time of manufacture are coated with a listed encapsulating nonconductive compound.

680-25(b)(3) Exception: This is a new exception that permits liquidtight flexible metal conduit or liquidtight flexible nonmetallic conduit to connect a transformer for pool lights. The intent is to permit some flexibility of the conduit when servicing or replacing the transformer. Any one length of flexible conduit is not permitted to be more than 6 feet. If more than one section of flexible conduit is used in a circuit, the total combined length of the flexible conduit sections is not permitted to exceed 10 feet.

680-25(c): In the case of a one-family dwelling, a motor associated with a pool is permitted to have the circuit run in nonmetallic-sheathed cable for the portion of the run inside of the dwelling. Now when such a motor is supplied from another building or structure associated with a one-family dwelling, it is permitted to have the circuit run as nonmetallic-sheathed cable.

680-25(d): If a panelboard or disconnecting means supplies pool equipment and is supplied from a separately derived system such as a transformer, the equipment grounding conductor run from the separately derived system to the panelboard is not permitted to be smaller than the conductor size determined from *Table 250-66* based upon the largest ungrounded conductor size or not smaller than AWG number 8. This is the case when the panelboard is supplied directly from the separately derived system and the first overcurrent device is located in the panelboard, but the grounding electrode conductor and main bonding jumper are located at the separately derived system. The previous edition of the Code based this conductor size on the minimum equipment grounding conductor size from *Table 250-95* but not smaller than AWG number 12.

680-38: An emergency shutoff switch is now required for the recirculation and jet system of a spa or hot tub. It is to be clearly marked and located within sight of the spa or hot tub, but not closer than 5 feet. This applies to all installations except single-family dwellings.

680-41(a)(1): Receptacles are not permitted to be located within 5 feet of the inside wall of a spa or hot tub installed indoors. The change is that this distance is now required to be measured horizontally from the inside edge of the spa or hot tub as illustrated in Figure 5.20.

680-57: New requirements have been added for the installation of electric signs in fountains. The signs are to have no moving parts and they are to be located at least 5 feet from the outer edge of the fountain to minimize potential contact with persons. All circuits supplying the sign shall have **personnel** ground-fault circuit-interrupter protection. A disconnect and bonding that meets the requirements of *Article 600* shall also be provided.

680-62(a): The previous edition of the Code required that all therapeutic tubs be ground-fault circuit-interrupter protected. This was not practical for fixed therapeutic tubs installed in commercial facilities and powered from a 3-phase circuit operating at more than 250 volts or requiring circuits that supply greater than 50 amperes. This new section does not require GFCI protection for field-assembled therapeutic tubs with a heater drawing more than 50 amperes, or operating at more than 250 volts, or supplied by 3-phase power.

680-70: All 125-volt, single-phase receptacles located within 5 feet from the inside edge of a hydromassage bathtub are required to be ground-fault circuit-interrupter protected. The change is that the 5-foot distance is measured horizontally from the inside edge of the hydromassage bathtub, not the shortest distance.

680-72: A new section was added that requires electrical equipment for hydromassage bathtubs to be accessible without the necessity of damaging the building structure or finish.

680-73: This is a new section requiring bonding of metal parts and equipment that are a part of a hydromassage bathtub, including metal piping and drains. The bonding conductor is required to be solid copper and not smaller than AWG number 8. The conductor can be bare, insulated, or covered. If equipment is listed as having a system of double insulation and does not require grounding, then that equipment is not required to be bonded.

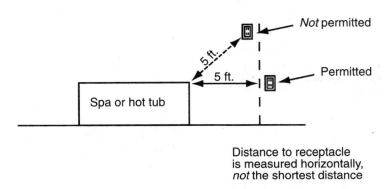

Section 680-41(a)(1)
Spa or hot tub installed indoors

Distance to receptacle
is measured horizontally,
not the shortest distance

**Figure 5.20   Receptacles are not permitted to be located closer than
5 feet measured horizontally from the inside edge of a spa or hot tub.**

WORKSHEET NO. 5
GROUNDING AND BONDING

These questions are considered important to understanding the application of the *National Electrical Code®* to electrical wiring, and they are questions frequently asked by electricians and electrical inspectors. People working in the electrical trade must continue to study the Code to improve their understanding and ability to apply the Code properly.

DIRECTIONS: Answer the questions and provide the Code reference or references where the necessary information leading to the answer is found. Space is provided for brief notes and calculations. An electronic calculator will be helpful in making calculations. You will keep this worksheet; therefore, you must put the answer and Code reference on the answer sheet as well.

1. The only available grounding electrode for the dwelling shown in Figure 5.21 is a driven ground rod. If the 100-ampere service is supplied with AWG number 2 aluminum conductors, the minimum size bare copper grounding electrode conductor permitted is AWG number _____ . Assume the ground rod will be driven just below the surface of the earth, the clamp is suitable for direct burial, and the grounding electrode conductor is located where it is not exposed to physical abuse.

    Code reference _____

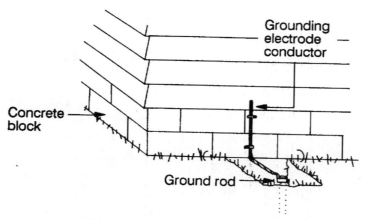

Figure 5.21   **The grounding electrode wire for a service entrance is fastened to the outside surface of the building.**

2. A commercial building is served by a 400-ampere, 120/208-volt, 3-phase wye electrical system, and the service entrance conductors are 500-kcmil copper with type THWN insulation. The 3/4-inch galvanized steel–reinforcing bars in the foundation are welded together and used as the grounding electrode for the service. The minimum size copper grounding electrode conductor permitted to run to the reinforcing bars is AWG number _____

   Code reference _____

3. A building is provided with a 200-ampere, single-phase service using AWG number 4/0 aluminum service entrance conductors. If the only practical grounding electrode available is a 5/8-inch diameter galvanized steel rod driven 8 feet into the earth, the minimum size copper grounding electrode conductor permitted is AWG number _____.

   Code reference _____

4. The reinforcing steel rods with a conductive coating in the foundation of a building are welded together and have been made available as a grounding electrode for the service. The building is also supplied by a metal underground water pipe. Are both required to serve as grounding electrodes for the service? (Yes or No)

   Code reference _____

5. All 3-phase, 480-volt electrical services are required to have a grounded circuit conductor bonded to a grounding electrode. (True or False)

   Code reference _____

6. A 480-volt delta service entrance has two parallel sets of 900-kcmil copper type THWN service entrance conductors with each set in a separate service conduit as shown in Figure 5.22. The metal water line is used as a grounding electrode and is supplemented with a driven ground rod. The copper grounding electrode conductor to the ground rod is AWG number 6. The copper main bonding jumper to the service equipment enclosure is 250 kcmil, and the copper bonding jumper to the service conduits is AWG number 2/0. The minimum permitted size of copper grounding electrode conductor to the water pipe is _____.

Code reference _____

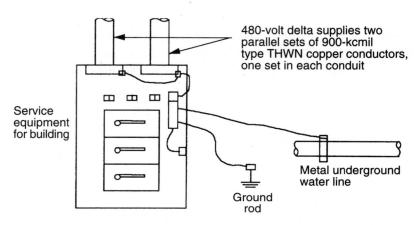

Figure 5.22   Determine the minimum permitted size copper grounding electrode conductor from the service equipment to the metal water pipe.

7. Is an insulated aluminum grounding electrode conductor permitted to be run as direct burial without raceway protection from a building to an 8-foot ground rod driven into the earth on the outside of a building? (Yes or No)

Code reference _____

8. A second ground rod was driven at a service entrance to a building because the electrician did not think there was less than 25 ohms resistance to earth for the first ground rod. The minimum separation required between these two ground rods is _____ feet.

Code reference _____

9. Refer to the 120/240-volt, single-phase service entrance shown in Figure 5.23. A bonding bushing is installed at the point where the service raceway enters the service panel. The conduit and fittings for this service entrance are metal. Is a bonding bushing also required in the meter socket at the location shown? (Yes or No)

    Code reference _____

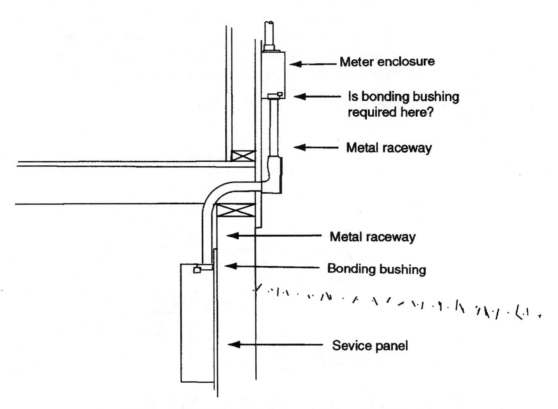

Figure 5.23 **The metal conduit between the meter enclosure and the service entrance enclosure is bonded using a bonding bushing in the service entrance enclosure.**

10. When installing a lightning surge arrester on a service, the lead wires are not to be cut to length, and the leads are to be installed with good square bent corners. (True or False)

    Code reference _____

11. A dwelling contains a metal water piping system; however, the water pipe entering the building from the ground is nonmetallic. Is the interior metal water piping system required to be bonded in some manner to the service raceway enclosure, grounding electrode conductor, or the grounding electrode? (Yes or No)

    Code reference _____

12. A single-phase, 3-wire subpanel will be installed at another location in the same building, and supplied power directly from the service entrance. The wires are run in metal conduit. Is the neutral terminal block at the subpanel permitted to be bonded to the subpanel enclosure? (Yes or No)

    Code reference _____

13. Two enclosures rated as service equipment make up the service for a building as shown in Figure 5.24. The grounded circuit conductor is run to each service enclosure. The 100-ampere panelboard is bonded to the grounding electrode conductor with a tap as indicated. The minimum size bare copper grounding electrode conductor tap permitted for the 100-ampere panelboard is AWG number _____. Some inspection jurisdictions may require that this grounding electrode conductor be routed differently than shown in Figure 5.24.

    Code reference _____

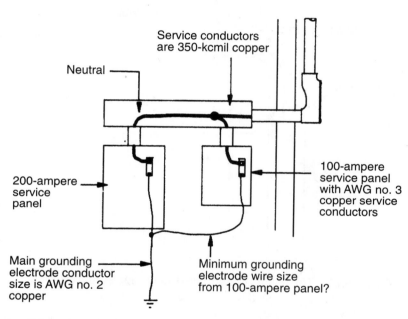

Figure 5.24   A grounding electrode conductor tap connects the 100-ampere panelboard to the main grounding electrode conductor for the service.

14. An AWG number 4 copper wire with black insulation may be used for equipment grounding provided the entire exposed wire is marked with green tape or painted or made bare at each end and at any point where it is visible and accessible. (True or False)

    Code reference _____

15. Each living unit of an apartment building has a panelboard to supply electrical circuits in that living unit. These panelboards (subpanels) are fed from the building service entrance. A single-phase, 3-wire, 120/240-volt electric range supplied from an apartment subpanel is permitted to be wired with a service entrance cable containing only two insulated wires and a bare neutral wire within the outer cable jacket. (True or False)

    Code reference _____

16. A lug installed in a disconnect switch for terminating an equipment grounding wire is permitted to be held in place by a bolt or screw used also to mount the disconnect to the building surface. (True or False)

    Code reference _____

17. A 3-foot length of flexible metal conduit connects a motor to a controller in a dry, dustfree location, as shown in Figure 5.25. There is no other length of flexible metal conduit, liquidtight flexible metal conduit, or flexible metal tubing in the circuit between the motor and the grounding bus of the service panel. The circuit overcurrent protection is rated at 30 amperes, and the flexible metal conduit is 1/2-inch trade size. Is a separate equipment grounding wire required between the controller and the motor? (Yes or No)

    Code reference _____

- Motor started

Fittings approved for grounding

3-foot length of flexible metal conduit

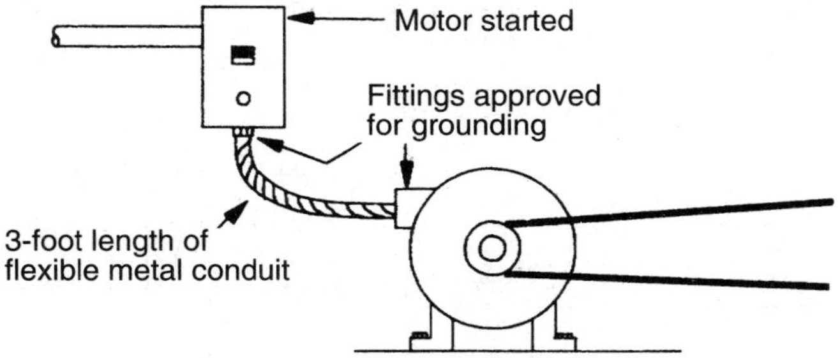

Figure 5.25   The electric motor supplied from a circuit with 30-ampere overcurrent protection is grounded using a 3-foot section of flexible metal conduit.

18. A lug for terminating an equipment grounding wire in an enclosure has a flat surface against the enclosure. Is it necessary to remove the paint in the area between the lug and the metal enclosure before mounting the lug? (Yes or No)

    Code reference _____

19. A motor control center in a building is fed under the floor in rigid nonmetallic (PVC) conduit from a disconnect at the main service. The feeder is protected with 200-ampere time-delay fuses, and the wires are AWG number 3/0 copper with type THHN insulation. The minimum size copper equipment grounding wire permitted for this feeder is AWG number _____.

    Code reference _____

20. Two separate circuits are run in the same nonmetallic conduit. One circuit consists of AWG number 8 copper wire protected at 50 amperes, and the other of AWG number 3 copper protected at 100 amperes. The minimum size copper equipment grounding wire for this conduit is AWG number _____.

    Code reference _____

21. Two parallel sets of conductors are run to a panelboard from the service equipment through two rigid nonmetallic conduits with one set of the parallel conductors in each conduit as shown in Figure 5.26. The feeder to the panelboard is provided with 600-ampere overcurrent protection located in the service entrance panel. A copper equipment grounding conductor is routed with each set of parallel conductors. The minimum permitted size of equipment grounding conductor run in each conduit is _____.

   Code reference _____

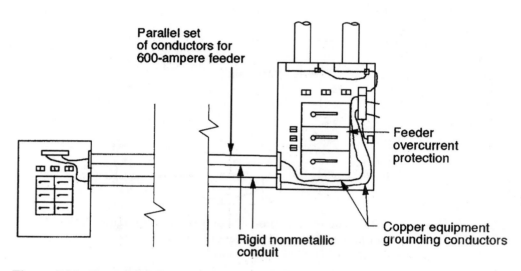

Figure 5.26 Determine the minimum permitted size of copper equipment grounding conductor in each rigid nonmetallic conduit for this 600-ampere feeder consisting of two parallel sets of conductors.

22. Is electrical metallic tubing permitted to serve as an equipment grounding conductor for a branch circuit or feeder? (Yes or No)

   Code reference _____

23. An electrical device with exposed noncurrent-carrying metal parts is installed at a distance from a building with two insulated circuit conductors. Is it permitted to ground the metal frame of this equipment to an 8-foot ground rod driven at the equipment in place of running an equipment grounding conductor back to the single-phase, 120/240-volt electrical service panel in the building? (Yes or No)

   Code reference _____

24. A permanent swimming pool has a diving board with an exposed metal frame. Is this metal frame required to be bonded to other metal associated with the pool and to the common bonding grid of the pool? (Yes or No)

    Code reference _____

25. Rigid nonmetallic conduit is used to connect the forming shell of a wet-niche, low-voltage swimming pool light to the transformer enclosure, as shown in Figure 5.27. The minimum size copper bonding conductor permitted to be run inside the conduit to the forming shell is AWG number _____.

    Code reference _____

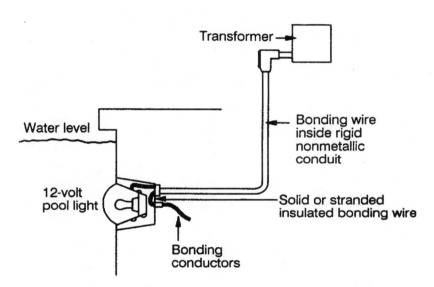

Figure 5.27   Determine the minimum size copper bonding wire run inside the conduit to bond the forming shell for this wet-niche swimming pool lighting fixture.

ANSWER SHEET   Name _____

No. 5   GROUNDING AND BONDING

Answer	**Code reference**
1. _____	_____
2. _____	_____
3. _____	_____
4. _____	_____
5. _____	_____
6. _____	_____
7. _____	_____
8. _____	_____
9. _____	_____
10. _____	_____
11. _____	_____
12. _____	_____
13. _____	_____
14. _____	_____
15. _____	_____
16. _____	_____
17. _____	_____
18. _____	_____
19. _____	_____
20. _____	_____
21. _____	_____
22. _____	_____
23. _____	_____
24. _____	_____
25. _____	_____

UNIT 6

Overcurrent Protection

OBJECTIVES

Upon completion of this unit, the student will be able to:

- name two types of overcurrent conditions.
- explain two ways in which conductors and equipment are damaged during overcurrent conditions.
- name two types of electrical fault conditions.
- explain interrupting rating as related to electrical equipment.
- name the two types of overcurrent conditions from which an overcurrent device protects.
- determine the voltage drop along a length of conductor.
- determine the minimum size wire that will limit voltage drop to a specific value for a given length of run and load.
- answer wiring installation questions relating to *Articles 240, 364, 384, 490, 550, 551, 552, 605,* and *645.*
- state at least five significant changes that occurred from the 1996 to the 1999 Code for *Articles 240, 364, 384, 490, 550, 551, 552, 605,* or *645.*

CODE DISCUSSION

The main emphasis of this unit is overcurrent protection. Several other articles are covered in which overcurrent protection is of particular importance. These additional topics are busways, office furnishings, mobile homes, recreational vehicles, and computers.

Article 240 is the subject of overcurrent protection. The fine print note to *Section 240-1* describes the purpose of overcurrent protection. Rules on overcurrent protection for specific equipment are placed in the articles covering that specific equipment. Standard ratings of overcurrent protective devices are listed in *Section 240-6.* There is a circuit-breaker or a fuse of each of these sizes available. There are not necessarily fuses or circuit-breakers available in all of these sizes. Manufacturers also have overcurrent protective devices available in other rated sizes. Information on the use and the ratings of fuses is given in *Parts E* and *F.* For circuit-breakers, this information is given in *Part G.* The maximum permitted rating of overcurrent protective device for a specific size of wire is determined from *Section 240-3.* There are particular circumstances where the overcurrent device rating for a particular conductor may be permitted to be higher than the allowable ampacity of the conductor as determined by the tables in *Article 310.* These situations are covered in *Parts A* and *B* of *Article 240.*

Part B of *Article 240* states which conductors shall have overcurrent protection and the location of this overcurrent protection in the circuit. Taps to feeders are cases in which the overcurrent protection is not at the point where the conductor receives the supply of electricity. There are many cases where it is only practical to install a limited length of conductor before an overcurrent device can be installed. These limited lengths of conductor are called taps, and they end at the point where an overcurrent device is sized adequate to handle the load and to protect the tap conductor from overload. The rules for taps are found in *Section 240-21*. Figure 6.1 shows a typical example of a tap conductor.

Article 364 covers busways, which are factory-assembled metal enclosures containing electrical conductors, usually in the form of copper or aluminum bars insulated from the metal enclosure. Busway is frequently used in commercial and industrial work areas by attaching to the ceiling. Circuit-breaker or fusible tap boxes are attached to the busway, and conduit or bus drop cable extends down to a workstation, machine, or electrical panel. The rules for use, installation, and overcurrent protection of busway are covered in this article.

Article 384 covers the use, installation, and protection of switchboards and panelboards. Definitions for switchboard and panelboard are given in *Article 100*. A panelboard contains fuses and/or circuit-breakers, and the inside is accessible only from the front. A load center is a type of panelboard. A switchboard is generally a freestanding switching device containing main and/or feeder overcurrent devices and electrical instruments. Usually these devices are accessible from the back and the front. The main electrical supply to a large building generally runs to a switchboard and then to panelboards. When applying this article, it is also important to consider the working space requirements of *Section 110-26* and the requirements for the installation of cabinets and cutout boxes in *Article 373*. Other important related installation requirements are the mounting of electrical equipment in *Section 110-13* and electrical connections in *Section 110-14*.

Article 490 deals with installations where the conductors operate at high voltage, which is defined as more than 600 volts. All of the requirements for high-voltage installations are not in this article. The emphasis in this article is on specifications for high-voltage equipment such as circuit-breakers, power fuses, power switchgear and industrial control assemblies, mobile and portable equipment, and electrode-type boilers.

Article 550 begins with some definitions in *Section 550-2* that are important for understanding the intent of the provisions of this article. *Part B* applies to the wiring within the

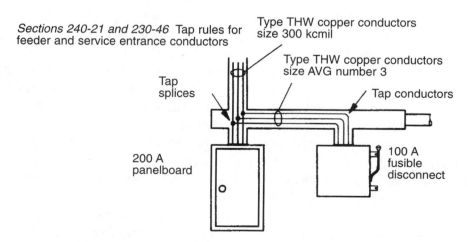

Figure 6.1   A conductor of a smaller size is permitted to be tapped from a conductor when Code rules are followed.

mobile home itself, the electrical disconnecting means, and the distribution equipment for the mobile home. It is important to understand that the mobile home is movable; therefore, the service drop or service lateral would logically connect to service equipment that is not mobile. The service equipment or disconnecting means for a mobile home is required by *Section 550-23(a)* to be located adjacent to the mobile home, and not on or within the mobile home. This is illustrated in Figure 6.2. *Section 550-13* gives a method to determine the minimum permitted rating of the electrical power supply to the mobile home. The minimum permitted size of power supply cord or feeder to the mobile home is given in *Section 550-5*. *Part C* covers services for mobile homes and feeders in a mobile home park. The service for a mobile home in a mobile home park is permitted to be remote from the mobile home, but a disconnecting means for the mobile home is required to be installed according to the rules of *Section 550-23*.

Article 551 contains definitions in *Section 551-2* that are important for the understanding of the article. *Part B* covers the installation requirements for low-voltage wiring in a recreational vehicle. Low voltage for recreational vehicle applications is defined in *Section 551-2* as 24 volts or less. *Part C* deals with the installation of wiring intended to operate from a combination of power sources. For example, the circuits may be supplied from a battery, another type of direct current supply, or from a 120-volt alternating current source. *Part D* applies to the installation of power sources, such as an engine-driven generator or batteries. When there are multiple power sources, such as a generator and a power supply cord, a transfer switch shall be installed to prevent interconnection of the power sources if such a transfer device is not supplied as part of the generator.

Part E covers the installation of wiring to be supplied from a nominal 120-volt or 120/240-volt electrical system. *Section 551-41* provides the requirements for the spacing and type of receptacle outlets. The receptacle outlet next to the lavatory in the bath is required to be ground-fault circuit-interrupter protected. Also, all receptacle outlets installed within 6 feet of the lavatory in the bathroom that are intended to serve the counter top are required to be ground-fault circuit-interrupter protected. This provision of *Section 551-41(c)(2)* is illustrated in Figure 6.3. *Section 551-42* covers the minimum branch circuit requirements for a recreational vehicle. The type and installation of the distribution panelboard including working clearances are covered in *Section 551-45*. The means of connecting the recreational vehicle to

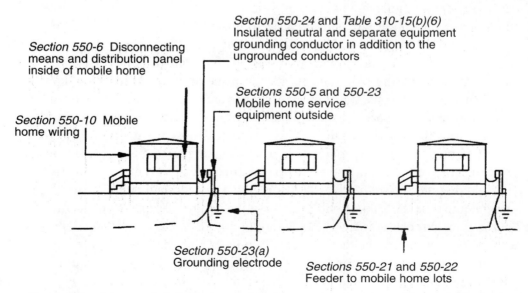

Figure 6.2   Service equipment for a mobile home is required to be located adjacent to the mobile home, and not on or within the mobile home.

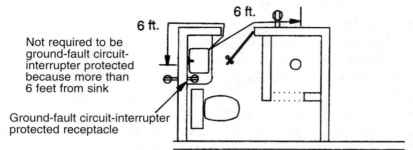

Not required to be ground-fault
circuit-interrupter protected
because not intended to serve
the counter top

Not required to be
ground-fault circuit-
interrupter protected
because more than
6 feet from sink

Ground-fault circuit-interrupter
protected receptacle

6 ft.   6 ft.

Section 551-41(c)(2) Receptacles within 6 feet of the
bathroom sink intended to serve counter tops are
required to be ground-fault circuit-interrupter protected

**Figure 6.3   Receptacle outlets installed within 6 feet of the lavatory in the
bathroom of a recreational vehicle are required to be ground-fault circuit-
interrupter protected unless not intended to serve the counter top.**

an external power supply are covered in *Section 551-46. Section 551-47* describes the wiring methods permitted to be used in a recreational vehicle for 120-volt or 120/240-volt electrical systems. The methods of grounding equipment are covered in *Section 551-55.* Part G provides requirements for determining the minimum size and installation of equipment at each recreational vehicle site and the feeders supplying the sites.

Article 552 specifies the rules for wiring on park trailers. A park trailer is defined in *Section 552-2* as a unit that is built on a single chassis which is mounted on wheels, and the unit gross area does not exceed 400 square feet when set up for occupancy. These units are ones that are intended for seasonal use, and they are not intended to have commercial uses or to be occupied as a permanent dwelling. A park trailer has differences that are not addressed in *Article 550* on mobile homes or in *Article 551* on recreational vehicles. This article does cover the wiring of lighting required to transport the trailer on public roads and highways. Electrical wiring operating from an alternating current source of 120 volts or 120/240 volts is specified in *Part D.* The minimum permitted size of distribution panelboard and power supply cord or feeder for the park trailer can be determined by using the method of *Article 220, Part B,* or the size determination can be made by using *Section 552-47.*

Article 605 provides the requirements for the installation of wiring that is part of relocatable partitions in office areas. It is popular for office areas to consist of large open areas with relocatable partitions and working surfaces and equipment. The means of electrically connecting office furnishings wiring to the building electrical system are covered in this article.

Article 645 on electronic computer and data processing equipment is concerned with the permanently installed equipment, and not with equipment such as remote terminals and stand-alone desktop computers. *Section 645-10* requires that for computing equipment and the air-conditioning equipment, a disconnecting means or the control for the disconnecting means be located at the principal operator exit. The battery of an uninterruptible power supply within a computer/data processing room, according to *Section 645-11,* shall be installed such that the disconnecting means at the operator exit will disconnect the load from the batteries.

Grounding of computers, data processing equipment, and related electronic equipment can introduce operational problems. *Section 645-15* requires that this equipment be grounded. It is not permitted to interrupt the equipment grounding conductor, but there are methods to help reduce the chances of problems due to grounding. One problem may be due to grounding loops where connected equipment is grounded to different points of an electrical system. This can be through the equipment grounding conductor for the individual pieces of equipment, or it can be by means of physical contact with structural or other metal intentionally or unintentionally connected to the equipment grounding of the electrical system in the building. It is sometimes a tedious task to eliminate grounding loops that can be affecting the operation of electronic equipment. Effective bonding and grounding of all components of an electronic system back to the same equipment grounding bus helps minimize problems.

Section 250-6(d) states that currents that introduce noise in such a system are not considered objectionable currents, as defined in *Section 250-6*. *Section 250-146(d)* permits electronic equipment to be supplied using receptacles with the equipment grounding terminal separately grounded with an insulated equipment grounding conductor. This separates the electronic equipment ground from the electrical wiring system ground, such as the metal conduit. This type of receptacle is identified with an orange triangle. *Section 250-96(b),* permits the enclosure of electronic equipment supplied with permanent wiring to be insulated from metal raceways, but an equipment grounding conductor is required to be connected to the equipment. This is illustrated in Figure 6.4. A grounding electrode is permitted to be connected to the equipment, but this grounding electrode is not permitted to take the place of an equipment grounding conductor.

OVERCURRENT PROTECTION FUNDAMENTALS

The purpose of overcurrent protection is to open a circuit if the temperature in the wiring or equipment rises to an excessive or dangerous level. This is discussed in the fine print note to *Section 240-1.*

Every circuit shall be protected by two levels of overcurrent protection: overload and short-circuits or ground-faults. Usually one overcurrent device provides both levels of protection, but in the case of a motor circuit, this protection is often provided at two locations. In a typical motor branch circuit, short-circuit and ground-fault protection is provided by the fuse or circuit-breaker, while overload protection is provided by the thermal unit in the motor starter.

Overloads

An overload is a gradual rise in current level above the current rating of the wires in the circuit or in the equipment. Heat is produced as the current flows through the wires. The two

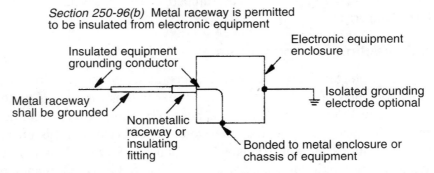

Figure 6.4 Electronic equipment is permitted to be insulated from metal raceways, but an equipment grounding conductor is required to be run to an equipment grounding bus of the electrical system supplying the equipment.

important variables for conductor heating are current level and time, in addition to resistance. Equation 1.16 discussed in Unit 1 may be used to determine the heat produced as electrical current flows in a conductor for a given length of time.

$$\text{Heat} = \text{Current}^2 \times \text{Resistance} \times \text{Time}$$

> Heat is in joules
> Current is in amperes
> Resistance is in ohms
> Time is in seconds

Faults

A high level of current generally flows during a fault. Common faults are short-circuits between wires and faults from an ungrounded wire to ground. This condition is called a **ground-fault.** If arcing occurs, then the fault is also referred to as an **arcing fault,** such as the case when a wire touches the conduit causing an **arcing ground-fault.** If two conductors with voltage between them are accidentally connected together, arcing generally will not occur. If the conductors are connected together so that arcing does not occur, this is termed a **bolted fault.**

How Damage Occurs

When a fault occurs, the amount of current that flows depends on the impedance (resistance) of the fault—the impedance of the wires, transformer, and electrical system feeding the fault. The circuit voltage between the faulted wires is also important in determining the amount of fault current that will flow. The current flow during a fault is often in the thousands of amperes.

High current flow during a fault will cause strong magnetic forces that can move wires and deform electrical parts if the parts are not adequately secured and braced to withstand these forces. Arcing will occur when parts with a voltage between them touch together. This may be a secondary effect after the magnetic forces deform the parts.

Interrupting Ratings of Fuses and Circuit-Breakers

The responsibility of the electrician is to know when excessive short-circuit currents can occur. Standard circuit-breakers are only required to withstand 5,000 amperes of short-circuit current. However, most are rated at 10,000 amperes as stamped on the circuit-breaker. Fuses are required to have an interrupting rating of at least 10,000 amperes. Fuses are commonly available having withstand ratings up to 200,000 amperes. Some fuses have a withstand rating as high as 300,000 amperes.

Overcurrent Protection for Conductors

Wires shall be protected for overcurrent in accordance with the allowable ampacity listed in the ampacity tables of *Article 310*. There are exceptions for circuits protected at not more than 800 amperes. In this case, the next standard size larger overcurrent device rating shall be permitted to be used, even though the overcurrent device rating is greater than the listed allowable ampacity of the wire. This rule is found in *Section 240-3(b)*. The wire is required to have an ampere rating not smaller than the load to be served. Standard ratings of overcurrent devices are found in *Section 240-6*.

A conductor tapped from a branch circuit or feeder with an overcurrent device rated higher than the allowable ampacity of the tap conductor is permitted by *Section 240-3(e)* with specific details for the particular situation covered in *Section 240-21*. The tap conductor is protected for short-circuits and ground-faults by the feeder or branch circuit overcurrent device. The tap conductor is protected from overloads by an overcurrent device at the load end of the tap conductor. *Section 240-21(b)(1)* permits a tap conductor to have an ampere rating not smaller than one-tenth of the rating of the feeder overcurrent device; it shall have an ampacity not smaller than the load to be served; it shall be protected from physical damage or be enclosed in raceway; and it shall be terminated at an overcurrent device with a rating not larger than the ampacity of the tap conductor, which is no more than 10 feet from the tap point. Example 6.1 illustrates the 10-foot tap rule of *Section 240-21(b)* as shown in Figure 6.5.

Example 6.1   Copper feeder conductors, type THW, size AWG number 4/0 are in conduit. The overcurrent protection for the feeder is a 200-ampere circuit-breaker. The tap load is a resistance electric heater with a full load current of 37 amperes. Determine the minimum permitted size of type THWN copper tap conductor assuming terminations are rated 75°C.

Answer: The tap conductor shall be rated at a minimum of 46 amperes to supply the electric heater, as required by *Section 424-3(b)(1)*. The tap to the feeder conductors is not more than 10 feet in total length from the branch circuit overcurrent device for the electric heater. The minimum permitted rating of the branch circuit overcurrent protective device shall be not less than 46 amperes. From *Section 240-6,* the minimum permitted standard rating of branch circuit overcurrent protection is 50 amperes. *Section 240-21(b)(1)(d)* requires that the tap conductor in this case have an ampacity not less than one-tenth the rating of the feeder overcurrent protection, which would be 20 amperes. In this particular case, the minimum permitted size of copper, type THWN tap conductor is found from *Table 310-16* based on the 50-ampere overcurrent device, *Section 240-21(b)(1)(a)(2)*. The minimum permitted size of type THWN copper conductor is AWG number 8.

$$37 \text{ A} \times 1.25 = 46 \text{ A}$$

$$\frac{200 \text{ A}}{10} = 20 \text{ A}$$

Example 6.2   A tap not more than 10 feet in length and ending at a 30-ampere overcurrent device is made from type THWN copper, 500-kcmil feeder conductors that have

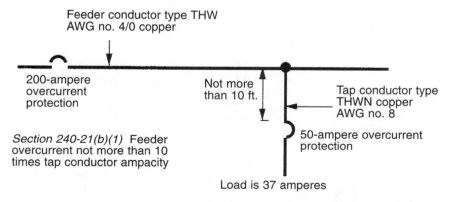

Load is 37 amperes

Figure 6.5   A tap to a feeder where the tap conductor is not more than 10 feet in length.

400-ampere overcurrent protection. The tap is shown in Figure 6.6. Determine the minimum permitted size of the copper, type THWN tap conductors assuming terminations rated 75°C.

Answer: Based on the load, the tap conductor would be required to be sized only at a minimum of 30 amperes. One-tenth of the rating of the 400-ampere feeder overcurrent protection is 40 amperes. The minimum tap conductor size permitted in this case is AWG number 8 copper, type THWN based on 40 amperes.

Section 240-21(b)(2) allows the tap conductor to extend more than 10 feet but not beyond 25 feet. However, in this case, an additional restriction is imposed. The ampacity of the tap conductor shall not be less than one-third the rating of the feeder overcurrent device. The 25-foot tap rule of *Section 240-21(b)(2)* is illustrated in Figure 6.7.

Example 6.3   A load is served by means of a tap from 300-kcmil, type THWN copper feeder conductors protected from overcurrent at 300 amperes, as shown in Figure 6.7. The tap conductors are copper type THWN in conduit, and the total length of the tap from the tap point to 70-ampere branch circuit overcurrent protection is more than 10 feet but not more than 25 feet. Determine the minimum size of type THWN copper tap conductor permitted assuming conductor terminations are rated 75°C.

Answer: The minimum permitted size of conductor is determined based on 100 amperes, which is one-third the rating of the 300-ampere feeder overcurrent protection. *Section 240-21(b)(2)*. The tap ends at an overcurrent protective device of only 70 amperes. In this case, the minimum permitted size of type THWN copper tap conductor is size AWG number 3 from *Table 310-16*.

$$\text{Feeder overcurrent protection:} \quad \frac{300 \text{ A}}{3} = 100 \text{ A}$$

VOLTAGE DROP

The Code does not have a requirement on the amount of voltage drop permitted on branch circuits and feeders. It is suggested, however, that a maximum of 5 percent voltage

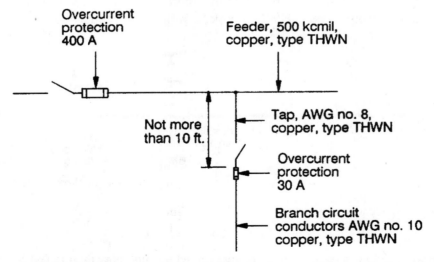

Figure 6.6   The 10-foot tap conductor shall have an ampacity not less than one-tenth the rating of the feeder overcurrent protection.

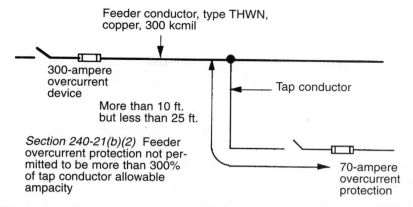

Figure 6.7 A tap to a feeder where the tap conductor is more than 10 feet but not beyond 25 feet in length.

drop be permitted on a feeder plus branch circuit to the most distant point of power use. This is in the form of a fine print note in *Section 215-2* and *Section 210-19(a) FPN 4*. Many tables, charts, and formulas are available to size wires that limit voltage drop to within a specified limit.

Voltage drop is the extent to which the line voltage at the source end of a branch circuit or feeder is reduced as the load current flows along the conductors. The way to measure voltage drop is to operate the load, then measure the voltage at the source end of the branch circuit or feeder and at the load end. Excessive voltage drop results in inefficient operation of electrical equipment, sometimes malfunctioning of control systems and premature motor failure. Equation 6.1 can be used to determine the expected amount of voltage drop for a single-phase circuit. Equation 6.2 is used for a 3-phase circuit. Resistivity **K** of the wire for typical operating temperatures is approximately 12 for copper and 19 for aluminum. Resistivity of conductors at different temperatures is given in *Table 1* of Unit 1. This does not take inductive reactance into effect. The Circular Mil area of the wires can be found in *Table 8* of *Chapter 9*. The percent voltage drops in Equations 6.1 and 6.2 are given in decimal.

Single-phase:

$$\text{Voltage Drop} = \frac{2 \times K \times \text{Amps} \times \text{One-Way Length}}{(\text{Circular Mil Area of Wire})} \qquad \text{Eq. 6.1}$$

3-phase:

$$\text{Voltage Drop} = \frac{1.73 \times K \times \text{Amps} \times \text{One-Way Length}}{(\text{Circular Mil Area of Wire})} \qquad \text{Eq. 6.2}$$

Example 6.4 A single-phase feeder consists of AWG number 3 type THWN copper wire in conduit with a length of 258 feet. The typical maximum load on the feeder is 80 amperes. The voltage at the supply end of the feeder is 240 volts. Determine the amount of voltage drop on the feeder wire.

Answer: This is a single-phase circuit; therefore, use Equation 6.1 and solve for the amount of voltage drop.

$$\text{Voltage drop} = \frac{2 \times 12 \times 80 \text{ A} \times 258 \text{ ft.}}{52,620 \text{ cm}} = 9.4 \text{ V}$$

Equation 6.3 is a common method for determining the minimum size wire needed for an application to maintain a specific level of voltage drop for a particular load over a particular distance. It is actually the same as Equation 6.1, except it has been rearranged to solve for the

Circular Mil area of the wire. Equation 6.4 is for a 3-phase circuit or feeder. Percent voltage drop is in decimal.

Single-phase:
$$\text{Circular Mil Area} = \frac{2 \times K \times \text{Amps} \times \text{One-Way Length}}{\% \text{ Voltage Drop} \times \text{Supply Voltage}} \qquad \text{Eq. 6.3}$$

3-phase:
$$\text{Circular Mil Area} = \frac{1.73 \times K \times \text{Amps} \times \text{One-Way Length}}{\% \text{ Voltage Drop} \times \text{Supply Voltage}} \qquad \text{Eq. 6.4}$$

Example 6.5   Refer to the feeder described in Example 6.3. Determine the minimum copper wire size required to limit voltage drop to 2 percent.

Answer: Determine the Circular Mil area of the wire by using Equation 6.3. From *Table 8, Chapter 9*, select AWG number 1/0 copper wire. The Circular Mil area of an AWG number 1/0 conductor is greater than the 103,200 Circular Mils determined by using Equation 6.3.

$$\text{Circular Mil Area} = \frac{2 \times 12 \times 80 \text{ A} \times 258 \text{ ft.}}{0.02 \times 240 \text{ V}} = 103,200 \text{ cm}$$

Alternate Voltage Drop Calculation Method

This alternate method of sizing a circuit wire or feeder for a specified maximum value of voltage drop uses *Table 8, Chapter 9* in the Code. It is not necessary to choose the type of conductor material in advance or to choose a value of conductor resistivity (K). Resistivity of the material is built into the method at 75°C because the resistance values of *Table 8, Chapter 9* are used to determine the wire size.

Equation 6.5 applies for single-phase, and Equation 6.6 applies for 3-phase feeder and branch circuit voltage drop calculations for conductors not larger than AWO number 4/0. For larger conductors, these formulas will also work as long as the power factor of the circuit is close to unity. When the power factor is low, inductive reactance also will have to be considered in the voltage drop calculation when the wire size is larger than AWG number 4/0. Equations 6.5 and 6.6 are used to determine the minimum size of conductor permitted to limit voltage drop to the specified level.

Single-phase:
$$\text{Resistance}_{(1,000 \text{ ft})} = \frac{5 \times \% \text{ Volt Drop} \times \text{Circuit Voltage}}{\text{Current} \times \text{One-Way Length}} \qquad \text{Eq. 6.5}$$

3-phase:
$$\text{Resistance}_{(1,000 \text{ ft})} = \frac{5.8 \times \% \text{ Volt Drop} \times \text{Circuit Voltage}}{\text{Current} \times \text{One-Way Length}} \qquad \text{Eq. 6.6}$$

These equations will determine the maximum permitted resistance per 1,000 feet to prevent voltage drop from exceeding the value specified. It is not necessary to know if the wire is copper or aluminum until time to select the wire size. *Table 8, Chapter 9* contains resistance values for both copper and aluminum wires. Select the wire size from *Table 8* using the value of resistance determined by the equation. To determine a wire size to limit voltage drop using this method, first determine the following information about the application:

The maximum percent voltage drop desired _____ %
(leave this as percent, not decimal)

The circuit voltage _____ V

Maximum branch circuit or feeder current   _____ A

The one-way length of the circuit or feeder   _____ ft.

Is the circuit or feeder single-phase or 3-phase?   _____

Example 6.6   A 150-ampere feeder has a one-way length of 235 feet, and the design current level is 150 amperes. The feeder is single-phase, 240-volt. Select the minimum size wire to limit voltage drop to 2.5 percent.

Answer: Use Equation 6.5 to determine the resistance of the wire on a 1,000-foot basis, which will limit the voltage drop to 2.5 percent. The value obtained is 0.085 ohms. From the *Table 8* resistance column for bare copper wire, select AWG number 3/0.

$$\text{Resistance}_{(1000\ ft)} = \frac{5 \times 2.5\% \times 240\ V}{150\ A \times 235\ ft.} = 0.085\ ohm$$

It is also necessary to check this with *Table 310-16* to determine the minimum size wire permitted, based on the load. This check reveals that AWG number 3/0 is the minimum permitted conductor size. If the wire is overhead triplex aluminum, then the minimum wire size permitted is from the aluminum resistance column of *Table 8, Chapter 9.* Also check *Table 310-20* for the minimum permitted. The minimum size aluminum wire to maintain 2.5 percent voltage drop for this example is 250 kcmil.

For wire sizes larger than AWG number 4/0, skin effect and inductive reactance frequently become significant and must be considered. *Table 9, Chapter 9* of the Code contains impedance values of wires in different types of raceway where the power factor is not unity. If a motor load is being supplied by a branch circuit or feeder, the voltage drop can be estimated using the impedance values from *Table 9.* If the power factor of the circuit or feeder is approximately 0.85, then impedance values are given in *Table 9.*

Values of impedance, rather than resistance, cannot be substituted into the equations where the value of conductor resistivity **K** is used. Some manufacturers' literature provides values of **K** to take impedance into consideration. Equations 6.5 and 6.6 can be used when the power factor is less than one. Just look up the wire size from *Table 9* instead of *Table 8.* For example, assume the single-phase feeder of Example 6.5 is run in steel conduit, and the power factor of the 150-ampere load is 0.85. Look up the wire size from the steel conduit column of *Table 9* for effective impedance **Z** of 0.85 power factor. The impedance determined by Equation 6.5 in Example 6.6 was 0.085 ohm. Therefore, choose AWG number 4/0 copper wire with an effective impedance of 0.080 ohm.

MAJOR CHANGES TO THE 1999 CODE

These are the changes to the 1999 *National Electrical Code*® for electrical systems that correspond to the Code sections studied in this unit. The following analysis explains the significance of only the changes from the 1996 to the 1999 Code, and this analysis is not intended to be used in place of the Code. Refer to the actual section of the 1999 Code for the exact wording and meaning of each section discussed. Changes are indicated in the Code with a vertical line in the margin. If material has been deleted or moved to another part of the Code, the location of the deletion is indicated with a dark dot in the margin.

Article 240 **Overcurrent Protection**

240-1: This article now recognizes a category of installation known as **supervised industrial installations,** and there is a new *Part H* that deals with supervised industrial installations. *Section 240-91* defines supervised industrial installations.

240-3(d): This new paragraph in *Article 240* was part of the footnote at the bottom of *Table 310-16* and *Table 310-17.* This is the requirement that unless specified in specific parts of the Code, an AWG number 14 copper conductor is not permitted to be protected with an overcurrent device rated more than 15 amperes. For an AWG number 12 copper conductor, the maximum is 20 amperes, and for an AWG number 10 copper conductor, the maximum is 30 amperes. This rule did not change; just the location of the rule changed.

240-3(e): A new paragraph was added defining **tap conductor.** Basically the definition states that the tap conductor is connected to another conductor with a higher ampere rating, and the overcurrent device for that main conductor has a higher rating than would otherwise be permitted for the tap conductor.

240-6(a): This section lists the standard ratings of fuses and circuit-breakers. A new sentence was added stating that nonstandard ratings of fuses and circuit-breakers are permitted to be used. Apparently some interpretations were that only the standard ratings were permitted to be used.

240-6(b): This is a new paragraph, but it was covered by default in the previous edition of the Code. In the case of an adjustable-trip circuit-breaker, if the adjustment is not designed to be inaccessible, then the maximum rating of the circuit-breaker must be used.

240-21: This section specifies the rules when a conductor of lower ampacity is tapped without overcurrent protection from a conductor with a higher rating of overcurrent protection. This section was reorganized to separate the rules that deal with conductors tapped from a feeder, from the rules that deal with conductors tapped from the secondary winding of a transformer. This section also specifies the rules for installing taps to conductors. A new sentence was added that prohibits a conductor being tapped to a tap. This was the intent, but it was never definitively stated in the Code. This is illustrated in Figure 6.8.

240-21(c)(1): This paragraph is the same wording as in *Section 240-3(f).* It states that in the case of a single-phase, 2-wire transformer secondary and in the case of a 3-phase, 3-wire single-voltage secondary, the secondary conductors are permitted to be protected with the

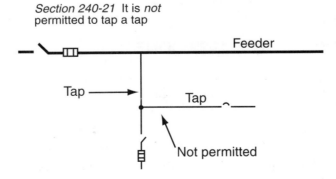

Section 240-21 It is *not* permitted to tap a tap

Feeder

Tap

Tap

Not permitted

Figure 6.8   A conductor is permitted to be tapped from another conductor with a higher ampacity and overcurrent device rating than normally would be permitted for the tap, but it is not permitted to add a second tap to the first tap conductor.

overcurrent device on the primary side of the transformer. In this case, the secondary conductors are protected according to their ampacity and, therefore, are not tap conductors.

240-21(c)(2): This is the 10-foot tap rule where the conductors are tapped from the secondary of a transformer. In the previous edition of the Code, the transformer 10-foot tap rule and the feeder 10-foot tap rule were the same. There is a change in the transformer 10-foot tap rule. There is no longer a limit to the ampere rating of the tap conductors on the secondary side of a transformer with respect to the overcurrent protection on the primary side of the transformer. The previous edition of the Code did not permit the secondary conductors to have an ampere rating less than one-tenth the rating of the primary overcurrent device times the ratio of the primary voltage divided by the secondary voltage.

240-24(b): The previous edition of the Code in what was *Exception 2* permitted all overcurrent devices serving circuits in guest rooms of hotels and motels to be accessible only to authorized management personnel. Now only the service overcurrent device and feeder overcurrent device where the feeder is supplying more than one occupancy are permitted to be not accessible to the occupants. The individual circuit overcurrent devices are required to be accessible to the occupants.

240-33: In the previous edition of the Code, an exception in this section permitted circuit-breaker panelboards to be mounted in other than the vertical position as long as the individual circuit-breakers operated either rotationally, horizontally, or vertically with on being in the up position. This was permitted only if vertical mounting was **impractical.** It is no longer required to show that vertical mounting is impractical. The installer has the option of mounting the panelboard vertically or horizontally as long as the position of the individual circuit-breakers is in compliance with *Section 240-81.*

A new sentence was added to *Section 240-33* that specifies that enclosures for overcurrent devices must be mounted in the vertical position unless such mounting is impractical. The new sentence deals with the installation of plug-in units for busway. Plug-in units are permitted to be installed in busway corresponding to the orientation of the busway. If a section of busway runs horizontally for a distance and then makes a turn so it now runs vertically, this new sentence permits a busway plug-in unit to be installed on the vertical section of busway.

240-86(b): This entire section is new, except most of this section dealing with circuit-breaker series interrupting ratings was derived from *Section 240-83(c)* in the previous edition of the Code. The requirement that is new is paragraph (b). Figure 6.9 illustrates a case where circuit-breaker series rating is used. The available short-circuit current at various points in an electrical system was determined by calculations beyond the scope of the Code. The present Code rule requires the **series combination interrupting rating** required at a switchboard, panelboard, or motor control center to be displayed at that location whenever the available fault current is greater than the interrupting rating of the circuit-breaker. Assume in Figure 6.9 that the circuit-breaker has a 10,000-ampere interrupting rating, but the fault current available is greater than 10,000 at that location. The available fault current must be indicated at that location, which in Figure 6.9 is the subpanel. This installation is acceptable if the circuit-breaker in the subpanel and the overcurrent device ahead of it in the main panel have a series combined interrupting rating greater than the fault current at the subpanel.

The change in this edition of the Code deals with the amount of motor load in the facility and where in the circuit the motor load is connected. Motors convert electrical power to mechanical power by creating a magnetic field in the motor windings. When power is

Section 240-86(b) If circuit-breakers are installed
as a series rated system, motor load connected between
series rated circuit-breakers is not permitted to exceed
one percent of the down line circuit-breaker interrupting rating

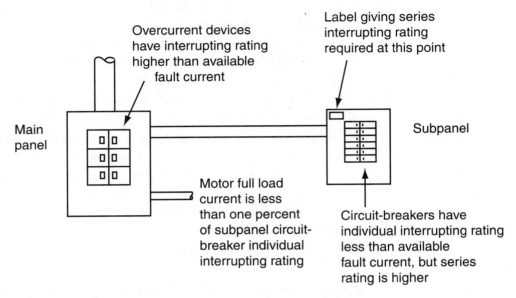

Figure 6.9   If the available fault current at the subpanel exceeds the interrupting rating of the individual circuit-breaker interrupting ratings, then the fault current available must be marked on the panelboard as the series combination interrupting rating required as long as the motor full-load current installed ahead of the panelboard does not exceed one percent of the panelboard circuit-breaker interrupter rating.

suddenly interrupted to an electric motor, the magnetic field around the windings collapses and a current surge is induced into the windings and sent back to the power source. This surge of current from the deenergized motors can add to the fault current coming into the building from the power system resulting in a total fault current level greater than expected. The change in the Code points this out as a potential problem when motors are installed after the first overcurrent device and ahead of the second circuit-breaker. This is the case in Figure 6.9 where the motor load is supplied from the main panel. If the full-load current of all the motors is greater than one percent of the interrupting rating of the second circuit-breaker, then this equipment is not permitted to be installed as a series rated system. The alternative is to install a subpanel with overcurrent devices that have fault current ratings equal to or greater than the available fault current. For the example in Figure 6.9, the circuit-breaker has an interrupting rating of 10,000 amperes. One percent of 10,000 amperes is 100 amperes (0.01 × 10,000 A = 100 A). The total full-load current of all the motors installed off the main panel and before the subpanel cannot exceed 100 amperes or a series rated system is not permitted to be installed.

240-90: This is a new section stating the general purpose of the new *Part H* dealing with what is known as supervised industrial installations. There are some differences in overcurrent protection requirements for these large power-using manufacturing and process control facilities.

240-91: This is a definition of a supervised industrial installation. Only qualified personnel will monitor and service the electrical system. The facility has at least one electrical system that operates at more than 150 volts to ground and more than 300 volts between conductors. The calculated load must be greater than 2,500 kilovolt-amperes. At 480 volts, 3-phase, this is 3,010 amperes. Offices and other nonmanufacturing functions are not permitted to make up any part of the 2,500-kilovolt-ampere load.

240-92(a): This section specifies the rules for overcurrent protection for conductors. One option is to install overcurrent protection according to the rules in *Section 240-21.*

240-92(b): The length of the secondary conductor from a transformer to the overcurrent device does not exceed 50 feet, and the overcurrent device on the primary side of the transformer is limited to 150 percent of the ampacity of the secondary conductor times the ratio of the secondary to primary voltage. This first option is illustrated in Figure 6.10. The length of conductor from the secondary of the transformer is permitted to be 75 feet if overcurrent protection is provided that will prevent damage to conductors due to any short-circuit or ground-fault condition. This type of protection is required to be determined under the supervision of an engineer. In addition, the conductor must end at a single overcurrent device or up to six overcurrent devices where the single or combined rating does not exceed the ampacity of the secondary conductors. There is even an additional option where overcurrent devices can have an even higher combined rating if engineering calculations show that overloading of the secondary conductors will not likely occur. The secondary conductors also must be suitably protected from physical damage.

240-92(c): In this case, the conductors are completely outside except at the point where they enter the building. The conductors must be suitably protected from physical damage, but the length of the conductors is not limited. The conductors must end in up to six overcurrent devices grouped at one location with a combined rating not exceeding the ampacity of the conductors.

240-100(a): The overcurrent protection requirements for feeders and branch circuits operating at over 600 volts were combined into one section. There is a new requirement that an overcurrent device must be located at the point where the feeder or branch circuit

Section 240-92(b) Special secondary overcurrent
protection rules for supervised industrial installations

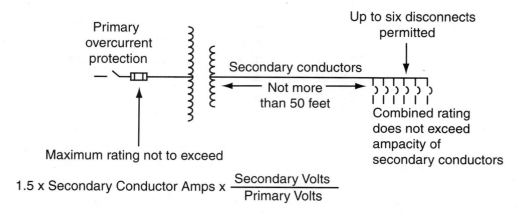

Figure 6.10   For a supervised industrial installation where at least one electrical system operates at more than 300 volts between conductors and more than 150 volts to ground, and the total manufacturing load is not less than 2,500 kilovolt-amperes, special overcurrent rules apply to the secondary conductors of the transformer.

receives the supply of power. There is a provision that allows the overcurrent device to be located elsewhere in the circuit or feeder.

240-100(a)(1) and (2): These two paragraphs specify the type of overcurrent protection that is acceptable for circuits and feeders operating at more than 600 volts. This was *Section 710-20* in the previous edition of the Code. There was no change in the overcurrent protection requirement. The individual requirements for fuses and circuit-breakers were in *Section 710-21*, and those requirements are now found in the new *Article 490* dealing with equipment operating at more than 600 volts.

240-101(a): A new sentence was added permitting conductor overcurrent protection for fire pump installations to be determined according to *Section 495-4(b)*.

Article 364   Busways

364-6(b): The change is for nonindustrial buildings when a riser busway passes vertically through two or more floors. Such installations now require curbs to be place around floor openings for riser busways to prevent the entrance of liquids that are spilled onto the floor. In order to contain liquid spills and prevent the liquid from traveling from floor to floor and into the busway system, the curbs are required to be a minimum of 4 inches in height and within 12 inches of the floor penetration.

364-8(c): This new paragraph places requirements on cord-and-cable assemblies that are used as branches from trolley-type busway that is used to supply movable equipment. Approved extra hard usage or hard usage cord-and-cable assemblies and listed bus drop cable is required when connecting movable equipment to a trolley busway. In the previous edition of the Code, the type of cord-and-cable assemblies permitted for these types of installation was not clear.

Article 384   Switchboards and Panelboards

384-14: Panelboards are classified either as **lighting and appliance branch circuit panelboards** or as a **power panelboard.** What is new is that the Code now uses the term **power panelboard.** It is important to know the difference because the overcurrent protection rules are different for the two types.

384-14(a): This section defines what is considered to be a **lighting and appliance branch circuit panelboard.** This is important because in *Section 384-16*, there is a requirement that a lighting and appliance branch circuit panelboard must have overcurrent protection either in the panelboard or on the feeder supplying the panelboard with a rating not greater than the rating of the panelboard. If the panelbaord does not meet this definition, then it is considered to be a power panelboard. Power panelboards are required to be sized adequate to handle the calculated load, but they are not required to be protected by an overcurrent device limited to the rating of the panelboard. The definition of lighting and appliance branch circuit panelboard in the previous edition of the Code was not completely clear, and there was a lack of uniformity of application in the field. It is assumed panelboards supplying lighting and appliance branch circuits are likely to have personnel working in them who have limited knowledge of panelboard loading. By requiring the panelboard to have overcurrent protection limited to the rating of the panelboard, the likelihood of causing a fire due to panelboard overloading is limited.

   The definition now states that a lighting and appliance branch circuit panelboard is one with more than 10 percent of the overcurrent devices supplying lighting or appliance branch circuits. A new sentence was added to the Code defining a lighting and appliance branch circuit. If the circuit has a neutral conductor and the overcurrent device is rated not more than 30 amperes, it is a lighting or appliance branch circuit. The actual difference in the definition is that the previous edition of the Code only required a neutral terminal in the panelboard while the new definition requires the neutral to be used in the circuits. This is illustrated in Figure 6.11. According to this definition, if a panelboard has all 2-pole or 3-pole circuits, none of which use a neutral, but the neutral terminal is in the panelboard, it is no longer considered a lighting and appliance branch circuit panelboard.

384-14(b): This is a new definition of a **power panelboard.** In the trade, if a panelboard did not meet the definition of a lighting and appliance branch circuit panelboard, it was considered to be a power panelboard. If 10 percent or fewer of the overcurrent devices supply circuits rated 30 amperes or less where the neutral is used, then it is considered to be a power panelboard. For the purpose of determining the percentage of circuits that are considered to be of the lighting and appliance type, the Code considers a 3-pole overcurrent device to be three devices and a 2-pole device to be two devices. For example, if a 30-space panelboard has a neutral bus and only three of the overcurrent devices are rated 30 amperes or less using the neutral, then it is considered to be a power panelboard. Assume that in the same panelboard there are several other circuits that do not use the neutral, but the overcurrent devices are rated 30 amperes or less. According to the new definition, it is still a power panelboard, but the previous edition of the Code would have considered it to have been a lighting and appliance branch circuit panelboard.

384-16(b): This is a new section stating the overcurrent protection requirements for power panelboards. If a panelboard meets the definition of a power panelboard, its supply includes a neutral conductor, and more than 10 percent of the overcurrent devices are rated 30 amperes or less, then the panelboard is required to be protected on the supply side with an overcurrent device having a rating not greater than the rating of the power panelboard. A main in the panelboard or an overcurrent device protecting the feeder is considered adequate protection. This is not a change from the previous edition of the

Section 384-14(a) Lighting and appliance branch circuit panelboard has more than 10 percent of circuits 30 amperes or less using a neutral conductor

Supply is 120/240-volt, 3-wire, single-phase

Classified as power panelboard

20 space

Classified as lighting and appliance panelboard

20 space

Assume neutral is not used with 20- and 30-ampere, 2-pole circuit-breakers

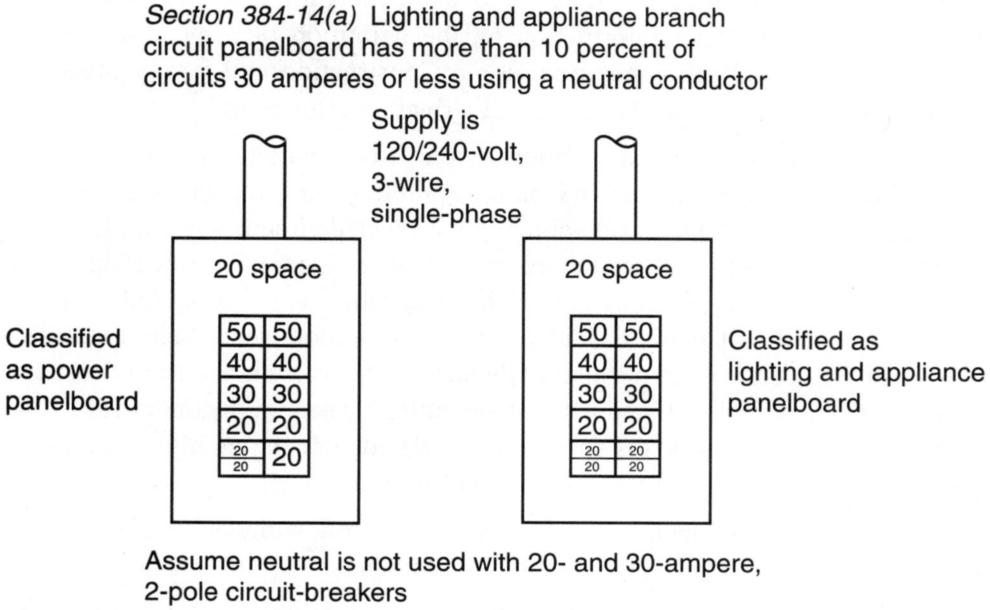

Figure 6.11 A panelboard is classified as a lighting and appliance branch circuit panelboard if more than 10 percent of the circuits are rated 30 amperes or less and the circuits have a neutral conductor.

Code except that two main overcurrent devices in the power panelboard are no longer permitted because the previous edition of the Code considered this to be a lighting and appliance branch circuit panelboard. The actual change with respect to overcurrent protection for panelboards is in the new exception. If a power panelboard is supplied with a neutral conductor and more than 10 percent of the overcurrent devices are rated 30 amperes or less, a single overcurrent device is not required to protect the panelboard if the panelboard is used as service equipment and meets the six disconnect requirements of *Section 230-71*. The neutral is available in the panelboard, but not used with the circuits rated 30 amperes or less.

384-16(g): This section of the Code permitted back-feeding plug-in type overcurrent devices or main lug assemblies, provided additional fasteners were installed to prevent these devices from being disconnected from their mounting means. This requirement now only applies to those conductors that are field-installed.

Article 490   **Equipment over 600 Volts Nominal**

This is a new article made up primarily of material from *Article 710* in the previous edition of the Code. The material in that article was moved to various locations in the Code with the majority of the material moved to this article. Some of the sections were reworded to eliminate as many of the exceptions as possible, but the material remained essentially the same as in the previous edition of the Code.

Article 550   **Mobile Homes, Manufactured Homes, and Mobile Home Parks**

550-2: The definition of a manufactured home was changed. All manufactured homes are required to be labeled, indicating that they are manufactured. The previous edition of the Code defined a manufactured home as being built on a permanent foundation. A manufactured home is now defined as one that can be with or without a permanent foundation.

550-2 FPN 1: This new fine print note reminds installers to check with the local building codes for the definition of the term **permanent foundation.**

550-2 FPN 2: This is a new fine print note. Field and local enforcement agency personnel who would like more in-depth information on the definition of a manufactured home are referred to Part 3280, Manufactured Home Construction and Safety Standards of the United States Department of Housing and Urban Development.

550-8(g): This section covers the installation of a heat-tape outlet. To increase protection for this portion of the wiring system, it is now required that this receptacle outlet, if installed, shall be connected to the load side of a GFCI protected circuit. The GFCI protection may come from the receptacle in the bathroom; however, this GFCI protection is not permitted to be on a small-appliance circuit. If this heat-tape outlet is installed on the underside of the mobile home, it is not considered to be an outdoor receptacle, regardless of distance from the outside edge of the mobile home. This is a change from the previous edition of the Code, which allowed a heat-tape outlet located on the underside of the mobile home and not more than 3 foot from the outside edge of the mobile home to be considered an outside receptacle. This is illustrated in Figure 6.12.

550-10(f): Metallic and nonmetallic surface raceway is now included in the list of permitted raceways for mobile homes and manufactured homes.

550-23(b): The conditions under which service equipment may be installed in or on a manufactured home are covered in this section. These conditions were found in *Section*

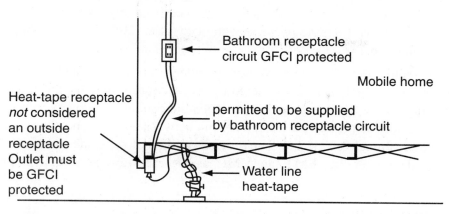

Section 550-8(g) Receptacle under a mobile home to supply water line heat-tape

Bathroom receptacle circuit GFCI protected

Mobile home

Heat-tape receptacle *not* considered an outside receptacle Outlet must be GFCI protected

permitted to be supplied by bathroom receptacle circuit

Water line heat-tape

Figure 6.12   A receptacle outlet installed under a mobile home for a water line heat-tape is required to be ground-fault circuit-interrupter protected and is permitted to be supplied from the bathroom receptacle circuit.

550-23(a) Exception 2 in the previous edition of the Code. Now the service equipment may only be installed in or on a manufactured home where the building is attached to a permanent foundation that complies with the local building codes.

Article 551   **Recreational Vehicles and Recreational Vehicle Parks**

551-30(e): New text was added to this section to concerning recreational vehicle generators that are mounted below the vehicle and not in a compartment. For these types of generator installations, a panelboard or junction box with a receptacle is required to be placed within 18 inches from the point the supply conductors from the generator enter the vehicle. If the generator first termination point is a junction box with a blank cover, the junction box is required to be mounted to any part of the generator supporting structure, or to the inside or bottom side of the vehicle floor within 18 inches of any point directly above the generator.

551-41(c)(2): Where 125-volt, single-phase, 15- or 20-ampere receptacles are installed, are intended to serve the lavatory counter top, and are mounted within 6 feet of the lavatory or sink, the receptacles are required to be ground-fault circuit-interrupter protected. The previous edition of the Code included all receptacles within 6 feet of the lavatory or sink and did not specify only those serving the counter top.

551-41(c)(2) Exception 3: This new exception permits receptacles to be installed within 6 feet of a sink or lavatory that are not ground-fault circuit-interrupter protected if they are a part of an expandable room where the receptacles are deenergized when the room is retracted.

551-47(a): It is now permitted to install ENT, liquidtight flexible nonmetallic conduit, and surface metal raceways in recreational vehicles.

551-47(g): When installing ENT through holes of framing members, additional protection for tubing is required when the separation distance from the inside and outside surfaces of the framing member to the ENT is less than 1 1/4 inches.

551-72: The type of electrical distribution system permitted to be used to supply sites in a recreational vehicle park is covered in this section. In the previous edition of the Code, a

120/240-volt, 3-wire system was required. This electrical system is still required for all 50-ampere receptacles at recreational vehicle sites. However, for sites with 20 and 30 amperes, 120-volt receptacles may now be derived from any grounded electrical system that delivers 120-volt, single-phase power, such as a 120/208-volt, 3-phase system as shown in Figure 6.13.

551-80(b): The requirements for the protection of underground cables and conductors installed for recreational vehicle sites are covered by this section. Where not exposed to physical damage, liquidtight flexible metal and nonmetallic conduit are now permitted to protect direct burial conductors and cables that are run between a trench and equipment or enclosure at a recreational vehicle site. The previous edition of the Code recognized only rigid metal, rigid nonmetallic, EMT, and IMC as the raceways permitted to provide this protection.

Article 552 **Park Trailers**

552-41(d): This section covers the installation of a ground-fault circuit-interrupter protected heat-tape receptacle outlet for a park trailer. The outlet is not permitted to be installed on a small-appliance branch circuit. If installed on the underside of the park trailer, it is not considered to be an outdoor receptacle.

552-43(a): The Code now permits the use of field-installed cord cap to supply power to a park trailer.

552-43(c): Liquidtight flexible nonmetallic conduit is now permitted to be installed from the disconnecting means in the park trailer to the underside of the trailer. In the previous

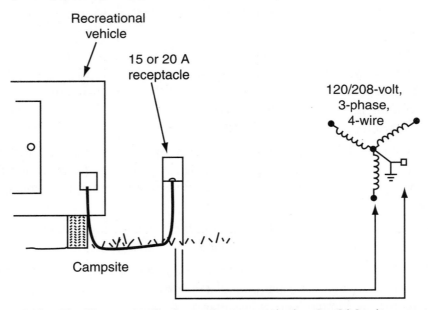

Section 551-72 Campsite receptacles rated 20 or 30 amperes, 125 volts are permitted to be derived from any electrical system providing 120 volts

Figure 6.13 **The 50-ampere, 3-wire outlets at recreational vehicle sites are to be derived from a single-phase, 120/240-volt electrical system, but the 20- and 30-ampere, 125-volt receptacles are permitted to be derived from any electrical system that supplies 120-volt power, such as a 120/208-volt, 3-phase electrical system.**

edition of the Code, only metal raceway and rigid nonmetallic conduit were permitted for this application.

552-46(a) Exception: This new exception permits more than five 125-volt branch circuits rated 15 or 20 amperes to be installed in a park trailer provided they are equipped with an energy management system rated at 30 amperes maximum.

552-48(a): It is now permitted to install ENT, liquidtight flexible nonmetallic conduit, and surface metal raceways in park trailers.

552-48(g): When installing ENT through holes in framing members, additional protection is required when the 1 1/4-inch separation distance from the inside and outside surfaces of the framing member cannot be maintained.

Article 605 **Office Furnishings**

No significant changes occurred in this article.

Article 645 **Information Technology Equipment**

In several sections of this article, the references to **computer rooms** or **data processing systems** have been replaced with the term **information technology equipment room.** No change of intent occurred with this change; these changes are an editorial correction from the previous edition of the Code when the title of this article was modified to its present form.

645-5(d)(5)(c): In the previous edition of the Code, this was the third exception to this section that permitted the installation of other than type DP cables, such as type TC, for use under elevated floors of an information technology equipment room. All type TC rated cables contain at least two or more conductors; therefore, single conductors such as equipment grounding cannot be identified as type TC. A new sentence was added to permit single conductors to be used for equipment grounding. If this single conductor is insulated green and is AWG number 4 or larger, the conductor is permitted to be marked **for CT use** or **for use in cable trays.**

645-15: There is a new requirement for the grounding of information technology equipment. If a signal reference structure or grid has been installed, this structure must be bonded to the equipment grounding system provided for the equipment.

WORKSHEET NO. 6
OVERCURRENT PROTECTION

These questions are considered important to understanding the application of the *National Electrical Code®* to electrical wiring, and they are questions frequently asked by electricians and electrical inspectors. People working in the electrical trade must continue to study the Code to improve their understanding and ability to apply the Code properly.

DIRECTIONS:　Answer the questions and provide the Code reference or references where the necessary information leading to the answer is found. Space is provided for brief notes and calculations. An electronic calculator will be helpful in making calculations. You will keep this worksheet; therefore, you must put the answer and Code reference on the answer sheet as well.

1.　Is 225 amperes a standard rating for a fuse or circuit-breaker? (Yes or No)

　　Code reference _____

2.　An electrician installs an electrical panel containing fuseholders of the screw-shell type for general purpose 15-ampere circuits in a building associated with a dwelling. Does the Code require the electrician to install type S fuses as shown in Figure 6.14? (Yes or No)

　　Code reference _____

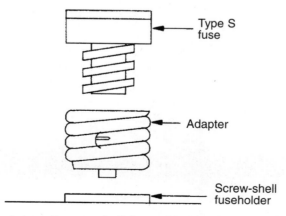

Type S fuse

Adapter

Screw-shell fuseholder

Figure 6.14　A type S screw-shell fuse with an adapter for a screw-shell fuseholder.

3. A busway is suspended from the ceiling of an industrial area. A 40-foot addition to the existing 400-ampere busway is to be installed. If 225-ampere busway is of ample size to serve the intended load, is it permitted to add this 225-ampere section to the existing busway without overcurrent protection for the 225-ampere section, other than the 400-ampere fuses protecting the entire busway? (Yes or No)

    Code reference _____

4. A disconnect switch is intended to break current at fault levels. The disconnect shall, for the system voltage and current available at the line terminals of the equipment, have:
    A. a continuous current rating not less than the available short-circuit current.
    B. an adjustable current trip mechanism.
    C. a low voltage cutout.
    D. a sufficient interrupting rating.

    Code reference _____

5. Two 100-ampere, 120/208-volt, 3-phase lighting panelboards are mounted next to each other, and each is supplied with a set of AWG number 1 type THWN ungrounded copper conductors in the same conduit. The conductors have been sized based on an 80-ampere load and a 70 percent derating factor. All conductor terminations are rated at 75°C. Also, assume that the load served does not produce nonlinear or harmonic currents on the neutral conductor. Is an AWG number 4/0, type THWN copper conductor permitted to serve as a common neutral for both feeders, as shown in Figure 6.15? (Yes or No)

    Code reference _____

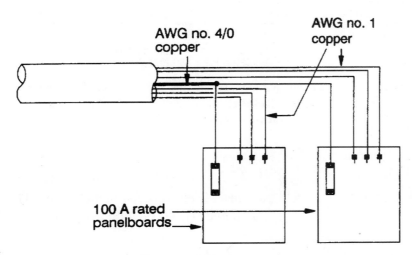

Figure 6.15   A common neutral for two 120/208-volt, 3-phase, 4-wire feeders.

6. The maximum number of single-pole circuit-breakers permitted in a panelboard is _____

   Code reference _____

7. A 3-phase, 120/208-volt electrical system supplies a building, and the service equipment consists of an 18-space circuit-breaker panelboard. The building load calculated according to *Article 220* is 130 amperes, but the panelboard is rated 200 amperes, and the three ungrounded service entrance conductors are copper type THHN, AWG number 4/0. A neutral conductor enters the panel, and it is copper type THHN, AWG number 1/0. There is no single main overcurrent device in the panelboard. There are six 3-pole circuit-breakers, one rated 100 amperes, three rated 50 amperes, one rated 30 amperes, and one rated 20 amperes, as shown in Figure 6.16. The neutral conductor is not run as a part of the 20- or 30-ampere, 3-phase circuits. This service equipment is classified as a (lighting and appliance branch circuit panelboard, power panelboard).

   Code reference _____

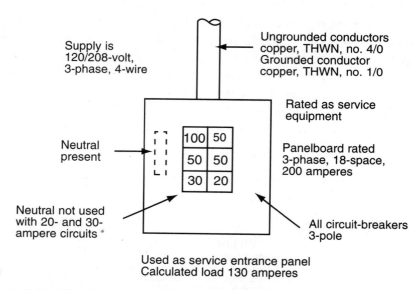

Figure 6.16   The 18-space panelboard for this 120/208-volt, 3-phase service contains six 3-pole circuit-breakers; there is a neutral present in the panel, but the neutral is not used as a circuit conductor for the 20-ampere or 30-ampere circuits.

8. A recreational vehicle park has 52 recreational vehicle sites. The minimum number of sites required to be equipped with a 30-ampere, 125-volt receptacle is _____.

   Code reference _____

9. A fusible disconnect switch, as shown in Figure 6.17, receives the supply of electricity for a branch circuit at the bottom of the disconnect switch with the load attached to the top. The disconnect only accepts cartridge fuses. Is this installation, as shown, permitted by the Code? (Yes or No)

Code reference _____

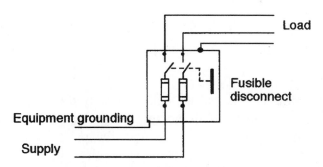

Figure 6.17   A fusible disconnect switch is wired with the supply directly to the fuses.

10. An electrician has a fusible disconnect switch rated 100 amperes and 250 volts. The disconnect only accepts cartridge fuses. Is this disconnect permitted to be used on a 100-ampere, 480-volt circuit? (Yes or No)

Code reference _____

11. Which of the screw-shell fuses shown in Figure 6.18 is rated 15 amperes? (A or B)

Code reference _____

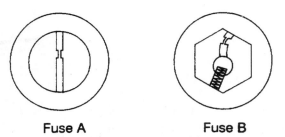

Fuse A   Fuse B

Figure 6.18   The tops of two screw-shell fuses are shown.

12. A small office data processing computer system has a nameplate rating of 28 amperes at 240 volts. Conductor terminations and insulation are rated 75°C. The minimum size copper type THHN wire permitted for this circuit is AWG number _____.

Code reference _____

13. A feeder consists of 500-kcmil copper type THWN conductors protected from overcurrent at 400 amperes. A tap is run from a pull box to a disconnect switch fused at 40 amperes to serve a 32-ampere continuous load. All terminations are 75°C rated. The total length of the tap is 7 feet. The minimum size copper type THHN tap conductor permitted is AWG number _____.

Code reference _____

14. A mobile home is installed as a single-family dwelling on the owners' property, and not in a mobile home park. Is the electrical service to the mobile home required to comply with the provisions of *Article 550,* assuming that electrical inspection is required in the area where the mobile home is to be installed? (Yes or No)

Code reference _____

15. A mobile home is supplied with a 50-ampere rated, 4-wire cord from a service on a pole adjacent to the mobile home. If a 15-ampere, 120-volt duplex receptacle is installed in the box on the pole, as shown in Figure 6.19, is it required to be protected by a ground-fault circuit-interrupter? (Yes or No)

Code reference _____

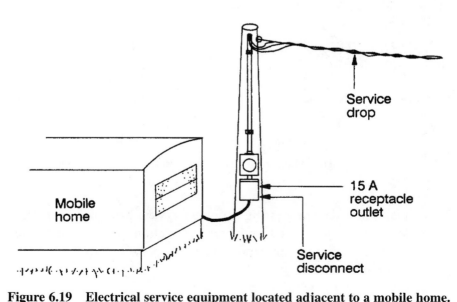

Figure 6.19　Electrical service equipment located adjacent to a mobile home.

16. A building is served with a 120/240-volt, 3-wire single-phase electrical system. A 2-pole knife blade disconnect switch is used as the disconnect switch for a 240-volt single-phase water pump. Are type S screw-shell fuses permitted for use, as shown in Figure 6.20? (Yes or No)

    Code reference _____

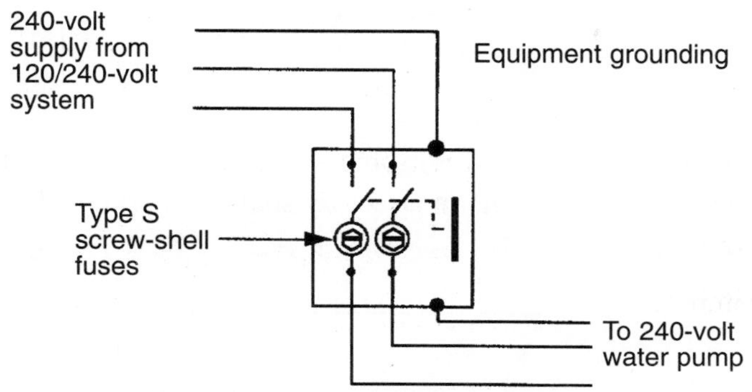

240-volt
supply from
120/240-volt
system

Equipment grounding

Type S
screw-shell
fuses

To 240-volt
water pump

Figure 6.20   A disconnect switch with screw-shell fuses for a 240-volt circuit.

17. In a recreational vehicle park, all sites with electrical power are required to have a 20-ampere, 125-volt receptacle outlet. (True or False)

    Code reference _____

18. A 300-ampere service entrance consists of a panelboard with a single main overcurrent device rated at 200 amperes and a 100-ampere fused disconnect switch. The panelboard and the disconnect switch are both rated as suitable for use as service equipment, and they are tapped from the 500-kcmil, type THWN aluminum service entrance conductors as shown in Figure 6.21. All conductor terminations are 75°C rated. The minimum size type THWN aluminum tap conductor permitted for the 100-ampere disconnect is AWG number _____.

    Code reference _____

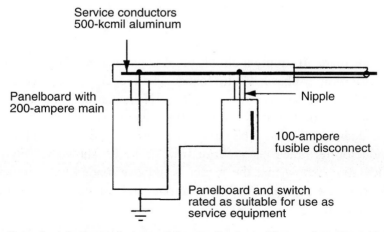

Service conductors
500-kcmil aluminum

Panelboard with
200-ampere main

Nipple

100-ampere
fusible disconnect

Panelboard and switch
rated as suitable for use as
service equipment

Figure 6.21   Two electrical panels tapped from a feeder, with taps less than 10 feet in length.

19. A feeder protected by a 200-ampere circuit-breaker supplies two panelboards for lighting and receptacle loads. Is it permitted to install 100-ampere rated panelboards tapped directly to this feeder without a main overcurrent protective device on the supply side of each panelboard? (Yes or No)

   Code reference _____

20. Listed 20-ampere rated circuit-breakers used as switches for 277-volt fluorescent lighting circuits in a factory shall be marked _____ on the circuit-breaker.

   Code reference _____

21. Listed extension cords AWG number 16 and larger without integral overcurrent protection and adequately sized for the load to be served are plugged into a 20-ampere 120-volt circuit. Is such a listed extension cord set considered to be protected by the circuit overcurrent device? (Yes or No)

   Code reference _____

22. A disconnecting means located near the principal operator exit doors is required for a group of data processing equipment located in a data processing room as well as the air-conditioning equipment for the room. (True or False)

   Code reference _____

23. A tap is made to supply a circuit-breaker panelboard from a 400-ampere, 120/208-volt, 3-phase, 4-wire feeder. The feeder ungrounded conductors are 600 kcmil copper with type THWN insulation. The distance from the tap point on the feeder to the 100-ampere main circuit-breaker in the panelboard is 25 feet. The tap conductor is copper, type THWN, and all conductor terminations are rated at 75°C, as shown in Figure 6.22. Calculations show that the feeder is not overloaded with the additional load from the panelboard. The minimum size tap conductor permitted in this case is _____.

   Code reference _____

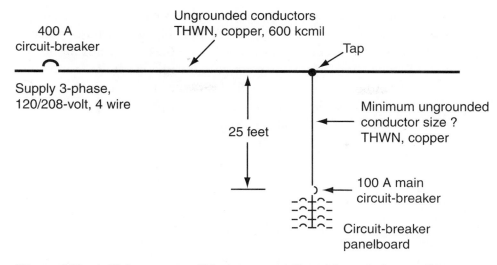

Figure 6.22   A 25-foot tap to a 100-ampere panelboard is made from a 400-ampere feeder with copper conductors with 75°C insulation and terminations.

24. The owner of a recreational vehicle park wants to charge one rate for a 20-ampere circuit and another rate for a 30-ampere circuit at a campsite. The wire is AWG number 10 copper, and the receptacle outlet is rated 30 amperes. For the owner's convenience, is the parallel installation shown in Figure 6.23 permitted? (Yes or No)

Code reference _____

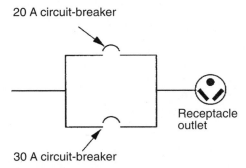

Figure 6.23   Two circuit-breakers are connected in parallel.

25. An electric heater circuit is protected with 60-ampere fuses, and the heater is controlled with a contactor. A line voltage thermostat operates the coil of the contactor, as shown in Figure 6.24. All wires are in conduit. All conductor terminations are 75°C rated, except the terminations at the thermostat are only 60°C rated. Determine the minimum size of the copper type THWN Class 1 control circuit wire permitted if it is only protected with the 60-ampere heater circuit fuses.

Code reference _____

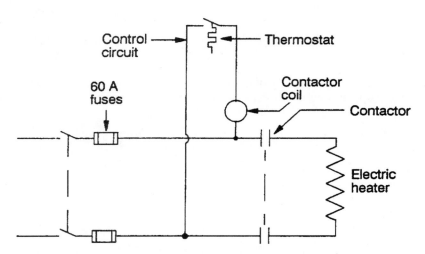

Figure 6.24   A thermostat controls a contactor to operate a resistance-type electric heater.

ANSWER SHEET   Name _____

No. 6   OVERCURRENT PROTECTION

Answer	**Code reference**
1. _____	_____
2. _____	_____
3. _____	_____
4. _____	_____
5. _____	_____
6. _____	_____
7. _____	_____
8. _____	_____
9. _____	_____
10. _____	_____
11. _____	_____
12. _____	_____
13. _____	_____
14. _____	_____
15. _____	_____
16. _____	_____
17. _____	_____
18. _____	_____
19. _____	_____
20. _____	_____
21. _____	_____
22. _____	_____
23. _____	_____
24. _____	_____
25. _____	_____

UNIT 7

Motor-Circuit Wiring

OBJECTIVES

Upon completion of this unit, the student will be able to:

- read and explain the use of the information on a motor nameplate.
- determine the minimum size conductor permitted for a motor branch circuit.
- determine the maximum rating of the overcurrent device for the motor branch circuit short-circuit and ground-fault protection.
- select the type and rating of disconnect required for a motor circuit.
- select the type and rating of the controller for a motor.
- determine the maximum permitted rating of motor overload protection.
- determine the minimum size conductor permitted for a feeder supplying several motors.
- explain when overcurrent protection for the motor-control circuit is needed in addition to the motor branch circuit overcurrent device.
- explain the meaning of NEMA enclosure type for motor controllers.
- diagram a control circuit for a magnetic motor starter.
- answer wiring installation questions relating to *Articles 430, 440, 455, 460, 470, 675, 685,* and *Examples D8* through *D10* in *Appendix D.*
- state three significant changes that occurred from the 1996 to the 1999 Code for *Articles 430, 440, 455, 460, 470, 675, 685,* or *Examples D8* through *D10* in *Appendix D.*

CODE DISCUSSION

Sizing and installation of components of an electric motor circuit are covered in this unit. An electric motor is a device to convert electrical energy into mechanical energy or power. An electric motor is a perfect servant. It tries to power any load to which it is connected. For this reason, overcurrent protection must be provided that will prevent self-destruction of the motor and possible fire or an electrical shock hazard. The task of sizing overcurrent protection for a motor is difficult because of the current characteristics of the motor. The motor will typically draw four to six times as much current during starting as it will when running under constant full-load. Figure 7.1 shows the typical motor current draw during starting, compared with the full-load running current. The real challenge for overcurrent protection is to protect for overloads during running and yet not experience opening of the overcurrent device during the starting in-rush current.

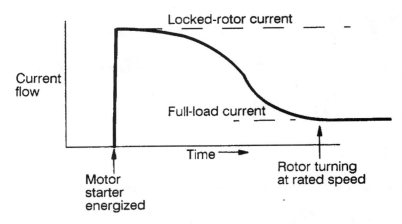

Figure 7.1 When a motor is started, there is an in-rush of current until the motor obtains running speed.

Article 430 deals with motors, motor branch circuits and feeders, conductors and their protection, motor overload protection, motor-control circuits, motor controllers, and motor-control centers. *Diagram 430-1* in the Code shows a motor circuit and can be used as an index to find the section of the article dealing with the various circuit components. *Part A* of this article provides general information about various motors and general wiring requirements for the motor circuit. There are several tables at the end of the article, which will be used frequently when sizing components for a motor circuit. *Tables 430-147* through *430-150* give the motor full-load current for various types of motors. *Section 430-6(a)(1)* requires that these values are used except when sizing the motor running overcurrent protection, and unless the nameplate full-load current for the motor is a higher value than given in the table. For a motor-operated appliance with both the horsepower and full-load current listed on the nameplate, the full-load current is to be used for sizing conductors and other circuit components, *Section 430-6(a)(2)*.

The rules for determining the minimum permitted size of single motor branch circuit conductors are found in *Section 430-22*. The minimum wire size for a feeder serving a specific group of motors is determined according to *Section 430-24*. The wire size for a motor circuit is determined directly from *Table 310-16* once the ampacity has been determined in *Part B*. The former footnote to *Table 310-16*, which states that the overcurrent protection for an AWG number 14 copper wire is not to exceed 15 amperes, 20 amperes for an AWG number 12, and 30 amperes for an AWG number 10, is now *Section 240-3(d)* and does not apply in the case of a motor circuit. Other references in the Code that clarify that overcurrent protection for motor branch circuits is to be sized according to the provisions of *Article 430* are *Section 430-1* and *Section 240-3(g)*.

Part C discusses the type of overload protection permitted for an electric motor and the sizing of the overload protection. *Part D* explains the short-circuit and ground-fault protection for the motor branch circuit. Frequently, a single overcurrent device does not protect for overloads, ground-faults, and short-circuits. *Part E* deals with the overcurrent protection of motor feeders. This would be a conductor supplying power to more than one motor branch circuit. *Table 430-152* is used when determining the maximum permitted rating of motor branch circuit short-circuit and ground-fault protection.

The electrical current to many motors is controlled by a motor-control circuit. The wiring to a start-stop station for a magnetic motor starter is a Class 1 control circuit. *Part F* covers the wiring and overcurrent of motor-control circuits. *Part G* covers controllers for motors. This part provides requirements to what is permitted to be a motor controller and the ratings of controllers. *Part H* deals with motor-control centers. A motor-control center is an assembly of one to several sections in which there is a bus on which motor-control units are attached. The requirements for the motor-control center are in this part of *Article 430*, but installation requirements are in *Article 384* on switchboards and panelboards. *Part J* covers the type and ratings of disconnecting means for motors and controllers.

Article 440 deals with the branch circuits and motors associated with air-conditioning and refrigeration equipment. These are special cases because of frequent use of hermetically sealed motor compressors, and the use of multimotor branch circuits. Branch circuit selection current is covered in *Section 440-4(c)*. The minimum permitted size of the branch circuit conductors is found in *Article 440, Part D*. The minimum permitted rating of the disconnecting means for the refrigerant motor compressor is determined from *Article 440, Part B*. *Section 440-22(c)* states that if the maximum permitted rating of branch circuit short-circuit and ground-fault protective device rating is marked on the nameplate, that value shall not be exceeded. Overload protection for the motor compressor and for the branch circuit conductors shall not exceed the value required in *Article 440, Part F*. The installation of room air conditioners is covered in *Part G*.

Article 455 deals with the installation of phase converters. A phase converter is an electrical device that permits a 3-phase electric motor or other 3-phase equipment to be operated from a single-phase electrical supply. The sizing of components of the circuit or feeder for a phase converter installation is difficult because of the 1.73-to-1 current ratio between the single-phase and 3-phase portions of the circuit. It theoretically takes 17.3 amperes flowing on the single-phase input conductors to a phase converter to supply 10 amperes at 3-phase to a load connected to the phase converter. A phase converter circuit is illustrated in Figure 7.2. An electric motor is permitted to be operated at the nameplate rated horsepower, but the motor starting torque will be greatly reduced when operated from a phase converter. Motors powering hard starting loads should not be operated from a phase converter unless recommended by the manufacturer.

Phase converters may be a static type with no moving parts. A static phase converter usually serves only one 3-phase load, and it is sized specifically for that load. Most static phase converters consist primarily of capacitors. A transformer can be used to provide an output voltage different from the input voltage. Solid-state static phase converters rectify the single-phase alternating current supply to direct current and then invert the direct current into

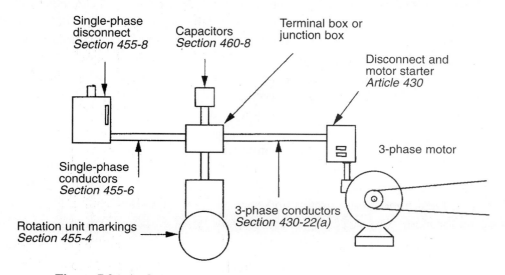

Figure 7.2   A phase converter permits a 3-phase motor or load to be operated from a single-phase electrical supply.

3-phase alternating current. This type of phase converter also can control the output frequency so the speed of induction motors supplied can be made variable.

Rotary phase converters are capable of supplying one or more loads. If an installation consists of several 3-phase motors, then it may be possible to supply the 3-phase power from a single-phase supply with one rotary phase converter. The basic components of a rotary phase converter are a rotating unit that is actually a 3-phase motor, a capacitor bank, and if the output voltage is different than the input voltage, a transformer is included. With this type of unit, it is necessary to start the phase converter and bring the rotating unit to full speed before any 3-phase load is applied.

When sizing the conductors and overcurrent protection for a phase converter installation, it is necessary to know the "rated single-phase input full-load amperes" from the nameplate or the actual load to be served. *Section 455-6(a)(1)* requires that the minimum single-phase input conductor ampacity not be less than 1.25 times the rated single-phase input full-load current as marked on the nameplate. It is not uncommon for a phase converter to be oversized for the load to be served. In this case, *Section 455-6(a)(2)* permits the single-phase input conductors to have an ampacity not less than 2.5 times the full-load current of the 3-phase load supplied by the phase converter.

Overcurrent protection is required for the single-phase input conductors and the phase converter. That overcurrent protection is located at the supply end of the single-phase input conductors. *Section 455-7(a)* requires that the overcurrent protection not be more than 1.25 times the rated single-phase input full-load current as marked on the phase converter nameplate. *Section 455-7* permits the calculated value to be rounded up to the next standard rating of overcurrent device as given in *Section 240-6* when the calculated value does not correspond with a standard rating. *Section 455-7(b)* permits the maximum rating of overcurrent protection to be sized not more than 2.5 times the sum of the 3-phase loads supplied by the phase converter.

A disconnecting means is required to be located within sight of the phase converter. *Section 455-8* permits that disconnecting means to be a switch rated in horsepower, a circuit-breaker, or a molded case switch. If only nonmotor loads are served and a switch is used as the disconnecting means, the switch is not required to have a horsepower rating. The disconnecting means is required to have an ampere rating not less than 1.15 times the rated single-phase input full-load current as marked on the phase converter nameplate. *Section 455-8(c)(1)* permits a circuit-breaker or a molded case switch to be sized not less than 2.5 times the sum of the 3-phase full-load currents of the loads supplied. *Section 455-8(c)(2)* explains how to determine the horsepower rating of the load served by the phase converter. The disconnecting means must be capable of opening the circuit under full-load or locked-rotor conditions. The horsepower rating is selected from either *Table 430-151A* or *Table 430-151B* by using a calculated equivalent locked-rotor current. That equivalent locked-rotor current is 2.0 times the locked-rotor current of the largest motor served, plus the full-load current of all other motors served, plus the full-load current of all nonmotor loads served.

Example 7.1 A 20-kVA rotary phase converter with a rated single-phase input full-load current of 105 amperes is connected to a 240-volt single-phase electrical system to supply several 230-volt, 3-phase design B motors. The electric motors supplied are 10, 3, and 2 horsepower. Determine the minimum horsepower rating of switch permitted to supply this specific motor load.

Answer: Add the locked-rotor current of the 10-horsepower, 230-volt, 3-phase design B motor (*Table 430-151B*) to the full-load currents of the 3- and 2-horsepower motors and then multiply the sum of these numbers by 2.0.

$$\text{Equivalent locked-rotor current} = 2.0 \times (162 \text{ A} + 9.6 \text{ A} + 6.8 \text{ A}) = 357 \text{ A}$$

Now find a single-phase horsepower rating from *Table 430-151A* that has a locked-rotor current equal to or greater than 357 amperes. Note that there is no single-phase switch shown in *Table 430-151A* with a horsepower rating high enough to handle the equivalent locked-rotor current of 357 amperes. Manufacturers do have single-phase switches with horsepower ratings higher than listed in *Table 430-151A*. If a single-phase switch is not available, a 3-phase switch can be used as the phase converter disconnecting means. In this case, one of the switch poles is not used. It is important that the 3-phase switch has a locked-rotor current rating equal to or higher than the calculated equivalent locked-rotor current for the load. In this example, the equivalent locked-rotor current was 357 amperes. Look down the 230-volt 3-phase column of *Table 430-151B* until a locked-rotor current value is found that is equal to or greater than 357 amperes. A switch rated 20-horsepower 3-phase is not adequate because it only has a locked-rotor current rating of 290 amperes. The 25-horsepower 3-phase switch has a locked-rotor current rating of 365 amperes, which is larger than 357 amperes. Therefore, a 25-horsepower 3-phase switch can be used as the disconnecting means for the phase converter.

Article 460 deals with the installation of capacitors wired as part of electrical circuits, and not as a component part of electrical equipment. Typical examples would be capacitors added for power factor correction or auxiliary capacitors added to motor circuits. An important aspect of this article is the discharging of the capacitor when it is deenergized. A capacitor will store a charge when power is shut off. If the charge is not removed from the capacitor, it can in some cases become a serious electrical hazard even though power has been disconnected. Sizing the capacitor supply wires, overcurrent protection, disconnects, markings, and grounding are covered in this article.

Article 470 covers the installation of resistors and reactors, which are sometimes used as a part of a motor or other equipment circuit. The main points of this article are the prevention of physical damage to the components and the prevention of overheating of wiring. Resistors and reactors are sometimes used in motor controllers for soft starting to reduce initial motor in-rush current.

Article 675 covers the installation of electrically driven or controlled irrigation machines. These are devices consisting of a water pipe with periodic sprinkler nozzles to irrigate crop land. These devices are frequently propelled through the field with electric motor-driven wheels at regular intervals along the machine. Some irrigation machines rotate about a central pivot point, while others laterally move across the field. A typical center pivot irrigation machine is shown in Figure 7.3. The equivalent current rating of *Section 675-7* is used to determine the size of conductors, overcurrent protection, disconnecting means, and controller for the irrigation machine. The drive motors for some irrigation machines, such as the center pivot type, operate intermittently. Therefore, a duty cycle is permitted to be applied when determining the equivalent current rating of the machine. *Section 675-22* provides the value to use for the duty cycle for the center pivot machine. *Section 675-8(b)* requires a disconnecting means at the point where electrical power connects to the machine or within 50 feet of that point. *Section 675-15* requires a grounding electrode system for lightning protection.

Article 685 applies to integrated electrical systems in industrial applications where equipment must be shut down in an orderly manner for a particular reason. In the event of a malfunction in a component machine of a continuous process, automatic shutdown of that particular machine may cause great economic loss or even endanger human life. A malfunction

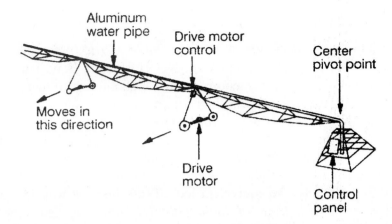

Figure 7.3   An electrically driven center pivot irrigation machine that is driven with motors at each support and pivots about a stationary center point.

signal to the required on-duty maintenance personnel is permitted in place of automatic shutdown.

Appendix D, Example D8 is an excellent example of an installation of three motors served by a common feeder. Each motor is on a separate branch circuit. This example gives a clear understanding of the meaning of the various rules in *Article 430*. It is highly recommended that the person learning about motor-circuit and feeder installations work through this example. This example shows the method of determining the maximum permitted rating of the branch circuit short-circuit and ground-fault protective device. *Section 430-52(c)(1) Exception 1* permits rounding up to the next standard rating of overcurrent device listed in *Section 240-6*. The full-load current of the motor as determined from *Table 430-148* or *Table 430-150* is multiplied by the factor found in *Table 430-152*. It is permitted to round up to the next higher standard overcurrent device rating.

Example D9 shows how to determine the minimum permitted ampacity of the feeder conductors supplying several elevator motor branch circuits. *Example D10* deals with branch circuits and a feeder supplying elevators with adjustable speed drives. There is a branch circuit conductor supplying the transformer and a conductor with a different ampacity between the transformer and the drive unit. Finally, it is shown how the individual circuit loads are combined to determine the required ampacity of the feeder.

MOTOR CIRCUITS AND CALCULATIONS

Information is provided in the Code that is necessary to size and install the components of a motor circuit. It is also necessary to determine information from the specific motor to be installed. The type of environment must be known to choose the proper enclosures for the motor and other equipment of the circuit.

Motor Nameplate and Other Information

A motor circuit is wired to fit the specifications of a specific motor. Information for determining the size or rating of the various parts is found on the motor nameplate, and the full-load current for single-phase and 3-phase motors from *Table 430-148* and *Table 430-150*. A motor nameplate is shown in Figure 7.4. The most important information needed for wiring the circuit is: (1) horsepower; (2) phase; (3) voltage; (4) full-load current; (5) temperature rise

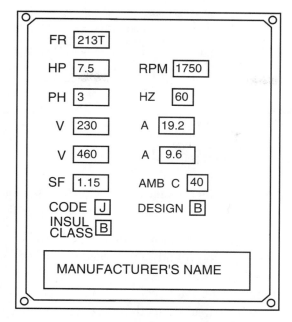

Figure 7.4   The motor nameplate contains essential information for sizing components of a motor circuit.

above ambient or service factor; (6) code letter; and (7) design letter. Also, it is important to know about the physical environment, the location of the installation, the type of controller desired, and the type of load that will be powered (easy starting or hard starting load). The motor nameplate of Figure 7.4 shows ambient temperature, and not temperature rise above ambient. Ambient temperature is the maximum environmental temperature in which the motor is to be operated. If the surrounding temperature is higher than the ambient temperature marked on the nameplate, the motor will be in danger of overheating if operated at the nameplate horsepower rating.

The in-rush current of a motor is affected by the design of the motor. The National Electrical Manufacturers Association (NEMA) has established specifications and designations for motor design. Design letters are used to group motors into categories of similar operating characteristics. These characteristics include slip at 100 percent load, locked-rotor current, and torques at various speeds. The most common type of motor in use is the design B motor. Motors are available that are designed especially as high-efficiency motors. These are designated as design E motors. The design E motor has a higher in-rush current than the more common design B motor. In the case where a single disconnecting means serves more than one motor, the equivalent locked-rotor current may be required to be determined. *Table 430-151B* is used to determine the locked-rotor current for 3-phase motors for this calculation. There is a separate locked-rotor current column for design E motors in *Table 430-151B*. Motors operating at low speeds or high torque may have full-load currents in excess of the values listed in *Table 430-148* or *Table 430-150*. In these cases, the nameplate current shall be used if it is higher than the values given in the tables.

Motor Circuit

A typical motor circuit is diagrammed in Figure 7.5. There is a chart in *Section 430-3* that gives the location in the article where necessary information is found for sizing and wiring the circuit. The following components shall be sized or specified for a motor branch circuit:

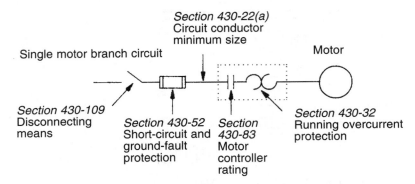

Figure 7.5   A typical single motor branch circuit with a motor starter.

1. Branch circuit disconnecting means minimum rating, *Part J*
2. Branch circuit short-circuit and ground-fault protection rating, *Part D*
3. Branch circuit conductors minimum size, *Part B*
4. Motor controller minimum size, *Part G*
5. Motor and branch circuit overload protection maximum rating, *Part C*
6. Motor-control circuit conductor minimum size, *Part F*
7. Motor-control circuit overcurrent protection maximum rating, *Part F*
8. Motor feeder conductor minimum size, *Part B*
9. Motor feeder short-circuit and ground-fault protection maximum rating, *Part E*
10. Motor-control center, *Part H*
11. Grounding, *Part M*

Methods of Controlling Motors

The motor controller directly controls the flow of electrical current to the motor. The definition is found in *Article 100*. Several methods are permitted to control a motor, as the following list indicates:

*   For portable motors rated 1/3 horsepower and less, the cord-and-plug, as stated in *Section 430-81(c)*
*   For stationary motors not over 1/8 horsepower, normally operating continuously, the branch circuit protective device, as stated in *Section 430-81(b)*
*   For stationary motors rated not more than 2 horsepower, a snap switch, as stated in *Section 430-83(c)(1)*
*   For a stationary motor, an inverse-time circuit-breaker rated in amperes, as stated in *Section 430-83(a)(2)*
*   For a stationary motor, a fusible switch rated in horsepower, as stated in *Section 430-90* and *Section 430-83*
*   A manual motor starter rated in horsepower, as stated in *Section 430-83(a)(1)*
*   A magnetic motor starter rated in horsepower, as stated in *Section 430-83(a)(1)*

The National Electrical Manufacturers Association (NEMA) has established a size numbering system for horsepower ratings of electric motor starters. The NEMA sizes are shown in Table 7.1. A 3-pole motor starter may be used to control a single-phase motor. If only 3-phase horsepower is listed on a 3-pole motor starter to be used for a single-phase motor, divide the 3-phase horsepower rating for the desired voltage by 2 to obtain the maximum single-phase horsepower rating permitted for that motor starter. A single-phase motor with the same

Table 7.1   Motor horsepower and voltage ratings for NEMA size motor starters.

NEMA size	Single-phase 120 V	240 V	Three-phase 208 V	240 V	480 V
00	1/3	1	1½	1½	2
0	1	2	3	3	5
1	2	3	7½	7½	10
1P	3	5	—	—	—
2	3	7½	10	15	25
3	7½	15	25	30	50
4	—	25	40	50	100
5	—	50	75	100	200
6	—	—	150	200	400
7	—	—	—	300	600
8	—	—	—	450	900

horsepower and voltage rating as a 3-phase motor will draw 1.73 times as much current as the 3-phase motor.

Enclosure Types

Enclosure type ratings for different environmental conditions have been established by NEMA. Enclosure types for motor starters for applications other than hazardous locations are listed in *Table 430-91*. The following is a general description of the common NEMA enclosure types and their typical applications.

NEMA 1:   General purpose enclosure used in any location that is dry and free from dust and flying flammable materials.

NEMA 3:   Weather-resistant enclosure suitable for use outdoors. Not suitable for use in dusty locations. This type is no longer available for motor starters from some manufacturers.

NEMA 4:   Watertight and dusttight enclosure suitable for outdoor locations and inside wet locations. Water can be sprayed directly on the enclosure without leaking inside. Suitable for most agricultural locations, provided corrosion is not a problem.

NEMA 4X: Watertight, dusttight, and corrosion-resistant enclosure suitable for outside and inside wet, dusty, and corrosive areas. Suitable for agricultural buildings.

NEMA 7:   Explosion-proof enclosure suitable for installation in Class I areas containing hazardous vapors. Required to be rated for type of hazardous vapor such as gasoline, Group D.

NEMA 9:   Dust-ignition-proof enclosure suitable for installation in Class II hazardous areas such as grain elevators. Required to be rated for the type of dust such as grain dust, Group G. There is a tendency for manufacturers to build one enclosure rated as NEMA 7 and 9.

Overcurrent Device Ratings

Manufacturers' ratings of time-delay fuses not larger than 30 amperes are frequently used to provide overload protection for electric motors. Table 7.2 is a listing of some generally available fuses of sizes not listed in the Code as standard ratings. Refer to *Section 240-6* for the list of standard ratings of fuses and circuit-breakers. The standard ratings of fuses less than 15 amperes recognized by the Code are 1, 3, 6, and 10 amperes.

Table 7.2   Typical time-delay fuse ratings available up to 30 amperes.

¹⁄₁₀	¹⁵⁄₁₀₀	²⁄₁₀	⁴⁄₁₀	½	⁶⁄₁₀	⁸⁄₁₀
1	1⅛	1¼	1⁴⁄₁₀	1⁶⁄₁₀	1⁸⁄₁₀	
2	2¼	2½	2⁸⁄₁₀	3²⁄₁₀	3½	
4	4½	5	5⁶⁄₁₀	6¼	7	8
9	10	12	15	17½	20	25	30

Motor Overload Protection

Electric motors are required to be protected against overload, according to the rules in *Section 430-32*. Overload protection is usually provided as a device responsive to motor current or as a thermal protector integral with the motor. A device responsive to motor current could be a fuse, a circuit-breaker, an overload current sensor, or a thermal reset switch in the motor housing. An automatically resetting thermal switch placed in the windings will sense winding temperature directly.

The service factor or temperature rise must be known from the motor nameplate when selecting the proper size motor overload protection. These are indicators of the amount of overload a motor can withstand. If a motor has a service factor of 1.15 or greater, the manufacturer has designed extra overload capacity into the motor. In this case, the overload protection shall be permitted to be sized not greater than 125 percent of the **nameplate** full-load current. *Section 430-34* permits the maximum setting of running overcurrent protection to be increased if the size determined in *Section 430-32* is not sufficient to permit the motor to start.

Internal heat is damaging to motor winding insulation. A motor with a temperature rise of 40°C (104°F) or less has been designed to run relatively cool; therefore, it has greater overload capacity. The overload protection is permitted to be sized not greater than 125 percent of the nameplate full-load current. A service factor of less than 1.15 or a temperature rise of more than 40°C (104°F) indicates little overload capacity. The overload protection under these circumstances is permitted to be sized not greater than 115 percent of the nameplate full-load current.

A time-delay fuse is permitted to serve as motor overload protection. Screw-shell fuses or cartridge fuses are used for small motors. The fuse size is determined by selecting the proper multiplying factor, 1.15 or 1.25, based on the service factor or temperature rise marked on the motor nameplate. Time-delay fuse ratings 30 amperes and smaller are listed in Table 7.2.

Circuit-breakers are permitted to be used as running overload protection, but they are not generally available in sizes smaller than 15 amperes. If they are used for large motors, they will usually trip on starting if they are sized small enough to provide overload protection.

Magnetic and manual motor starters have an overload relay or trip mechanism activated by a heater sensitive to the motor current. The manufacturer of the motor starter provides a chart inside the motor starter listing the part number for thermal overload unit. The heaters are sized according to the actual full-load current listed on the motor nameplate. Find the thermal overload unit number from the manufacturer's list corresponding to the motor nameplate full-load current. A typical manufacturer's thermal overload unit selection chart is shown in Figure 7.6.

An example will help show how the thermal overload sensing unit chart is used. A 3-phase motor nameplate full-load current for 240-volt operation is 1.5 amperes. The proper thermal overload unit to use is a number JR 2.40 from Figure 7.6. The Code allows this heater to be sized at 125 percent of the motor nameplate full-load current provided the service factor is 1.15 or larger, or the temperature rise is not greater than 40°C. The manufacturer has considered this when setting up the chart in Figure 7.6. If the service factor is less than 1.15 or the temperature rise is greater than 40°C, then the heater will be 10 percent oversized. The motor

Motor Full-Load Current (AMP)	Thermal Unit No.	Maximum Fuse Rating (AMP)	Motor Full-Load Current (AMP)	Thermal Unit No.	Maximum Fuse Rating (AMP)
0.28–0.30	JR 0.44	0.6	2.33–2.51	JR 3.70	5
0.31–0.34	JR 0.51	0.6	2.52–2.99	JR 4.15	5.6
0.35–0.37	JR 0.57	0.6	3.00–3.42	JR 4.85	6.25
0.38–0.44	JR 0.63	0.8	3.43–3.75	JR 5.50	7
0.45–0.53	JR 0.71	1.0	3.76–3.98	JR 6.25	8
0.54–0.59	JR 0.81	1.125	3.99–4.48	JR 6.90	8
0.60–0.64	JR 0.92	1.25	4.49–4.93	JR 7.70	10
0.65–0.72	JR 1.03	1.4	4.94–5.21	JR 8.20	10
0.73–0.80	JR 1.16	1.6	5.22–5.84	JR 9.10	10
0.81–0.90	JR 1.30	1.8	5.85–6.67	JR 10.2	12
0.91–1.03	JR 1.45	2.0	6.68–7.54	JR 11.5	15
1.04–1.14	JR 1.67	2.25	7.55–8.14	JR 12.8	15
1.15–1.27	JR 1.88	2.5	8.15–8.72	JR 14.0	17.5
1.28–1.43	JR 2.10	2.8	8.73–9.66	JR 15.5	17.5
1.44–1.62	JR 2.40	3.2	9.67–10.5	JR 17.5	20
1.63–1.77	JR 2.65	3.5	10.6–11.3	JR 19.5	20
1.78–1.97	JR 3.00	4.0	11.4–12.7	JR 22	25
1.98–2.32	JR 3.30	4.0	12.8–14.1	JR 25	25

Figure 7.6   Chart for selecting thermal overload sensing units.

is then vulnerable to burnout. If the motor service factor is less than 1.15, multiply the motor full-load current on the nameplate by 0.9, and use this new value to size the thermal overload sensing unit. For the example of the 3-phase motor, multiplying 1.5 amperes by 0.9 gives 1.35 amperes. The thermal overload unit corresponding to 1.35 amperes from the table of Figure 7.6 is a thermal unit number JR 2.10.

Remote Control Circuit Wires

A magnetic motor starter is operated with an electric solenoid coil. A control circuit is installed to operate the solenoid. One or more operating devices may be on a control circuit. The control circuit wires are permitted to be smaller than the motor-circuit wires, and they are considered to be protected by the motor branch circuit short-circuit and ground-fault protection. There is a limit as to how high a rating is permitted for this branch circuit protection before the control circuit is no longer considered to be adequately protected. The rules for determining the minimum control circuit wire size permitted based on the branch circuit protection rating are given in *Section 430-72*. When the rating of the branch circuit protective device is too high, overcurrent protection shall be installed to protect the control circuit.

Motor-Circuit Examples

Some examples of motor-circuit component sizing will help illustrate the application of *Article 430*. Nameplate information is given for the motor in each example. The motor nameplate current, as stated in *Section 430-6(a)(1)*, is used to determine the rating of the running overcurrent protection. The current as found in *Tables 430-147, 430-148, 430-149,*

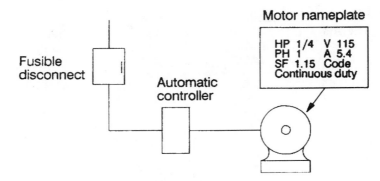

Figure 7.7   An automatically controlled, 1/4-horsepower, single-phase, 120-volt electric motor branch circuit.

and *430-150* is generally used to determine the size of branch circuit conductors and branch circuit short-circuit and ground-fault protection. Example 7.2 illustrates how components of a circuit are sized for a small horsepower motor.

Example 7.2   A 1/4-horsepower, 120-volt, single-phase electric motor is operated from an automatic controller, as shown in Figure 7.7. A fusible switch acts as the disconnect and contains fuses that act as both short-circuit and ground-fault protection, as well as running overload protection. The motor is not thermally protected, and it is not powering a hard starting load. Determine: (1) the minimum size copper, type THWN branch circuit conductor permitted assuming 75°C terminations, and (2) the maximum permitted rating of time-delay fuse to protect the motor from overloads and the branch circuit from short-circuits and ground-faults.

Answer: First, look up the motor full-load current from *Table 430-148*. The value of full-load current will be 5.8 amperes. The minimum permitted branch circuit wire size is determined according to *Section 430-22(a)*. The ampacity of the conductor is determined by multiplying the full-load current for the motor by 1.25, then the minimum wire size is found in *Table 310-16*. The smallest wire size permitted for a branch circuit is AWG number 14, even though the calculated value for the motor was 7.25 amperes.

$$1.25 \times 5.8 \text{ A} = 7.25 \text{ A}$$

A single set of fuses will serve as both branch circuit short-circuit and ground-fault protection, as well as overload protection for the motor. Fuses can provide both functions. The overload protection for the motor is a more restrictive requirement than short-circuit and ground-fault protection; therefore, size the fuses for the overload condition. Use the nameplate full-load current of 5.4 amperes to determine overload protection. The maximum overload rating selection is covered in *Section 430-32(c)(1)*. It is permitted to round up to next standard rating fuse if the size selected using *Section 430-32(c)(1)* is not adequate to start the motor, but it is not permitted to exceed 140 percent of the nameplate full-load current, as stated in *Section 430-34*. It is not permitted to select the 10-ampere standard rating of overcurrent device, as listed in *Section 240-6*, because it is larger than 140 percent. Select a set of fuses with a rating of 6.0 or 6.25 amperes from Table 7.2. If the 6.25-ampere fuse is not adequate to start the motor, then it is permitted to choose a 7-ampere fuse.

$$1.25 \times 5.4 \text{ A} = 6.75 \text{ A}$$

$$1.40 \times 5.4 \text{ A} = 7.56 \text{ A}$$

Example 7.3   A 10-horsepower, design B, 3-phase, 480-volt electric motor is controlled by a magnetic motor starter on a branch circuit with a fusible switch as the disconnecting means. The circuit is shown in Figure 7.8. The conductors are in conduit, and the motor is not powering a difficult starting load. Determine the following for the motor circuit:

1. The minimum permitted size copper, type THWN branch circuit conductor with 75°C terminations
2. The minimum permitted rating of the circuit disconnect
3. The minimum NEMA size motor starter permitted
4. The maximum permitted rating of time-delay fuses to be used for branch circuit short-circuit and ground-fault protection
5. The maximum permitted rating of motor overload device as found in the sample manufacturer chart of Figure 7.6
6. The minimum permitted size of type THWN control circuit wire when protected from overcurrent by the branch circuit fuses and assuming 75°C terminations

Answer: (1) Look up the motor full-load current rating of 14 amperes from *Table 430-150* in the Code. Next, determine the minimum permitted rating in amperes of the branch circuit wire. The wire size is determined from *Table 310-16. Section 240-3(d)*, which limits the overcurrent protection for AWG sizes 14, 12, and 10, does not apply in the case of motor circuits. The minimum wire size permitted is AWG number 14 copper.

$$1.25 \times 14\,\text{A} = 17.5\,\text{A}$$

(2) The disconnect is required to be rated in horsepower for the operating voltage of the motor. Electrical equipment typically has voltage ratings of 150, 250, and 600 volts. In the case of this motor, choose a 600-volt rated disconnect switch with a minimum 3-phase rating of 10 horsepower.

(3) The motor starter (controller) is required to have a minimum rating of 10-horsepower, 3-phase at 480 volts. Find the minimum permitted NEMA size 1 from Table 7.1.

(4) The maximum rating of the branch circuit short-circuit and ground-fault protection is determined using the information from the following Code sections:

- *Section 430-52* and *Table 430-152*
- *Section 430-52(c)(1), Exception 1,* which permits rounding up to the next standard rating of overcurrent device, as listed in *Section 240-6,* when the size is determined according to *Table 430-152*
- *Section 430-52(c)(1), Exception 2(b).* It is permitted to increase size if high motor starting current causes the overcurrent device to open, but the overcurrent device is not

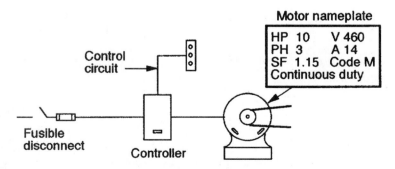

Figure 7.8   A 10-horsepower, design B, 3-phase, 480-volt electric motor controlled with a magnetic motor starter.

permitted to have a rating in excess of 225 percent of motor full-load current when time-delay fuses are used.

$$1.75 \times 14 \text{ A} = 24.5 \text{ A} \quad \text{maximum fuse ampacity}$$

$$2.25 \times 14 \text{ A} = 31.5 \text{ A} \quad \text{absolute maximum ampacity}$$

It is permitted to round the 24.5 amperes up to the next standard rating of fuse, which would be 25 amperes. If this fuse rating is too small to prevent fuse opening during difficult motor starting, then it is permitted to choose higher rating fuses, but it is not permitted to exceed 31.5 amperes. In this case, it would be permitted to use a 30-ampere fuse only if the 25-ampere fuse was not of a sufficiently high rating to carry the starting current.

(5) The motor overload protection for this motor is determined by using the nameplate full-load current of 14 amperes. In this case, the nameplate current and the current from *Table 430-150* are identical. Next, check the service factor (SF) or the temperature rise on the motor nameplate. This motor has a service factor of 1.15. This means that the overload protection is permitted to be sized at a maximum of 125 percent of the nameplate full-load current. The manufacturer has already taken the 125 percent into account; therefore, use the nameplate full-load current and look up the manufacturer's number for the overload thermal unit to be installed into the motor starter. Using the nameplate full-load current of 14 amperes, the manufacturer's thermal unit number is JR 25.

(6) The branch circuit fuses for this motor circuit are rated at 25 amperes. The control circuit is covered by *Section 430-72(b), Exception 2*. The maximum permitted branch circuit overcurrent device rating for various wire sizes is given in column C of *Table 430-72(b)*. If an AWG number 14 wire is used, the branch circuit fuses are permitted to be rated at 45 amperes. For this circuit, the fuses are 25 amperes, and the minimum permitted wire size is AWG number 14.

Example 7.4   A 3-phase, 3/4-horsepower, design B, 240-volt motor is operated by a motor starter from a circuit protected with an inverse-time circuit-breaker, as shown in Figure 7.9. Is the 15-ampere circuit-breaker permitted to act as short-circuit and ground-fault protection for the motor and controller?

Answer: The minimum branch circuit wire size permitted is AWG number 14, as determined by multiplying the 2.8-ampere full-load current for the motor by 1.25. The maximum permitted size of branch circuit short-circuit and ground-fault protective device rating is determined from *Section 430-52* and *Table 430-152*. The maximum permitted rating of circuit-breaker for the circuit is stated in *Section 430-52(c)(1), Exception 2(c)* as 400 percent of the motor full-load current. This is still smaller than the smallest circuit-breaker

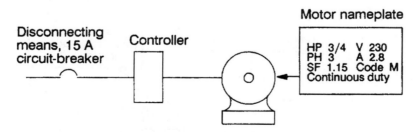

Figure 7.9   A circuit-breaker provides the branch circuit short-circuit and ground-fault protection for a 3/4-horsepower, 3-phase, 240-volt design B motor.

rating available and would tend to indicate that the circuit-breaker rating is still too high for the motor circuit. But this situation is covered in *Exception 1* to *Section 430-52(c)(1)*. Standard ratings for circuit-breakers smaller than 15 amperes are not available; therefore, this is the smallest standard circuit-breaker. The answer to the question is yes. The 15-ampere circuit-breaker is permitted to be used for this motor circuit.

$$2.5 \times 2.8 \text{ A} = 7.0 \text{ A}$$

$$4.0 \times 2.8 \text{ A} = 11.2 \text{ A}$$

Example 7.5   A specific fixed motor load consisting of 10-, 7.5-, and 5-horsepower, 3-phase, 240-volt design B motors is supplied by a single feeder conductor. Fuses are used as short-circuit and ground-fault protection for each motor branch circuit, and the ratings of these time-delay fuses for the branch circuits are as follows: 10-horsepower motor, 45-ampere; 7.5-horsepower motor, 30-ampere; and 5-horsepower motor, 25-ampere. The circuit is shown in Figure 7.10. Determine the following:

1. The minimum copper, type THWN feeder wire size permitted assuming 75°C terminations
2. The maximum feeder time-delay fuse size
3. The minimum tap wire sizes for each motor branch circuit assuming 75°C terminations

Answer: (1) Look up the motor full-load current from *Table 430-150*. Then determine the minimum size wire permitted as determined from *Section 430-24*. Look up the minimum wire size permitted from *Table 310-16*. The minimum feeder wire size is AWG number 4.

$$15.2 \text{ A} + 22 \text{ A} + 28 \text{ A} + 0.25 \times 28 \text{ A} = 72.2 \text{ A}$$

(2) The maximum permitted feeder time-delay fuse size is determined from *Section 430-62*. Start with the maximum branch circuit fuse rating permitted and add to it the full-load currents of the other motors supplied by the feeder. It is not permitted to exceed this value calculated; therefore, an 80-ampere time-delay fuse is the maximum permitted for the feeder. Another example of this procedure is given in *Example D8, Appendix D* of the Code.

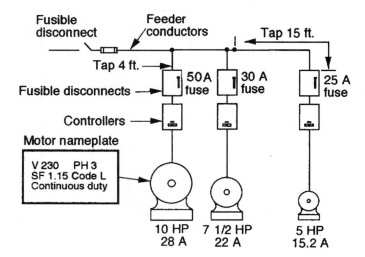

Figure 7.10   Three design B electric motor branch circuits are tapped from a feeder where these motors are the only loads on the feeder.

(3) The minimum branch circuit tap sizes permitted are determined using the tap rule of *Section 430-28*. With the 10-foot rule, the minimum tap conductor ampacity is one-tenth of the rating of the feeder overcurrent device. In this case, the feeder overcurrent device is a set of 80-ampere time-delay fuses. One-tenth of that rating is 8 amperes. In this case, the tap conductors will be simply sized adequate to serve the individual motor branch circuits.

10-horsepower motor: 28 A × 1.25 = 35 A, AWG number 10

7.5-horsepower motor: 22 A × 1.25 = 27.5 A, AWG number 10

For the 5-horsepower motor, the branch circuit tap is more than 10 feet; therefore, the 25-foot tap rule shall be used. The tap conductor shall have an ampacity of not less than one-third that of the feeder conductors, which is 85 amperes for an AWG number 4 copper type THWN conductor with 75°C terminations. Also, be sure to check the minimum size wire required to supply the motor. An AWG number 12 wire is the minimum permitted to supply the motor, but it is required to use an AWG number 10 wire to satisfy the minimum tap permitted.

5-horsepower motor: 15.2 A × 1.25 = 19 A, AWG number 12

$$\frac{85 \text{ A}}{3} = 28.3 \text{ A, AWG number 10}$$

MOTOR-CONTROL CIRCUIT WIRING

Motor-control circuit wiring can be confusing, but after the general concept is understood, it can be easily performed. Common types of control circuit wiring for a magnetic motor starter are the 3-wire control and the 2-wire control. A typical example of the 3-wire control circuit is a start-stop station that operates the solenoid of a magnetic motor starter. A 2-wire control is any device that opens and closes a switch to operate the motor starter. A pressure switch, limit switch, or thermostat are typical examples of a 2-wire control device. Ladder diagrams are frequently used to provide a means of visualizing the control circuit and how it works. Figure 7.11 shows a ladder diagram for a start-stop station operating a motor starter.

A schematic diagram of a magnetic motor starter operated with a start-stop station is shown in Figure 7.12. Compare the diagram of Figure 7.12 with Figure 7.11 to see how a

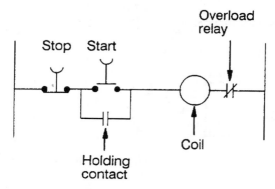

Figure 7.11 Ladder diagram for a start-stop station controlling a magnetic motor starter.

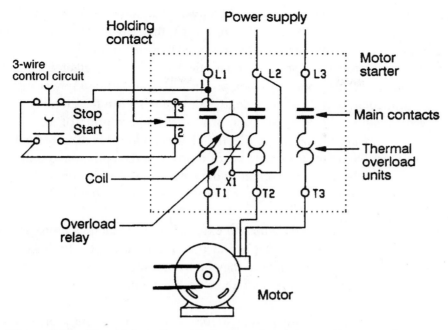

Figure 7.12   The wiring of a start-stop station to control a magnetic motor starter for a 3-phase motor.

ladder diagram represents the actual control circuit. With the ladder diagram, it is easy to see how the circuit works, but with Figure 7.12 it is easy to see how the wiring is installed.

A 2-wire control device is frequently used to open and close the motor-control circuit. Figure 7.13 shows a motor starter operated with a thermostat using power from two of the supply lines of the 3-phase source. Note that the thermostat in Figure 7.13 simply completes the circuit from line L1 to the coil. In the case of a 2-wire control circuit, the holding contact in

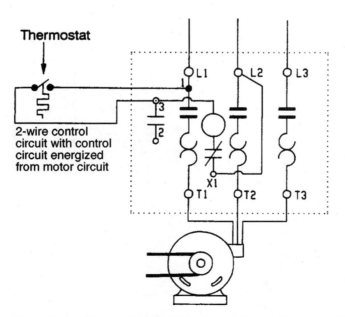

Figure 7.13   The wiring of a simple switch device such as a thermostat to control a magnetic motor starter.

the magnetic motor starter is not needed. Some physical action opens and closes the contacts of the 2-wire control device. In the case of Figure 7.13, that physical action is change in temperature. Other types of physical action are pressure, fluid level, flow rate, mechanical pressure, proximity, and many other physical quantities that can be detected by some type of sensor. It is important to remember that when power is restored after a power interruption, the motor will start immediately if the 2-wire control device is still in the closed position. This is the importance of *Section 455-22* in the case of a power interruption of a rotary phase converter circuit. It is necessary to make sure that the phase converter is restarted before the loads are started. Safety may be a factor in the case of a power interruption, and automatic restarting of loads may not be desirable. Note that in the case of the 3-wire control circuit of Figure 7.12, the motor will not restart after a power interruption because the holding contact is now open.

A 2-wire control circuit is used when an external device, such as a programmable controller, is used to operate the motor starter as illustrated in Figure 7.14. In this case, the power source to operate the motor starter solenoid may be from a source other than the motor branch circuit. It will be necessary to make sure there is no connection between the two power sources.

In the case of a control circuit power source separate from the motor branch circuit, *Section 430-74(a)* requires that all power sources be capable of being disconnected from the motor and the controller. It is permitted to have a separate disconnecting means for the motor branch circuit and the control circuit. The dotted line in Figure 7.14 shows the wire that must be removed to make sure the control circuit power source is separated from the motor branch circuit power source. The solenoid in the motor starter must match the control circuit voltage, and the control circuit must be separated from the motor power supply by removing a factory-installed wire inside the motor starter.

When a motor is operated at 480 volts, a control circuit is permitted to reduce the 480-volt supply to 120 volts by installing a control transformer inside or adjacent to the motor starter. Figure 7.15 shows a typical installation of a control transformer for a motor control circuit. Note in Figure 7.15 that a factory-installed wire inside the motor starter must be removed when a control transformer is supplying the control circuit power.

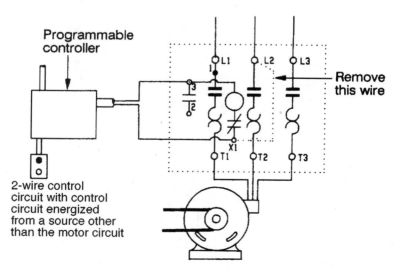

Figure 7.14   An external power source such as a programmable controller is used to control the motor at 120 volts.

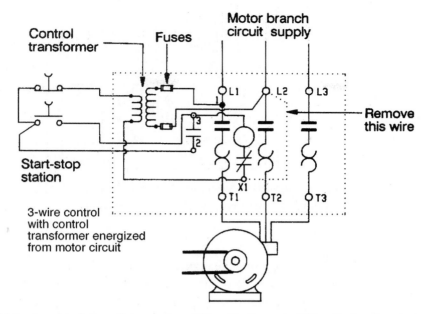

Figure 7.15 **A control transformer steps 480 volts down to 120 volts for the control circuit.**

MAJOR CHANGES TO THE 1999 CODE

These are the changes to the 1999 *National Electrical Code®* for electrical systems that correspond to the Code sections studied in this unit. The following analysis explains the significance of only the changes from the 1996 to the 1999 Code, and this analysis is not intended to be used in place of the Code. Refer to the actual section of the 1999 Code for the exact wording and meaning of each section discussed. Changes are indicated in the Code with a vertical line in the margin. If material has been deleted or moved to another part of the Code, the location of the deletion is indicated with a dark dot in the margin.

Article 430 **Motors, Motor Circuits, and Controllers**

430-6: When flexible cords are used in a motor circuit, the size is permitted to be determined from *Section 400-5*.

430-6(a)(1): This section states that motor current is to be determined from the ampere tables at the end of *Article 430*. The change is that it now states what that motor current is used to determine. It is used to size the conductors, switches, and branch circuit short-circuit and ground-fault protection.

430-6(a)(2): This is a section that was lifted from the previous section that states for the purpose of sizing motor running overcurrent protection, the nameplate rating of the motor is to be used.

430-6(a)(1) Exception 3: This is a new exception that specifies what current is to be used when sizing conductors, overcurrent devices, disconnects, and similar components for motor-operated appliances with both horsepower and full-load current marked on the nameplate. In these situations, the full-load current marked on the nameplate is to be used. This is illustrated in Figure 7.16.

Table 430-12(b): The dimensions for terminal housings for electric motors are found in this table. The minimum dimensions for cover openings of all motors with a diameter of 11 inches or less have changed. This is also the case for all ac motors that have a diameter

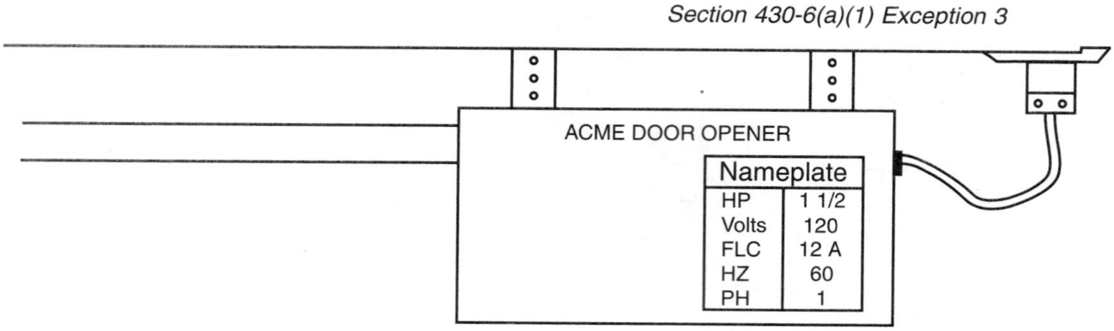

Figure 7.16   For a motor-operated appliance with both full-load current and horsepower marked on the nameplate, it is permitted to use the nameplate full-load current for sizing conductors and other circuit components.

greater than 11 inches. Cover openings for motors that have a diameter less than 11 inches and for all other ac motors have been reduced by approximately 40 percent. This 40 percent reduction applies only to cover openings. The minimum cubic inch capacities for all motors have not changed from the previous edition of the Code.

430-22(a): This section specifies that the circuit conductors for a single motor circuit shall not be less than 125 percent of the motor full-load current. It now states that this rule applies to continuous duty motors. Also, it is now clearly stated that the full-load current is determined by using *Section 430-6(a)(1)*. Neither of these modifications change the intent of this section.

430-22(b): This section specifies that conductor size for other than continuous duty motors be determined based upon percentages listed in *Table 430-22(b)*. This section was previously one of the exceptions. There is no change of intent.

430-52(c)(3) Exception 1: The exception, as in the past, permits a setting of an adjustable instantaneous-trip circuit-breaker to be increased to not more than 1,700 percent of a design E motor's full-load current when its setting is insufficient to handle the starting current. The change is that **energy efficient** design B motors are now recognized along with design E motors.

430-52(c)(4) Exception: As in the previous edition of the Code, this exception permits a single branch circuit short-circuit and ground-fault protective device sized to the full-load current of the winding drawing the largest value of current. Now the controllers used on each set of windings are required be sized for the highest horsepower rated winding of the motor, not just to the rating of each respective winding.

430-52(c)(6): This new paragraph recognizes a listed self-protected combination controller that provides short-circuit and ground-fault protection and overload protection and acts as a disconnect and controller. This paragraph permits these units to be installed and provide the short-circuit and ground-fault protection in place of other types of devices listed in *Table 430-152*.

430-52(c)(7): This paragraph was a last sentence to paragraph *(c)*. It was moved to a separate paragraph because it discusses a different type of device used to protect a combination motor starter. The only change of intent in this section is that **energy efficient** design B motors are treated the same as design E motors.

430-83: This section specifies the minimum rating of motor controllers. The section was rewritten to eliminate the many exceptions in the previous edition of the Code. The only change in this section was a rewording to eliminate a conflict that seemed to prohibit the use of circuit-breakers and molded case switches as controllers for design E motors. In *Section 430-83(a)(1)* a statement was added that made it clear that circuit-breakers and molded case switches are not required to be rated in horsepower.

430-87 Exception: The exception permits a group of motors under certain conditions to be operated by a single controller. The method of determining the minimum horsepower rating of the controller has been changed to be the same method as used to size a disconnecting means for the same group of motors as given in *Section 430-110(c)(1)*. The method in the previous edition of the Code simply required the summing of the individual motor horsepower ratings. Now, the minimum horsepower rating is determined by summing the full-load currents, summing the locked-rotor currents, and finding the equivalent horsepower rating.

430-92: The second paragraph of this section defined a motor-control center. That second paragraph was deleted because the definition is also contained in *Article 100*. The definition does not need to be in more than one place.

430-97(b): This section specifies the A, B, and C labeling arrangement of phase conductors in a motor-control center. A sentence was added specifying that in the case of a 4-wire delta electrical system with one phase with a higher voltage to ground than the other two conductors, the phase with the higher voltage to ground is required to be the B phase. Another sentence was also added that permits other phase arrangements for existing installations provided the conductors are labeled. For an existing installation where the phases were arranged in a manner other than that specified in this section, it is permitted to maintain the same arrangement of phases when additions or changes are made to the existing installation.

430-102(a): This section was changed to make it clear that **individual** disconnecting means is required for each controller. Each individual disconnecting means is required to be within sight of the controller to disconnect power from the controller.

430-102(b): This section requires a disconnecting means to be located in sight from the motor and driven machine. The word **separate** was added to this section to make it clear that a separate disconnecting means is required for each motor. But by adding the word **separate,** it implies that a separate disconnecting means is required for the motor in addition to the disconnecting means required for the controller. This additional disconnect for the motor is required in all cases unless the disconnect for the controller is capable of being locked in the open position, as permitted in the exception to this section.

430-109: This section specifies the type of device permitted to serve as a motor-circuit disconnecting means. The entire section, which consisted mostly of exceptions, was rewritten. The basic rule still remains that a disconnecting means is to be a listed switch rated in horsepower. Added as an acceptable means of providing a disconnecting means for a motor is a manual motor controller that is specifically marked **suitable as motor disconnect,** as illustrated in Figure 7.17. In general, manual motor controllers are not suitable as disconnects, but there are manual controllers now on the market that do meet disconnecting means requirements. A listed self-protected combination controller is now also recognized as an acceptable disconnecting means. This device has short-circuit and ground-fault protection, acts as a controller, provides overload protection, and has a disconnecting means that provides isolation from the supply by manually opening contacts.

430-109 FPN: This new fine print note was added to make it clear that a combination motor starter has a disconnecting means. This fine print note is a definition of a combination

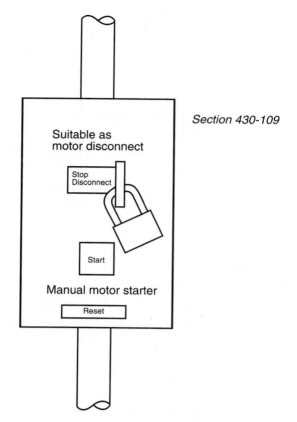

Figure 7.17   A manual motor starter is permitted to serve as a disconnecting means for the motor if it is marked "suitable as motor disconnect."

motor starter. The combination motor starter has a disconnecting means and a motor controller. It may also, but not necessarily, contain short-circuit and ground-fault protection and overload protection.

430-110(a) Exception: This is a new exception that deals with nonfused motor-circuit disconnecting switches that have a horsepower rating and an ampere rating. This section requires disconnecting means to have a current rating not less than 115 percent of the motor full-load current. In the case of a listed nonfused switch, 1.15 times the equivalent current listed for a motor of the maximum horsepower may be more than the current rating of the switch. That is permitted by this exception provided the motor does not exceed the horsepower rating of the switch at the rated voltage.

430-110(c)(1): The third paragraph of this section was revised to make the meaning clear. The paragraph specifies how to determine the equivalent horsepower rating where a disconnecting means supplies a combination load of one or more motors or nonmotor loads and where all of the loads will not operate at the same time. It is now clear that the maximum horsepower is based on the largest motor locked-rotor current and nonmotor load that can occur at any one time.

430-111(b)(2): A circuit-breaker acting both as controller and disconnecting means is permitted to be manually or power activated. Power activation of the circuit-breaker is new.

Article 440   **Air-Conditioning and Refrigerating Equipment**
No significant changes were made to this article.

Article 455   **Phase Converters**

455-6: This section specifies the minimum size of single-phase conductors supplying a phase converter. The section was completely rewritten to eliminate the exceptions. The intent of this section was not changed. The single-phase full-load input current is required to be provided on the phase converter nameplate. The single-phase input conductor size is to be not less than 125 percent of the single-phase input full-load current. A phase converter may have a rating larger than the load to be served. If that is the case, the single-phase input conductors are permitted to be sized based upon the load to be served. This method was the exception in the previous edition of the Code, and now it is paragraph *(2)*. The rewriting of this section did introduce two new terms that may be confusing. Paragraph *(1)* applies to **variable loads,** and the single-phase input conductors are required to have an ampacity not less than 125 percent of the nameplate full-load input current. The load does not necessarily have to be variable for this to apply. This rule applies when the phase converter is sized to the load, or when it is oversized but the load may increase in the future. The other term that applies in the case of paragraph *(2)* is **fixed load.** The more appropriate reference would be to a specific load less than the rating of the phase converter.

455-6(b): Two single-phase input conductors supply a phase converter, and the output consists of three conductors supplying 3-phase power to the load. Two of the output conductors connect directly to the input conductors. The third output conductor originates within the phase converter. This conductor is called the manufactured phase. The manufactured phase is defined in *Section 455-2*, and a restriction about using the manufactured phase is given in *Section 455-9*. Now there is a new requirement that the manufactured phase be identified at accessible locations with a distinctive marking. The identified manufactured phase is shown in Figure 7.18. This will serve as a reminder to the installer that this conductor is not permitted to be connected to supply single-phase loads such as the control circuit.

455-7: This section specifies the maximum permitted rating of the overcurrent protection for the single-phase input conductors and the phase converter. The general rule where the phase converter is matched to the load is that the maximum overcurrent device rating is not to exceed 125 percent of the phase converter nameplate full-load current rounded up to the next standard rating as given in *Section 240-6*. There was an exception in the previous edition of the Code where the load was less than the rating of the phase converter and the single-phase input conductors were smaller than 125 percent of the nameplate full-load input current; in that case, the maximum size of overcurrent device rating was not to exceed 250 percent of the full-load current of the 3-phase load rounded up to the next standard rating. That exception was eliminated, and it is now paragraph *(b)*. The same two new terms used in *Section 455-6*, **variable load** and **fixed load,** were also introduced into this section and will most likely create some confusion.

455-8: This section was rewritten to eliminate the exceptions. There was a change in this section that corrected an oversight in the past. In *Section 455-8(b),* a switch is permitted to be rated in amperes, not horsepower, when only nonmotor loads are served. In the previous edition of the Code, the switch was required to be rated in horsepower for all loads.

Article 460   **Capacitors**

460-8(c)(1) Exception: This exception, as in the past, allows for the installation of capacitors in motor circuits without installing additional disconnecting means to disconnect each

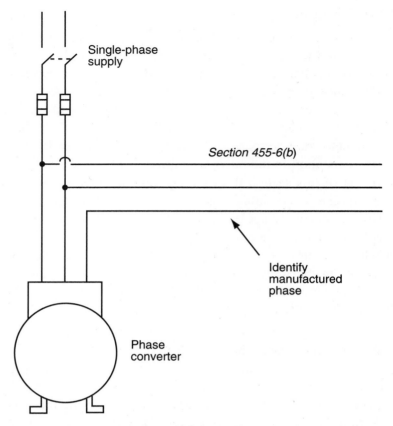

Single-phase
supply

Section 455-6(b)

Identify
manufactured
phase

Phase
converter

**Figure 7.18 The manufactured phase is required to be identified,
and it is not permitted to supply single-phase loads.**

ungrounded conductor from the circuit. Under current provisions of this exception, a separate disconnect is not required when the capacitor is connected on the load side of a **motor controller.** The previous edition of the Code allowed for the capacitor to be connected to the motor circuit on the load side of an **overload protective device.**

460-10 Exception: This exception requires that capacitor cases not be grounded where capacitors are mounted on a structure that operates at a voltage different than ground. The change made in this exception did not change its intent.

Article 470 **Resistors and Reactors**

470-19: A new exception was added that requires the grounding for resistor and reactor cases to be omitted when they are mounted on a structure that operates at other than ground potential.

Article 675 **Electrically Driven or Controlled Irrigation Machines**

There were no changes made to this article.

Article 685 **Integrated Electrical Systems**

685-2: This section lists other locations in the Code where the principle of an orderly shutdown is permitted. In the case of resistance heating elements in industrial installations under *Article 427,* an orderly shutdown as prescribed in this article is permitted in place of equipment ground-fault protection as stated in the *Exception* to *Section 427-22*. That reference was omitted in previous editions of the Code.

Appendix D **Examples D8, D9, and D10**

Examples were provided in *Chapter 9* to illustrate some of the more complex calculations. In the previous edition of the Code, the examples were *Part B* of *Chapter 9*. Examples are now in *Appendix D*.

Example D8: This is a sample calculation of three motor circuits and a feeder supplying all three motors. Two of the motors are of the wound-rotor type, which involves the sizing of secondary conductors from the rotor to a resistor bank. The problem is essentially the same as in the previous edition of the Code except that several of the values for the individual motors have changed slightly.

Examples D9 and D10: These are examples of feeder calculations supplying a group of elevator machines. These examples are the same as in the previous edition of the Code. The only change is their location, which is now in *Appendix D*.

WORKSHEET NO. 7
MOTOR-CIRCUIT WIRING

These questions are considered important to understanding the application of the *National Electrical Code®* to electrical wiring, and they are questions frequently asked by electricians and electrical inspectors. People working in the electrical trade must continue to study the Code to improve their understanding and ability to apply the Code properly.

DIRECTIONS: Answer the questions and provide the Code reference or references where the necessary information leading to the answer is found. Space is provided for brief notes and calculations. An electronic calculator will be helpful in making calculations. You will keep this worksheet; therefore, you must put the answer and Code reference on the answer sheet as well.

1.   A nonlocking circuit-breaker is the disconnect for a motor branch circuit, and it is 20 feet from the motor controller and in direct line of sight. The motor is 25 feet from the controller and in direct line of sight. But the motor is not in sight of the circuit-breaker. Does the Code permit this installation, as shown in Figure 7.19? (Yes or No)

Code reference _____

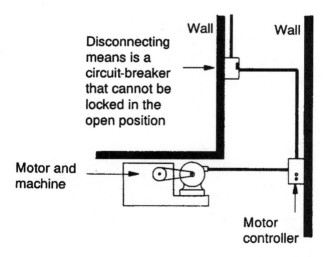

Figure 7.19   The motor and machine are in sight from the controller, but not from the disconnect.

2. The full-load current of a 200-volt, 3-phase, 15-horsepower electric motor is _____ amperes.

   Code reference _____

3. Is a 50-ampere inverse-time circuit-breaker permitted to serve as the controller for a thermally protected 5-horsepower, single-phase, 240-volt electric motor on a farm silo unloader? (Yes or No)

   Code reference _____

4. A single-phase, 1-horsepower, 230-volt electric motor is powering an easy starting load. The code letter on the design B motor is H, the service factor is 1.0, and the nameplate full-load current is 7 amperes. The branch circuit conductors are AWG number 14 copper with type THWN insulation. The motor is operated automatically controlled by a programmable controller. Using Figure 7.6, the maximum size of thermal units that could protect the motor is:
   A. JR 9.10.
   B. JR 10.2.
   C. JR 11.5.
   D. JR 12.8.

   Code reference _____

5. The full-load current if marked on a motor nameplate is generally used to size the:
   A. branch circuit conductors.
   B. running overcurrent protection.
   C. branch circuit short-circuit and ground-fault protection.
   D. feeder tap conductors.

   Code reference _____

6. A portable machine is powered with a 1/2-horsepower, single-phase, 120-volt electric motor. Is the attachment plug and receptacle permitted to serve as the controller? (Yes or No)

   Code reference _____

7. A 3-phase, 240-volt electric motor branch circuit is protected from short-circuits and ground-faults by a 40-ampere circuit-breaker in the supply panelboard. The motor is operated with a magnetic motor starter with a start-stop station located 15 feet from the motor starter. The wire to the start-stop station is AWG number 14 copper run in conduit. Conductor terminations are considered to be at a temperature of 60°C. Is this control circuit wire considered protected by the 40-ampere circuit-breaker if overcurrent protection is not installed specifically for the control circuit wires? (Yes or No)

   Code reference _____

8. Electrical equipment with several factory-installed electric motors, such as an air conditioner, shall be provided with an equipment nameplate stating, among other important information, the maximum ampere rating of the circuit short-circuit and ground-fault protective device. (True or False)

   Code reference _____

9. A tight-fitting box is constructed over an electric motor to protect it from the environment and muffle the noise. If ventilation holes are not provided, is this a violation of the Code? (Yes or No)

   Code reference _____

The following information applies to the next three questions.

A machine is powered by a 3-phase, 460-volt, 25-horsepower, design B electric motor with a service factor of 1.15 and code letter K. The motor is powering an easy starting load, and the nameplate full-load current is 33 amperes. The wire is copper, type THWN in conduit. The motor is shown in Figure 7.20. All conductor terminations are considered to be at a temperature not less than 75°C.

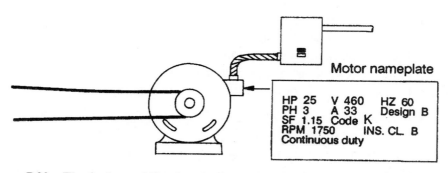

Figure 7.20   The 3-phase, 460-volt electric motor has a nameplate full-load current of 33 amperes, a service factor of 1.15, and is marked with code letter K.

10. The minimum size copper, type THWN wire permitted for the motor branch circuit of Figure 7.20 is AWG number _____.

    Code reference _____

11. The maximum rating inverse-time circuit-breaker permitted to serve as branch circuit short-circuit and ground-fault protection when the motor of Figure 7.20 starts the load without difficulty is _____ amperes.

    Code reference _____

12. If time-delay fuses are used to serve both as short-circuit and ground-fault protection and as running overload protection for the motor of Figure 7.20, the maximum ampere rating of the fuse permitted is _____ amperes. (The motor starts without difficulty.)

    Code reference _____

13. An open motor, with openings such that water falling from above will not enter, is permitted to be used in a farm animal feed preparation room where there is dust and flyings occasionally in the air. (True or False)

    Code reference _____

14. Two electric motors rated 5-horsepower, 3-phase, 240-volt have different design letters marked on the nameplate. One has a design letter E, and the other has the design letter B. Which motor will most likely have the lowest locked-rotor current, design E or design B?

    Code reference _____

15. If a continuous duty, 5-horsepower, design B electric motor has a temperature rise above ambient of 40°C and is powering an easy starting load, the running overload protection is not permitted to exceed _____ percent of the full-load current.

    Code reference _____

16. A single-phase, 3-horsepower, 230-volt electric motor is powering an easy starting load. The branch circuit conductors are AWG number 10 copper with type THWN insulation and 75°C terminations. The motor is protected from overloads by current-sensing relays in the motor starter. The code letter on the design B motor is H, the service factor is 1.0, and the nameplate full-load current is 16 amperes. The maximum standard rating time-delay fuse permitted for branch circuit short-circuit and ground-fault protection is _____ amperes.

    Code reference _____

17. A motor operates at 120 volts, and the controller is a magnetic motor starter with a 120-volt solenoid coil. The controller is activated with a start-stop station. Is the grounded circuit conductor permitted to be connected to the stop station rather than to the overload relay, as shown in Figure 7.21? (Yes or No)

Code reference _____

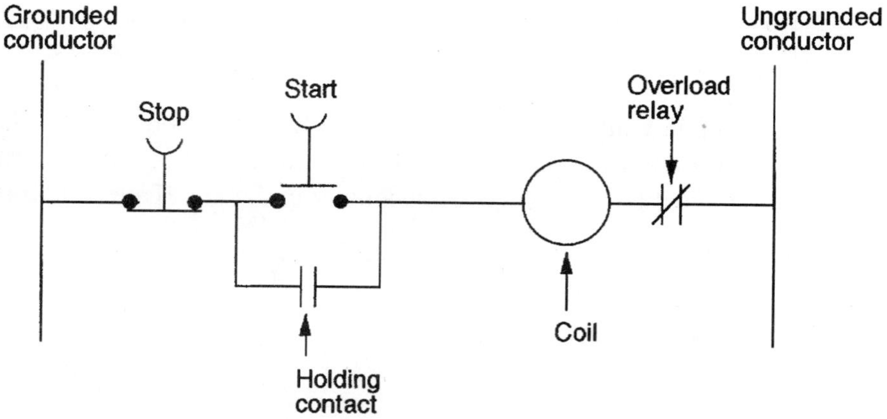

Figure 7.21 The ladder diagram of the start-stop station operating a motor starter shows the grounded circuit conductor connected to the stop station.

18. A motor branch circuit originates at a circuit-breaker in a panelboard with the circuit-breaker acting as the disconnecting means for the ungrounded conductors to a fusible switch. Is the fusible switch permitted to serve both as the controller and the disconnect for the motor, provided the switch breaks all ungrounded conductors? (Yes or No)

Code reference _____

19. Two electric motors on separate branch circuits not functionally associated are wired as shown in Figure 7.22. The start-stop stations are located between the motors, as shown in the figure. A disconnecting means is within sight of the controllers and the motors. Is this an acceptable installation? (Yes or No)

Code reference _____

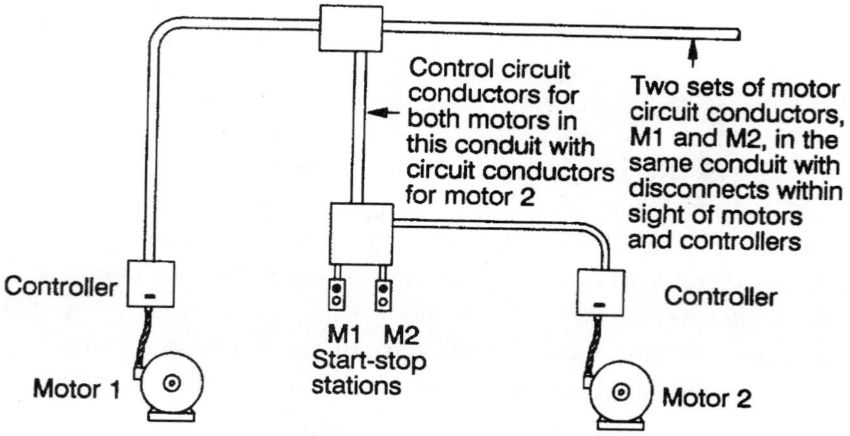

Figure 7.22 The two electric motors are on separate branch circuits, and they are not functionally associated.

20. A new hermetically sealed refrigeration unit consisting of more than one motor is required to state the branch circuit selection current on an equipment nameplate. (True or False)

   Code reference _____

21. A magnetic motor starter is permitted to receive a source of control circuit power from a source other than the motor branch circuit. (True or False)

   Code reference _____

22. Several machines form an integrated manufacturing process that must be shut down in an orderly manner and is permitted to have an alarm system, which alerts the operator on duty to activate orderly shutdown or take corrective action in the event of motor overload. This is in place of automatic individual motor shutdown. (True or False)

   Code reference _____

23. A copper type THWN feeder is supplying a group of motors located in a separate room, as shown in Figure 7.23. All conductor terminations are rated at 75°C. The motors are 3-phase, 230-volt, design B, and the sizes are 7.5-, 5-, 5-, 2-, and 1-horsepower. The minimum size feeder conductor permitted to supply this specific motor load where the wires are in conduit that is in free air is AWG number _____.

   Code reference _____

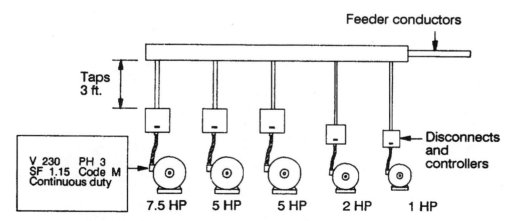

Figure 7.23   Five design B electric motors have branch circuits and are tapped from a feeder that supplies no other loads.

24. A 20-kVA rotary phase converter with a single-phase input full-load current rating of 105 amperes is powering only one 10-horsepower, design B, 240-volt, 3-phase electric motor from a 240-volt single-phase supply. No other load will be supplied. Determine the minimum permitted size of copper single-phase input conductors, assuming 75°C conductor insulation and terminations.

Code reference _____

25. A center pivot irrigation machine is propelled by ten 3-phase, 1-horsepower design B motors rated at 480 volts. The minimum equivalent locked-rotor current of the irrigation machine is _____ amperes.

Code reference _____

ANSWER SHEET   Name _____

No. 7   MOTOR-CIRCUIT WIRING

Answer   **Code reference**

1. _____   _____
2. _____   _____
3. _____   _____
4. _____   _____
5. _____   _____
6. _____   _____
7. _____   _____
8. _____   _____
9. _____   _____
10. _____   _____
11. _____   _____
12. _____   _____
13. _____   _____
14. _____   _____
15. _____   _____
16. _____   _____
17. _____   _____
18. _____   _____
19. _____   _____
20. _____   _____
21. _____   _____
22. _____   _____
23. _____   _____
24. _____   _____
25. _____   _____

UNIT 8

Transformers

OBJECTIVES

Upon completion of this unit, the student will be able to:
- define insulating transformer and autotransformer.
- determine the voltage of a transformer winding if turns ratio and the other winding voltage are given.
- determine the full-load current of a transformer winding if the voltage and kilovolt-amperes of the transformer are given for single-phase or 3-phase transformers.
- determine the minimum permitted kilovolt-amperes required for a specific application.
- draw the proper connections of the windings of a dual-voltage single-phase transformer for the desired voltage.
- explain the purpose of the taps on the primary winding of some transformers.
- state a specific example of the use of a boost and buck transformer.
- determine the maximum permitted overcurrent protection for a specific transformer.
- determine the minimum permitted primary and secondary conductors for a specific transformer application.
- determine the maximum distance permitted from a feeder tap to the transformer overcurrent protection.
- determine the maximum permitted input overcurrent protection for a boost or buck transformer application.
- explain how to ground the transformer and the secondary electrical system of the transformer.
- answer wiring installation questions relating to *Article 450*.
- state a significant change that occurred in Code *Article 450* from the 1996 to the 1999 edition.

CODE DISCUSSION

Article 450 deals with transformers and transformer vaults. *Section 450-1* gives the exceptions for transformers that are covered in other sections of the Code. *Section 450-3* deals with overcurrent protection of transformers. Some specific rules for autotransformer overcurrent protection are covered in *Section 450-4*. The remainder of *Part A* of this article covers general installation requirements. *Part B* covers requirements for specific types of transformers. *Part C* deals with transformer vaults that are enclosures of specific construction for the purpose of housing transformers.

Article 450 of the Code applies only to the transformer itself, and not to the conductors leading to or away from the transformer. The branch circuit, feeder, and tap conductors must be protected according to the rules of *Article 240*. Grounding must be accomplished according to the rules of *Article 250*. Here are Code sections helpful in working transformer circuit problems: *240-3(f), 240-21(b)(3), 240-21(c), 250-30, 384-16(e), 430-72(c), 600-21, 600-23, 600-31, 600-32, 680-5(a), 725-21(a)(1), 725-24 Exception 2, 725-41(a)(1),* and *725-51.* These are in addition to information found in *Article 450.*

TRANSFORMER FUNDAMENTALS

The purpose of a transformer is to change electrical voltage to a different value. For example, a large 480-volt 3-phase motor is powering a well pump. The motor is in a building, and one 120-volt circuit for a few lights and a receptacle outlet is needed. A transformer is used to lower the voltage from 480 to 120 for the lighting circuit. The controls for furnaces and air-conditioning units are often operated at 24 volts. A small transformer inside the equipment lowers the line voltage to 24 volts for the control circuit. Transformers are frequently used inside electronic equipment.

Types of Transformers

Transformers are of the dry type or oil-filled. Two to five percent of the electrical energy is lost in a transformer, mostly due to the resistance of the windings. Large transformers circulate oil through the windings to remove the heat. Dry transformers use air for cooling. Heat is moved from the windings to the case by conduction in smaller sizes of the dry type. Large dry-type transformers actually allow air to circulate through the windings. Oil-cooled transformers are used by the electric utility and for industrial or large commercial applications.

Common two-winding transformers are often called insulating transformers. The primary winding and the secondary winding are separate from each other, and they are not electrically connected. An autotransformer has the windings interconnected so that the primary and the secondary share the same winding. These transformers, therefore, have an electrically connected primary and secondary. A major advantage of the autotransformer over the insulating type is its lighter weight and compact size. An insulating transformer and an autotransformer are compared in Figure 8.1. A common application of an autotransformer is for electric discharge lighting ballasts.

A special type of autotransformer called a grounding autotransformer, or zig-zag transformer, is occasionally used to create a neutral wire or a ground for an ungrounded 480-volt 3-phase electrical system. These transformers are found occasionally in industrial wiring. The name zig-zag is derived from the shape of the schematic diagram. Standard insulating

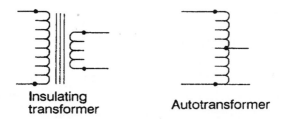

Figure 8.1   Two basic types of transformers are the insulating transformer and the autotransformer.

transformers can be used to make a zig-zag transformer. The wiring of these transformers is covered in *Section 450-5.*

Voltage and Turns Ratio

The input winding to a transformer is called the primary winding. The output winding is called the secondary winding. If there are more turns of wire on the primary winding than on the secondary winding, the output voltage will be lower than the input voltage.

It is important to know the ratio of the number of turns of wire on the primary winding as compared with the secondary winding. This is called the turns ratio of the transformer. The actual number of turns is not important, just the turns ratio. The turns ratio of a transformer can be determined with Equation 8.1 if the actual number of turns on the transformer windings is known.

$$\text{Turns Ratio} = \frac{\text{Number of Turns on the Primary}}{\text{Number of Turns on the Secondary}} \qquad \text{Eq. 8.1}$$

The step-down transformer of Figure 8.2 has 14 turns on the primary winding and 7 turns on the secondary winding; therefore, the turns ratio is 2 to 1, or just 2. The step-up transformer has 7 turns on the primary and 14 on the secondary; therefore, the turns ratio is 1 to 2, or 0.5. If the voltage of one winding and the turns ratio are known, the voltage of the other winding can be determined using either Equation 8.2 or Equation 8.3.

$$\text{Primary voltage} = \text{Secondary Voltage} \times \text{Turns Ratio} \qquad \text{Eq. 8.2}$$

$$\text{Secondary Voltage} = \frac{\text{Primary Voltage}}{\text{Turns Ratio}} \qquad \text{Eq. 8.3}$$

Transformer Ratings

Transformers are rated in volt-amperes (VA) or kilovolt-amperes (kVA). This means that the primary winding and the secondary winding are designed to withstand the VA or kVA ratings stamped on the transformer nameplate. The primary and secondary full-load current usually are not given. The installer must be able to calculate the primary and secondary full-load current from the nameplate information. If the volt-ampere rating is given along with the primary voltage, then the primary full-load current can be determined using Equation 8.4 or Equation 8.5 for the case of a single-phase transformer. Equation 8.6 is used for determining the full-load current of either the primary or the secondary winding of a 3-phase transformer.

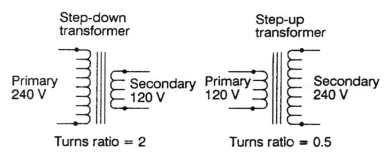

Figure 8.2   Schematic diagrams of step-down and step-up transformers.

Single-phase:

$$\text{Full-Load Current} = \frac{\text{VA}}{\text{Voltage}} \qquad \text{Eq. 8.4}$$

$$\text{Full-Load Current} = \frac{\text{kVA} \times 1{,}000}{\text{Voltage}} \qquad \text{Eq. 8.5}$$

3-phase:

$$\text{Full-Load Current} = \frac{\text{kVA} \times 1{,}000}{1.73 \times \text{Volts}} \qquad \text{Eq. 8.6}$$

An example will help to show how the previous equations are used to determine the full-load current of a transformer winding. It may be a good idea to write Equations 8.5 and 8.6 into a copy of the Code for easy reference when making transformer installations.

Example 8.1 A single-phase transformer is connected for a 480-volt primary and a 120-volt secondary. The transformer has a rating of 2 kVA. Determine the primary winding and the secondary winding full-load current of the transformer.

Answer: This is a single-phase transformer rated in kilovolt-amperes; therefore, Equation 8.5 is used to determine the full-load current for both windings. The full-load current of the primary winding of the transformer is 4.17 amperes, and for the secondary winding, the full-load current is 16.67 amperes.

$$\text{Primary Full-Load Current} = \frac{2 \text{ kVA} \times 1{,}000}{480 \text{ V}} = 4.17 \text{ A}$$

$$\text{Secondary Full-Load Current} = \frac{2 \text{ kVA} \times 1{,}000}{120 \text{ V}} = 16.67 \text{ A}$$

Connecting Transformer Windings

Transformer wiring diagrams are printed on the transformer nameplate, which may be affixed to the outside of the transformer or printed inside the cover to the wiring compartments. The lead wires or terminals are marked with Xs and Hs. The Hs are the primary leads, and the Xs are the secondary leads.

Some transformers have two primary and two secondary windings so they can be used for several applications. These are called dual-voltage transformers. Connections must be made correctly with dual-voltage transformers. If connected improperly, it is possible to create a short-circuit that will usually damage or destroy the transformer when it is energized.

Consider a dual-voltage transformer rated 240/480 volts on the primary and 120/240 volts on the secondary. Each of the two primary windings is, therefore, rated 240 volts. Each secondary winding is rated 120 volts. The transformer must be connected so each primary winding receives the proper voltage. Figure 8.3 shows the transformer with the primary windings connected in series with H1 and H4 connected to a 480-volt supply. The voltage across H1 and H2 is 240, and the voltage across H3 and H4 is 240. Each winding is receiving the proper voltage. With each primary winding receiving the proper 240 volts, each secondary winding will have an output of 120 volts. Connecting the secondary windings in series produces 240 volts across X1 and X4.

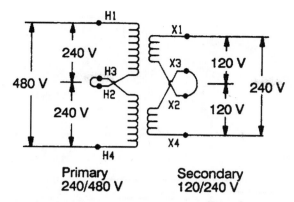

Figure 8.3 **The windings are connected in series to obtain the higher of the rated transformer voltages.**

Next consider a case where the primary voltage available is 480, but the desired output is 120 volts, single-phase. In this case, the primary windings are connected in series, while the secondary windings are connected in parallel, as shown in Figure 8.4.

Three-Phase Transformers

Changing the voltage of a 3-phase system can be done with a 3-phase transformer, or it can be done with single-phase transformers. Three-phase transformers are generally designed and constructed for specific voltages. For example, a transformer may have a 480-volt delta primary and a 120/208-volt wye secondary.

The 3-phase transformer has one core with three sets of windings. A primary and secondary winding are placed one on top of the other on each of the three legs of the core. Single-phase transformers can be used to form a 3-phase transformer bank. It is important that single-phase transformers are identical when connecting them to form a 3-phase system. They should be identical in voltage, kilovolt-amperes, impedance, manufacturer, and model number. Transformer impedance is the combined effect of resistance and inductance and is given in percent.

Connecting single-phase transformers to form a 3-phase bank must be done with extreme caution. The windings can only be connected in a certain way. Reversing a winding can damage the transformer. Figure 8.5 shows three individual single-phase transformers connected to step down from 480 volts delta to 240 volts delta. It may be advisable to obtain a 3-phase transformer rather than connecting single-phase transformers. To illustrate the complexity, standard

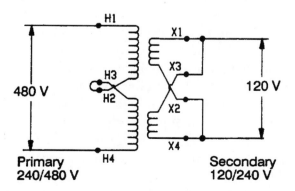

Figure 8.4 **The secondary windings are connected in parallel for an output of 120 volts.**

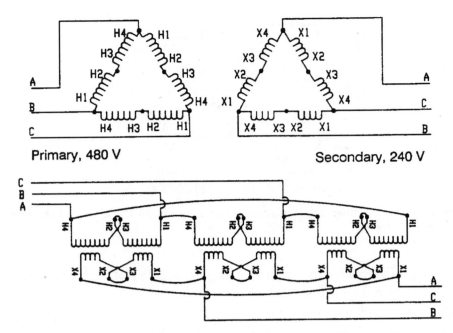

Primary, 480 V

Secondary, 240 V

Figure 8.5   Dual-voltage single-phase transformers with 240/480-volt primary windings and 120/240-volt secondary windings are shown connected to form a 480-volt delta to 240-volt delta 3-phase step-down transformer bank.

dual-voltage single-phase transformers are used to change 480-volt 3-phase delta to 120/208-volt wye, as shown in Figure 8.6.

Winding Taps

Transformers, except for small sizes, are often supplied with winding taps to compensate for abnormally low or high primary voltage. Assume, for example, that a transformer is rated 480 volts primary and 240 volts secondary. This means that 240 volts will be the output if the input is 480 volts. But what if the input is only 444 volts? The turns ratio for this transformer is 2-to-1; therefore, the output will be 222 volts. Equation 8.3 is used to determine the output, which would be 222 volts.

$$\text{Secondary Voltage} = \frac{444 \text{ Volts}}{2} = 222 \text{ Volts}$$

To get an output of 240 volts with an input of only 444 volts, the turns ratio will have to be changed to 1.85-to-1. The 1.85 was determined by dividing the 444 volts by 240 volts. The purpose of the tap connections, usually on the primary, is to easily change the transformer turns ratio. A typical single-phase transformer nameplate with primary taps is shown in Figure 8.7.

Consider an example in which the desired output voltage from a single-phase step-down transformer is 120/240 volts, but the available input is only 450 volts rather than 480 volts. A standard step-down transformer with a 2-to-1 turns ratio will give an output of 112.5/225 volts with a 450-volt input. By changing the primary taps, as shown in Figure 8.8, the turns ratio of the transformer is changed, and the output is now close to the desired 120/240 volts.

Winding taps each make a 2.5 percent change in the voltage. A transformer will often have two taps above normal voltage and four taps below normal voltage. A transformer usually comes preconnected for normal voltage. If an abnormal voltage is present, it is up to the installer to change the tap connections.

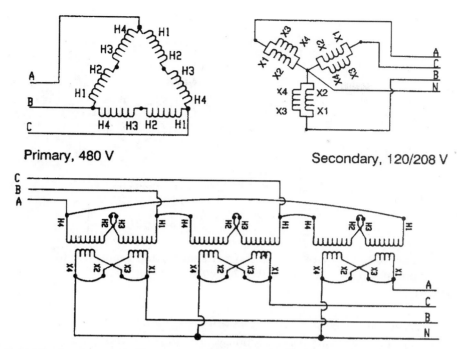

Primary, 480 V

Secondary, 120/208 V

Figure 8.6 Dual-voltage single-phase transformers with 240/480-volt primary windings and 120/240-volt secondary windings are shown connected to form a 480-volt delta to 120/208-volt wye 3-phase step-down transformer bank.

Input and Output Current

The primary kilovolt-amperes (kVA) of a transformer will be equal to the secondary kVA less any small losses. If the primary voltage is reduced from 240 to 120 volts, this is a voltage ratio reduction of 2-to-1. If the primary and secondary kVA are to remain equal, the current must be higher on the secondary than on the primary by a factor of two. For example, assume the

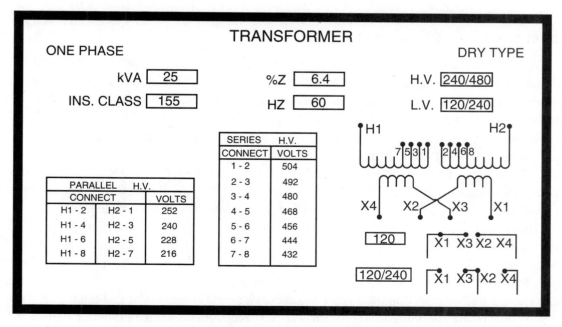

Figure 8.7 **Transformer nameplate showing primary taps.**

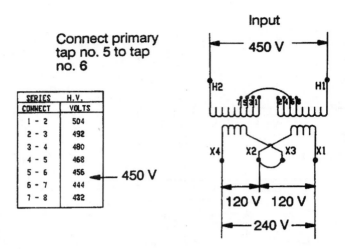

Figure 8.8   Single-phase transformer showing tap connection for an input of 450 volts, and a 3-wire 120/240-volt output.

240-volt primary current of the previous transformer is 5 amperes. The primary volt-amperes will be 1,200 VA. For the 120-volt secondary, 10 amperes must flow to keep the volt-amperes at 1,200 VA. If the primary and secondary voltages are known and if either the primary or secondary current is known, the other current may be determined using Equation 8.7 or Equation 8.8. These are useful equations especially when working with autotransformers.

Find primary current when secondary current is known:

$$\text{Primary Current} = \text{Secondary Current} \times \frac{\text{Secondary Voltage}}{\text{Primary Voltage}} \qquad \text{Eq. 8.7}$$

Find secondary current when primary current is known:

$$\text{Secondary Current} = \text{Primary Current} \times \frac{\text{Primary Voltage}}{\text{Secondary Voltage}} \qquad \text{Eq. 8.8}$$

Boost and Buck Transformers

A boost and buck transformer is an insulating transformer that can be connected as an autotransformer. The boost and buck transformer is used to make small adjustments in voltage, either up or down. For example, a machine has an electric motor that requires 208 volts, but the electrical supply is 240 volts. If ordering the machine with a 240-volt motor is expensive, a less costly solution to the problem may be to buck the voltage from 240 down to 208 with a boost and buck transformer.

Low voltage resulting from voltage drop can be corrected with a boost and buck transformer. This practice may not be energy-efficient, but it may be the best solution in unusual circumstances. Voltage drop on wires is wasted energy and should be avoided.

Boost and buck transformers for single-phase applications have a dual-voltage primary rated 120/240 volts. There is a choice of two sets of secondary voltages depending on the amount of boosting or bucking required: 12/24 volts and 16/32 volts. Three-phase applications from 380 to 500 volts require the use of a boost and buck transformer with a 240/480-volt primary and a 24/48-volt secondary. Figure 8.9 shows a single-phase boost and buck transformer connected to boost 208 volts to approximately 230 volts. Refer to manufacturer literature for other combinations of boosting and bucking.

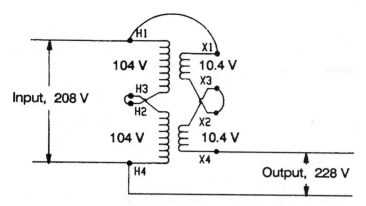

Figure 8.9   A boost and buck transformer is connected to boost a 208-volt supply to approximately 230 volts.

Boost and buck transformers may be connected for 3-phase applications, but all 3-phase combinations are not possible. The commonly used 3-phase boost and buck transformer connections are:

- Wye (4-wire)   to   wye (3- or 4-wire)
- Wye (3- or 4-wire)   to   open delta (3-wire)
- Delta (3-wire)   to   open delta (3-wire)

A confusing aspect of boost and buck transformers is the kVA rating of the transformer required to supply a load. For some applications, the kVA rating of the load may be several times larger than the kVA rating of the boost and buck transformer. A boost and buck transformer, when used as an autotransformer for boosting or bucking, can supply a load several times the kVA rating of the transformer. The maximum kVA rating of the load supplied depends on the full-load current rating of the transformer secondary winding and the operating voltage of the load. Each manufacturer supplies load current and kVA data for boost and buck transformers for all combinations of input and output voltages.

K-Rated Transformers

The K-rating marked on some transformers is an indication of the ability of the transformers to supply loads, which produce harmonic currents. A transformer designated K-1 is one that has not been modified to supply loads that produce nonlinear or harmonic currents. Standard transformer ratings are K-4, K-9, K-13, K-20, K-30, K-40, and K-50. The higher the K-rating number, the greater is the ability of the transformer to supply loads that have a higher percentage of harmonic current-producing equipment without overheating. It is most important to select a transformer with a K-rating for the specific harmonic frequencies present and their magnitude in relation to the total 60-hertz current. A number that is probably of little use when selecting transformer K-rating is the percent of total harmonic distortion (THD). Another value that can give an indication that harmonics are present but is of little use in determining proper transformer K-rating is the value of root-mean-square (rms) current. When viewed on an oscilloscope, the current sine wave will distort when large numbers and significant levels of harmonic currents are present. This will usually result in an increase in the rms line current. Transformer overheating due to harmonics is primarily the effect of specific frequencies of harmonics and their magnitudes.

Electronic equipment that draws current for only a portion of the cycle is the type that produces harmonic currents in the electrical system. Electronic dimmer switches and some other electronic controllers switch on for only a portion of the cycle. As a result, current flows to the loads in pulses. Other electronic equipment that produces nonlinear currents includes personal computers, video display terminals, copiers, fax machines, uninterruptable power supplies (UPS), variable speed drives, electronic high-efficiency ballasts, some medical electronic monitors, welders, mainframe computers, data processing equipment, and induction heating systems. Most electronic office equipment has switching-mode power supplies (SMPS) that draw current in pulses for only part of the cycle. These pulsing input currents produce harmonic currents in the electrical system. These harmonic currents can result in overheating of a transformer and other distribution equipment and wiring.

One way of dealing with loads that consist of a high percentage of harmonic-producing electronic equipment is to oversize the transformer for the load. But a transformer designed to supply 60-hertz loads does not perform the same when currents are at a higher frequency than 60 hertz. Core saturation can occur when a standard transformer is subjected to loads with a high percentage of harmonic currents. This can occur even when the transformer is supplying less than rated full-load current. Transformer heating is different for a given level of rms current at a different frequency.

There are industry recommendations for transformer K-ratings for particular loads. Dealing with loads that produce harmonic currents is sometimes complex, and even these K-rating recommendations may not be correct for all applications. A harmonic analysis may be necessary in some cases with the transformer specifically matched to the load. Here are some general recommendations for matching transformers to loads:

- Use nonrated K-1 transformers when the loads producing harmonic currents are less than 15 percent of the total load.
- Use K-4 rated transformers when the loads producing harmonic currents are 15 to 35 percent of the total load.
- Use K-13 rated transformers when the loads producing harmonic currents are 35 to 75 percent of the total load.
- Use K-20 rated transformers when the loads producing harmonic currents are 75 to 100 percent of the total load.
- Use K-30 and higher rated transformers for specific equipment where the load and transformer are matched for harmonic characteristics.

In the case where a transformer is supplying specific loads known to be producers of harmonic currents, some general guidelines can be used to help avoid transformer overheating. The transformer can be sized to the load kVA to be supplied when the proper K-rating is selected. Not only do different brands of the same equipment produce different harmonic currents, the identical equipment can produce different harmonic currents when supplied by different electrical systems. The harmonic current problem, when it exists, may be complex, sometimes requiring experienced personnel for analysis and design. The following is a general industry recommendation of approximate current K-ratings for different types of loads. Actual specifications for a specific installation made by an experienced engineer or technician may be different.

- K-4   welders and induction heaters
- K-4   electric discharge lighting
- K-4   solid state controls
- K-13   telecommunications equipment
- K-13   branch circuits in classrooms and in health care facilities
- K-20   mainframe computers and data processing equipment
- K-20   variable speed drives

On a balanced 3-phase wye electrical system, odd-numbered harmonics in multiples of the third harmonic (3rd, 9th, 15th, 21st, and so on) if present will not cancel and will increase the neutral current. Examples are the 208/120-volt and the 480/277-volt 4-wire 3-phase electrical systems serving line-to-neutral loads. When a 3-phase wye transformer is supplying balanced line-to-neutral loads where harmonics are present, it is possible for the neutral current to be as high as twice the level of line current. This can result in neutral conductor and conductor termination overheating. A K-rated transformer will not eliminate this type of problem. If the harmonic currents cannot be reduced, then it will be necessary to increase the size of the neutral bus and neutral conductor.

OVERCURRENT PROTECTION FOR TRANSFORMERS

Wiring a transformer circuit is one of the most difficult wiring tasks unless the installer understands transformer fundamentals. Rules for sizing overcurrent protection for a transformer operating at not more than 600 volts are covered in *Section 450-3(b)* and *Table 450-3(b)*. It must be noted that these rules apply only to the transformer itself, and not necessarily to the input and output circuit wires. If one or both of the windings operate at more than 600 volts, *Section 450-3(a)* and *Table 450-3(a)* will apply.

According to *Table 450-3(b)*, both the primary and secondary windings are permitted to be protected from overcurrent by one overcurrent device located on the primary side of the transformer and sized at not more than 125 percent of the transformer full-load current. If that primary overcurrent device has a rating greater than 125 percent of the transformer full-load primary current rating, then overcurrent protection not greater than 125 percent of the transformer secondary full-load current rating is required to protect the transformer secondary winding. Both situations are explained in the following two sections.

Overcurrent Protection Only on the Primary (600 volts or less)

A transformer is permitted to be protected by one overcurrent device on the primary side rated not more than 1.25 (125%) times the primary full-load current, as shown in Figure 8.10. This overcurrent device may be a set of fuses in a panelboard or fusible switch, or a circuit-breaker. The secondary winding is not required to be protected from overcurrent in this situation. When the secondary of the transformer is not single-voltage two-wire, the secondary conductors are required to be protected according to *Section 240-3(b)*, but the secondary winding is not required to be protected by other than the primary overcurrent device. *Section 240-21(c)(1)* applies to the secondary conductors, and not to the winding of the transformer.

Example 8.2 Consider the case of a 3-kVA transformer used to step down 480 to 120 volts to supply one 20-ampere single-phase 2-wire circuit. Determine the maximum permitted overcurrent device rating on the primary circuit to protect the transformer and the secondary circuit wires. There will be no secondary overcurrent protection as shown in Figure 8.10.

Answer: First, determine the primary and secondary full-load current rating of the 3-kVA transformer using Equation 8.5.

$$\text{Primary Current} = \frac{3 \text{ kVA} \times 1,000}{480 \text{ V}} = 6.25 \text{ A}$$

$$\text{Secondary Current} = \frac{3 \text{ kVA} \times 1,000}{120 \text{ V}} = 25 \text{ A}$$

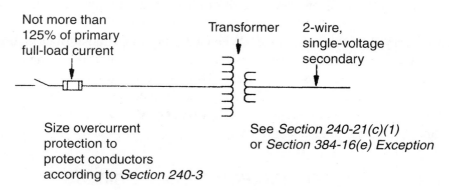

Table 450-3(b) Transformer protected only on the primary

Figure 8.10   A transformer may be protected with one overcurrent device on the primary sized at not more than 125 percent of the primary rated current.

Next, determine the maximum permitted overcurrent device rating for the primary using the rules of Code *Section 450-3(b)*. The primary overcurrent device rating is not to exceed 125 percent of the transformer primary full-load current.

$$\text{Overcurrent Rating} = 6.25 \text{ A} \times 1.25 = 7.8 \text{ A}$$

Circuit-breakers are not available in these small sizes; therefore, time-delay fuses will be used. The next standard fuse rating larger than 7.8 amperes is permitted to be used as long as it does not exceed 167 percent of the primary full-load current, according to *Table 450-3(b)*.

$$\text{Maximum Fuse Rating} = 6.25 \text{ A} \times 1.67 = 10.4 \text{ A}$$

A 10-ampere time-delay fuse is the maximum size permitted to protect the 3-kVA transformer. But this fuse is sized only to protect the transformer, and not to protect the circuit conductors. The primary circuit conductor is not permitted to be sized smaller than AWG number 14 copper; therefore, it is adequately protected with a 10-ampere fuse. If the primary fuse is sized at 10 amperes, determine how much current will flow on the secondary circuit wires in order to blow the primary fuse. Use Equation 8.8 to determine the current flow on the secondary circuit wires if 10 amperes flow on the primary circuit wires.

$$\text{Secondary Current} = 10 \text{ A} \times \frac{480 \text{ V}}{120 \text{ A}} = 40 \text{ A}$$

The purpose of installing this transformer was to get a 20-ampere, 120-volt circuit. It is possible to draw 40 amperes on the secondary before the primary fuse blows. The secondary wire would be required to be AWG number 8 copper if the primary fuse is 10 amperes. A more practical approach would be to install an AWG number 12 copper secondary wire that is adequate for the intended 20-ampere circuit. Then, determine the maximum permitted primary fuse size to prevent more than 20 amperes from flowing in the secondary circuit. Use Equation 8.7 to determine the maximum permitted primary fuse rating to limit the secondary current to 20 amperes.

$$\text{Primary Current} = 20 \text{ A} \times \frac{120 \text{ V}}{480 \text{ V}} = 5 \text{ A}$$

A 5-ampere time-delay fuse in the primary circuit will protect the transformer, and it will protect the secondary AWG number 12 copper circuit wires.

Primary and Secondary Overcurrent Protection (600 volts or less)

The overcurrent device protecting the primary of a transformer is permitted to be rated as large as 2.50 (250%) times the primary full-load current, provided the transformer secondary winding is protected. The transformer secondary overcurrent device rating is not permitted to be greater than 1.25 (125%) of the secondary full-load current. Protection for both the primary and the secondary of the transformer is illustrated in Figure 8.11.

Consider the 3-kVA, single-phase, 480-to-120-volt step-down transformer of Example 8.2, but this time overcurrent protection will be installed on both the primary and the secondary circuits. Assume the input wire is AWG number 14 copper with type THWN insulation, and the wire is protected by a 15-ampere circuit-breaker. The primary full-load current was determined to be 6.25 amperes, and for the secondary it was 25 amperes. In this case, the primary winding of the transformer is permitted to be protected at 250 percent of the primary full-load current.

$$\text{Maximum Primary Overcurrent Rating} = 2.5 \times 6.25 \text{ A} = 15.6 \text{ A}$$

The 15-ampere circuit-breaker is not too large to serve as overcurrent protection for the primary winding of this transformer. An overcurrent device installed to protect the transformer secondary winding at 125 percent of the secondary full-load current will satisfy the Code requirements for transformer protection, *Table 450-3(b)*.

$$\text{Maximum Secondary Overcurrent Rating} = 1.25 \times 25 \text{ A} = 31.3 \text{ A}$$

A 30-ampere fuse or circuit-breaker at the transformer secondary will meet the transformer secondary overcurrent protection requirement. But the purpose of installing the transformer was to provide one 20-ampere, 120-volt circuit. It would then make sense to provide a 20-ampere fuse or circuit-breaker at the transformer secondary and use AWG number 12 copper wire.

Overcurrent protection for the secondary winding of the transformer, when required, is permitted to consist of up to six separate overcurrent devices grouped together. The sum of the

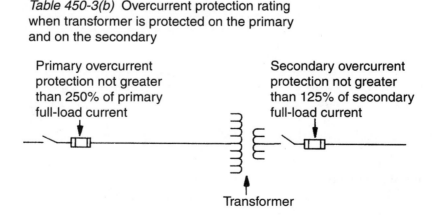

Table 450-3(b) Overcurrent protection rating when transformer is protected on the primary and on the secondary

Primary overcurrent protection not greater than 250% of primary full-load current

Secondary overcurrent protection not greater than 125% of secondary full-load current

Transformer

Figure 8.11   If the transformer is protected on the secondary with an overcurrent device rated not more than 125 percent of the secondary current, the primary overcurrent device is permitted to be sized as large as 250 percent of the primary current.

ratings of the secondary overcurrent protective device is not permitted to exceed 1.25 times the secondary full-load current rating. It is important to note that the location of the primary and the secondary winding overcurrent protective devices is not specified. This permits the overcurrent devices to be located away from the transformer provided the requirements for protecting the conductors in *Section 240-3* and *Section 240-21* are satisfied. For outside transformer taps, see *Section 240-21(c)(4)*.

Example 8.3   A 25-kVA single-phase transformer is supplied from a 480-volt panelboard with a circuit sized specifically to supply a 100-ampere, single-phase, 120/240-volt, 3-wire panelboard. The transformer will be protected from overcurrent in accordance with *Section 450-3(b)*. All conductor insulation and terminations are rated at 75°C. The total length of the circuit from the primary overcurrent device to the secondary overcurrent device is more than 25 feet. Determine the following:

1. The minimum permitted size of copper secondary circuit conductors to the 100 ampere panelboard
2. The minimum permitted size of copper primary circuit conductors supplying the transformer
3. The maximum permitted rating of the primary overcurrent device
4. The maximum permitted rating of the overcurrent device to protect the secondary conductors and panelboard

Answer: The purpose of this installation is to supply a 100-ampere panelboard at 120/240 volts. First, size the secondary feeder conductors supplying the 100-ampere panelboard. The minimum permitted copper conductor with 75°C insulation and terminations is AWG number 3 from *Table 310-16*.

Next determine the minimum permitted primary conductor size needed at 480 volts to supply 100 amperes at 240 volts. This can be determined using Equation 8.7. When 100 amperes is being drawn on the 240-volt secondary, 50 amperes will be flowing on the 480-volt primary conductors. The minimum permitted size of copper conductors with 75°C insulation and terminations is AWG number 8.

$$\text{Primary Current} = 100 \text{ A} \times \frac{240 \text{ V}}{480 \text{ V}} = 50 \text{ A}$$

Now use Equation 8.5 to determine the primary and secondary full-load current rating of the transformer at the desired voltages. The primary full-load current will be 52 amperes, and the secondary full-load current will be 104 amperes.

$$\text{Primary Full-Load Current} = \frac{25 \text{ kVA} \times 1,000}{480 \text{ V}} = 52 \text{ A}$$

$$\text{Secondary Full-Load Current} = \frac{25 \text{ kVA} \times 1,000}{240 \text{ A}} = 104 \text{ A}$$

The overcurrent protection for the transformer is sized using *Table 450-3(b)*. In this case, the transformer primary overcurrent protective device rating will not exceed 125 percent of the primary full-load current rating.

$$\text{Maximum Primary Overcurrent Rating} = 52 \text{ A} \times 1.25 = 65 \text{ A}$$

The transformer overcurrent protection is permitted to be 70 amperes according to *Note 1* of *Table 450-3(b)*. But the primary circuit conductor was sized at AWG number 8 copper, which was adequate to supply the load. *Section 240-3* requires that this conductor be protected in accordance with the 50-ampere allowable ampacity. The maximum permitted rating of primary overcurrent protective device is 50 amperes.

When the transformer is protected on the primary at not more than 125 percent of the transformer primary full-load current rating, then secondary winding overcurrent protection is not required. But, according to *Section 240-3(b)*, the secondary conductors are required to be protected. The panelboard is required to be provided with overcurrent protection on the secondary side of the transformer by *Section 384-16*. The conductors have an allowable ampacity of 100 amperes, and the panelboard has a maximum rating of 100 amperes; therefore, the secondary overcurrent device maximum rating is not permitted to exceed 100 amperes.

There are cases where the primary overcurrent device for the previous example may have a rating as large as 130 amperes, and the primary conductors may be larger than AWG number 8 copper. This could be the case where the transformer is tapped from a feeder. In this case, the secondary winding is required to be protected from overcurrent with a maximum rating of 125 percent of the transformer secondary full-load current, which in the case of Example 8.3 would be 150 amperes. This rule is found in *Table 450-3(b)*.

Boost and Buck Transformer Overcurrent Protection

The specific transformer, winding connections, and kVA rating for an application of voltage boosting or voltage bucking are determined with information from specific boost and buck transformer manufacturers. The overcurrent for the primary of the transformer is determined according to Code *Section 450-4*. The input conductor shall be protected from overcurrent at not more than 125 percent of the rated full-load input current of the transformer as it is installed.

The most practical method of approaching overcurrent protection for a boost or buck transformer application is to determine the kVA rating of the transformer for the specific application. Then, select the minimum size of transformer output wire to the load. Next, determine the transformer input current required to deliver the required output load current. Finally, determine the maximum permitted rating of the input circuit overcurrent protection.

Consider an example of a 208-volt, single-phase, 2-wire input circuit to a boost transformer with an output of 230 volts to supply a 3-horsepower, single-phase electric motor. The circuit is shown in Figure 8.12. Assume that the correct transformer was selected for the application.

Approach the circuit in this case as an ordinary motor branch circuit. Find the 17-ampere motor full-load current from *Table 430-148*. The minimum wire size permitted is AWG number 12, type THWN copper.

$$1.25 \times 17 \text{ A} = 21 \text{ A}$$

The motor full-load input current to the transformer is determined using Equation 8.7. In the case of autotransformers, the terms **input** and **output** are used instead of **primary** and **secondary.** Equation 8.9 is similar to Equation 8.7.

$$\textbf{Input Current} = \textbf{Output Current} \times \frac{\textbf{Output Voltage}}{\textbf{Input Voltage}} \qquad \textbf{Eq. 8.9}$$

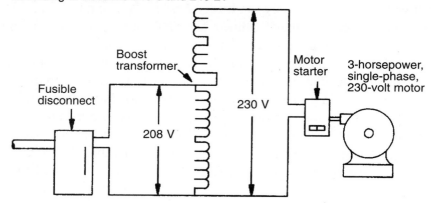

Figure 8.12 **A boost and buck transformer is used to boost the 208-volt supply to 230 volts for a 3-horsepower, single-phase motor.**

$$\text{Input Current} = 17 \text{ A} \times \frac{230 \text{ V}}{208 \text{ V}} = 18.8 \text{ A}$$

Now, treat this as a motor circuit to determine the minimum ampere rating of the input circuit wire.

$$18.8 \text{ A} \times 1.25 = 23.5 \text{ A}$$

In this case, the minimum size of input wire is AWG number 12, type THWN copper. The overcurrent device protecting the transformer on the input circuit is not permitted to be sized larger than 125 percent of the input full-load current of the transformer. In the case of this example, the next standard overcurrent device rating could be used, which would be 25 amperes. If an oversized boost transformer had been used, then the overcurrent device rating could have been sized larger.

TRANSFORMER TAP FROM A FEEDER

Electrical wires are required to be of sufficient ampacity to supply the load, and they shall be protected from overcurrent according to the wire ampere ratings, as given in ampacity tables of *Article 310.* Therefore, wire ampere rating must also be considered when sizing transformer overcurrent devices. Tap rules do apply to transformer installations where the wire ampacity is permitted to be of a value less than the overcurrent protection. There is a 10-foot tap rule, a 25-foot tap rule, and an outside tap rule with no length specified. The 25-foot tap rule is most common, and it is illustrated in Figure 8.13.

Whenever the overcurrent device on the primary side of the transformer will not limit the current on a secondary conductor to a value less than the allowable ampacity of the secondary conductors, as stated in *Section 240-3(f),* these would be considered to be tap conductors with the tap originating at the transformer secondary. If the primary conductors are tapped from a feeder, then the tap is sometimes considered to originate at the point where the primary conductor is tapped to the feeder.

One or more sets of conductors are permitted to originate at the secondary of a transformer in the case where the primary overcurrent protective device is sized at not more than

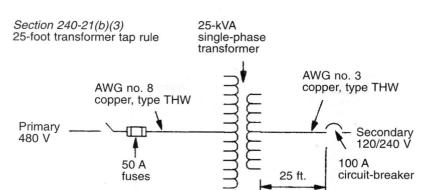

Section 240-21(b)(3)
25-foot transformer tap rule

25-kVA
single-phase
transformer

AWG no. 8
copper, type THW

AWG no. 3
copper, type THW

Primary
480 V

Secondary
120/240 V

50 A
fuses

100 A
circuit-breaker

25 ft.

Figure 8.13   The secondary overcurrent protection is permitted to be located 25 feet from the secondary winding when the primary circuit wire has overcurrent protection not exceeding the ampacity of the wire.

125 percent of the primary full-load current of the transformer rounded up to the next higher standard rating of overcurrent device. The length of that tap conductor is permitted to be 25 feet in the case of *Section 240-21(b)(3)*. If the allowable ampacity of the tap conductor is less than is permitted for the 25-foot tap, then the tap length is limited to a maximum of 10 feet according to *Section 240-21(c)(2)*.

Code *Section 240-21(b)(3)* applies to basically two situations. The first situation is where overcurrent protection is provided for the primary circuit wires and the primary windings, and the secondary of the transformer is not single-voltage, 2-wire. If the secondary was single-voltage, 2-wire, then overcurrent protection might not even be required on the secondary, as in the case of Example 8.2. This case is covered by paragraph *(c)* of *Section 240-21(b)(3)*. The primary circuit wires have overcurrent protection sized properly for the wire; therefore, the tap length begins at the secondary winding of the transformer, as shown in Figure 8.13.

The second case of the 25-foot tap rule is where the transformer primary circuit wire is protected with an overcurrent device of a rating larger than the ampacity of the wire. The distance from that primary overcurrent device to the transformer secondary overcurrent device is not permitted to be more than 25 feet. Another example is where the transformer primary circuit wires are tapped from a feeder. The distance from the tap point to the secondary overcurrent device is not permitted to be more than 25 feet, as shown in Figure 8.14. The overcurrent device protecting the feeder to which the primary conductor tap is made shall not be larger than 250 percent of the transformer primary full-load current. The primary tap wire is required to have an ampacity of at least 1/3 that of the feeder overcurrent protection. The secondary wire ampacity times the secondary-to-primary voltage ratio is required to be at least 1/3 that of the primary feeder overcurrent protection.

Example 8.4:   A 100-ampere, 120/208-volt, 3-phase panelboard is supplied from a 480-volt 3-phase system through a 37.5-kVA step-down transformer. Determine the following for the transformer circuit:

1. The minimum permitted secondary circuit wire size to supply the load
2. The maximum permitted rating of the secondary overcurrent protective device for the transformer secondary winding and secondary circuit wires
3. The maximum permitted distance from the secondary winding of the transformer to the secondary overcurrent device
4. The minimum primary circuit wire size required to supply the load on the secondary

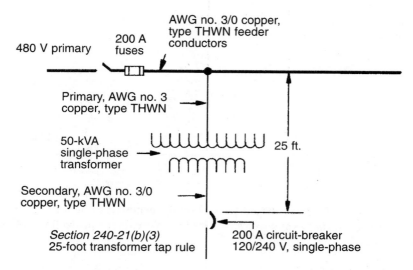

Figure 8.14   When the primary circuit wire is tapped from a feeder, the secondary winding overcurrent protection shall be located not more than 25 feet from the point of the primary tap.

5. The maximum permitted primary overcurrent device rating for the transformer primary winding and to protect the primary circuit wires at their ampacity

Answer: The first procedure is to determine the primary winding and the secondary winding full-load currents.

$$\text{Primary Full-Load Current} = \frac{37.5 \text{ kVA} \times 1,000}{1.73 \times 480 \text{ V}} = 45 \text{ A}$$

$$\text{Secondary Full-Load Current} = \frac{37.5 \text{ kVA} \times 1,000}{1.73 \times 208 \text{ V}} = 104$$

(1) The reason for installing the transformer is to supply a 100-ampere panelboard. Therefore, it would seem reasonable to size the secondary wires to handle the 100-ampere load. This would require the use of AWG number 3, type THWN copper wire.

(2) The AWG number 3, type THWN copper wire shall be protected from overcurrent with a device that has a rating not exceeding 100 amperes. Next, check to make sure the 100-ampere overcurrent device rating does not exceed the maximum value permitted for the transformer secondary winding. This value is 1.25 times the transformer secondary full-load current. The 100-ampere overcurrent device is permitted to protect both the secondary circuit wires and the transformer secondary winding.

$$1.25 \times 104 \text{ A} = 130 \text{ A}$$

(3) The primary circuit wires will be protected at their ampacity; therefore, according to *Section 240-21(b)(3)*, the length of the secondary circuit conductors to the overcurrent device shall not exceed 25 feet. A circuit-breaker or fuses may be provided for this purpose, or a main overcurrent device may be installed in the 100-ampere panelboard.

(4) If the secondary of the transformer will be permitted to carry only 100 amperes at 208 volts, it would seem reasonable that the primary circuit wires need only be large enough to deliver the 100 amperes to the secondary. Use Equation 8.7 to determine how much

current will flow on the primary when 100 amperes flow on the secondary. Select a wire size that will carry a minimum of 43 amperes. An AWG number 8, type THWN copper wire will carry 50 amperes.

$$\text{Primary Current} = 100 \text{ A} \times \frac{208 \text{ V}}{480 \text{ V}} = 43 \text{ A}$$

$$45 \text{ A} \times 2.5 = 112 \text{ A}$$

(5) The primary overcurrent device is not permitted to exceed 250 percent of the transformer primary winding full-load current. But if the primary circuit wire was selected to be AWG number 8, type THWN copper wire, then the maximum overcurrent device permitted to protect the wire is 50 amperes.

The tap for a transformer may include both the primary circuit wire and the secondary circuit wire, as shown in Figure 8.14. In this case, the tap begins at the feeder and continues through the transformer to the point where the secondary overcurrent device is located. An example will illustrate how the primary tap wire size is determined.

Example 8.5   A 50-kVA, single-phase transformer is installed to supply a 120/240-volt, 200-ampere panelboard. The secondary conductors are AWG number 3/0, type THWN copper, and end at the panelboard that has a 200-ampere main circuit-breaker. The primary circuit wires for the transformer are tapped to an AWG number 3/0, 480-volt, 3-phase feeder protected by 200-ampere fuses. The distance from the tap point at the 480-volt feeder to the 200-ampere secondary circuit panelboard is not more than 25 feet. This installation is shown in Figure 8.14. Answer the following questions about the installation:

1. Do the 200-ampere fuses protecting the 480-volt feeder have too high a rating to serve as overcurrent protection for the transformer primary winding?
2. What is the minimum size tap conductor permitted on the primary of the transformer?
3. Is the AWG number 3/0, type THWN copper wire of the secondary portion of the tap of sufficient ampacity for this installation?

Answer: The first step is to determine the primary full-load current of the transformer using Equation 8.5.

$$\text{Primary Full-Load Current} = \frac{50 \text{ kVA} \times 1,000}{480 \text{ V}} = 104 \text{ A}$$

(1) The secondary winding is protected from overcurrent with a rating not exceeding 125 percent of the transformer secondary full-load current; therefore, the primary winding is permitted to be protected at not more than 250 percent of the primary full-load current. The answer to the first question is no. The 200-ampere fuses protecting the feeder do not have too high a rating to also protect the primary winding of the transformer.

$$\text{Maximum Primary Overcurrent Rating} = 2.5 \times 104 \text{ A} = 260 \text{ A}$$

(2) The tap wire supplying the primary of the transformer shall be of sufficient ampacity to supply the 200-ampere secondary load, and it is not permitted to be smaller than 1/3 the ampacity of the 200-ampere feeder. First determine the minimum ampacity permitted for this tap to the primary feeder.

$$\text{Minimum Primary Tap Ampacity} = \frac{200 \text{ A}}{3} = 67 \text{ A}$$

Next, determine the primary circuit ampacity required to supply the secondary circuit 200-ampere load. This is determined using Equation 8.7. The minimum wire size permitted for the primary circuit of the transformer is AWG number 3, type THWN copper, which will carry 100 amperes.

$$\text{Primary Current} = 200 \text{ A} \times \frac{240 \text{ V}}{480 \text{ V}} = 100 \text{ A}$$

(3) The ampacity of the secondary circuit tap conductor is not permitted to be smaller than 1/3 the ampacity of the primary feeder and adjusted for the voltage ratio of the transformer. It was determined that 1/3 the ampacity of the feeder was 67 amperes. Use Equation 8.8 to determine the amperes on the secondary. The AWG number 3/0, type THWN copper wire on the secondary is adequate to meet this requirement. The secondary is adequately sized because the minimum permitted secondary wire rating to meet the tap rule is 134 amperes.

$$\text{Secondary Current} = 67 \text{ A} \times \frac{480 \text{ V}}{240 \text{ V}} = 134 \text{ A}$$

The 10-foot transformer tap rule is illustrated in Figure 8.15. It applies when there are two or more taps to the secondary of the transformer where the allowable ampacity of the secondary conductor is less than permitted for the 25-foot tap rule of *Section 240-21(b)(3)*. The distance from the secondary winding of the transformer to the tap overcurrent protection is not permitted to exceed 10 feet. In the example of Figure 8.15, three disconnects are tapped from the transformer secondary. The 10-foot transformer tap rule is found in Code *Section 240-21(c)(2)*. Equation 8.10 and Equation 8.11 can be used to determine the minimum permitted allowable ampacity for the tap conductors based on the primary overcurrent device rating.

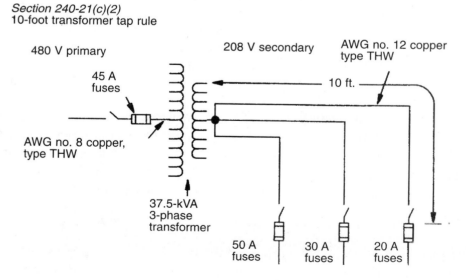

Figure 8.15 **When two or more circuits are tapped directly to the transformer secondary, the secondary overcurrent protection shall be located not more than 10 feet from the transformer secondary.**

The following two equations can be used to determine the maximum length of a transformer tap conductor when the overcurrent device rating on the primary side of the transformer is known. If the tap conductor allowable ampacity is not smaller than the value determined using Equation 8.10, then the length of the tap is permitted to be 25 feet. But if the allowable ampacity of the tap conductor is less than the value determined using Equation 8.10, then the maximum length of the tap is limited to 10 feet. Equation 8.11 is used to determine the minimum permitted allowable ampacity of a transformer tap conductor. Equation 8.10 is based on *Section 240-21(b)(3)*, and Equation 8.11 is based on *Section 240-21(c)(2)*.

25-foot tap rule:

$$\textbf{Tap conductor ampacity} = \frac{1}{3} \times \textbf{Primary O.C. device} \times \frac{V_P}{V_S} \qquad \textbf{Eq. 8.10}$$

10-foot tap rule:

$$\textbf{Tap conductor ampacity} = \frac{1}{10} \times \textbf{Primary O.C. device} \times \frac{V_P}{V_S} \qquad \textbf{Eq. 8.11}$$

TRANSFORMER GROUNDING

An electrical system derived from a transformer is required to have noncurrent-carrying metal parts and equipment grounded the same as any other part of the electrical system. In addition, the electrical system produced by the transformer also may be required to be grounded according to *Section 250-20(b)*. *Section 250-26* specifies the wire that shall be grounded. The method of grounding the grounded circuit conductor is specified in *Section 250-24* for services and separately derived systems. A separately derived system is defined in *Article 100*. A transformer installed at some point in an electrical system is considered to be a separately derived system if a derived grounded circuit conductor is not electrically connected to a grounded circuit conductor of the building electrical system. Further information about separately derived systems is found in *Section 250-20(d)*. Most transformer installations are considered to be separately derived systems, and the grounding is covered in *Section 250-30*. The rules of grounding are illustrated in Figure 8.16.

Separately derived systems do not necessarily originate from a transformer. An example of a separately derived system originating from a transformer would be any grounded electrical system where the transformer was supplied by an ungrounded electrical system. Assume that a single-phase, 120/240-volt, 3-wire electrical system was derived from a transformer that had the primary supplied by two conductors of an ungrounded delta, 3-wire, 480-volt, 3-phase electrical system. It is required to connect the grounded circuit conductor of the separately derived system to a grounding electrode, as specified in *Section 250-30*. This is illustrated in Figure 8.16. The grounded wire, usually the neutral, shall be connected to a grounding electrode. This grounding connection is permitted to be made at the transformer or at a service panel or disconnect switch supplied from the transformer. The grounding electrode shall be as close as practical to the point where the grounding electrode conductor connects to the grounded circuit conductor. Acceptable grounding electrodes are listed in *Section 250-30(a)(3)*. If neither of these grounding electrodes is available, then one described in *Section 250-50* or *250-52* shall be used. When a metal water piping system is used as the grounding electrode, as covered in *Section 250-30(a)(3)*, the connection shall be made at the nearest available point to the separately derived system.

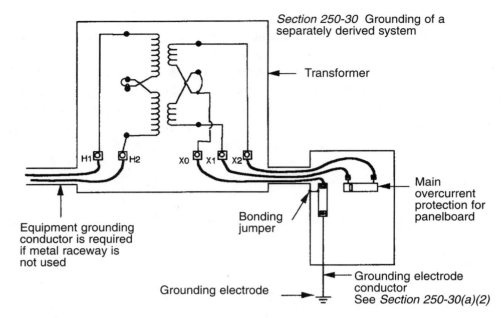

Figure 8.16   Grounding and bonding a separately derived system derived from a transformer are permitted to be done at a disconnect or panelboard.

The size of the grounding electrode conductor is based upon the size of the output conductor from the transformer, using *Table 250-66*. A bonding jumper is required between the grounding electrode conductor of the derived system and the metal enclosure of the transformer, as specified in *Section 250-30(a)(1)*. This connection is permitted to be made at a panelboard or disconnect supplied by the transformer.

MAJOR CHANGES TO THE 1999 CODE

These are the changes to the 1999 *National Electrical Code®* for electrical systems that correspond to the Code sections studied in this unit. The following analysis explains the significance of only the changes from the 1996 to the 1999 Code, and this analysis is not intended to be used in place of the Code. Refer to the actual section of the 1999 Code for the exact wording and meaning of each section discussed. Changes are indicated in the Code with a vertical line in the margin. If material has been deleted or moved to another part of the Code, the location of the deletion is indicated with a dark dot in the margin.

Article 450   **Transformers**

450-3: This section specifies the maximum rating permitted for transformer overcurrent protection. Over the years the rules changed, and section reorganization made it difficult to comprehend all of the conditions that applied in a particular situation. The entire section was reorganized with the overcurrent protection rules placed in two new tables. The tables are new, but the rules did not change from the previous edition of the Code. *Table 450-3(a)* specifies the maximum rating of overcurrent device for transformers that operate at more than 600 volts. *Table 450-3(b)* gives the maximum overcurrent device rating for transformers operating at 600 volts and less. The 600 volts referenced in this section is the voltage between any two conductors, and not necessarily the voltage as measured to ground.

An important aspect of new *Table 450-3(a)* and *Table 450-3(b)* is the notes to the tables. This section is easier to understand when one of the conditions described in the notes is to apply. The second, third, and fourth sentences of the first paragraph of the previous edition of the Code were moved to *Note 2* for each table. *Note 2* describes the conditions when a secondary overcurrent device is permitted to be replaced with up to six overcurrent devices grouped together in one location. The new words added to the beginning of *Note 2* make it clear that there are conditions where secondary overcurrent protection for a transformer is not required. The circumstances of those conditions are obvious from the two new tables.

Note 1 replaces an exception that appeared three times in this section in the previous edition of the Code. *Note 1* specifies the conditions when it is permitted for the calculated value of the overcurrent device to be rounded up to the next standard overcurrent device rating as given in *Section 240-6*. Those conditions can be determined with a quick glance at the tables.

Table 450-3(a) combines the tables for supervised locations and nonsupervised locations in the previous edition of the Code. Those tables only covered transformers with an impedance (usually listed as %Z) up to 10 percent. The general conditions of transformers operating at more than 600 volts and with an impedance of more than 10 percent are not included in *Table 450-3(a)* for supervised locations only. The description of what qualifies as a supervised location is now included as *Note 3* to *Table 450-3(a)*.

Note 4 to *Table 450-3(a)* applies to the case of electronically actuated fuses installed on the primary side of the transformer. The electronically actuated fuses were not previously permitted to be set at the same rating as circuit-breakers in supervised locations where secondary overcurrent protection is provided. The previous edition of the Code limited their setting to 300 percent of the primary rated current. Now they will be permitted to be set at 600 percent of primary current for a transformer with not more than 6 percent impedance and up to 400 percent of rated primary current when the impedance is more than 6 percent but not more than 10 percent.

Note 5 permits the secondary overcurrent protection to be omitted for a transformer operating at more than 600 volts when the transformer is equipped with coordinated thermal overload protection.

450-21(b): This section applies to dry-type transformers installed indoors and rated more than 112 1/2 kVA. If not installed in a fire-resistant transformer room, this section specifies the spacing requirements from combustible materials. The exceptions in the previous edition of the Code applied to transformers with a temperature rise of 80°C or higher. There is no longer a reference to transformer temperature rise. *Section 450-11* requires the temperature class for the insulation system to be marked on all dry-type transformers. The closest temperature class was Class 155, which is basically equivalent to a temperature rise of 80°C. This section now applies to dry-type transformers with an insulation system Class 155 or higher.

450-22: This section applies to dry-type transformers installed outdoors and rated more than 112 1/2 kVA. The reference to an 80°C rise was replaced with the reference to Class 155 insulation systems or higher.

450-23 FPN 1: The term **combustible material** is used in this section; therefore, a reference to combustible material and limited-combustible material was added to this fine print note.

450-26 Exception 4: An oil-insulated transformer rated not more than 75 kVA and supplied at not more than 600 volts that is an integral part of charged particle accelerating equipment is permitted to be installed in a building or room of noncombustible or fire-resistant construction. Installation in a vault is not required. It is also a requirement that suitable arrangements be made to prevent an oil fire from spreading to combustible materials in the building or room. This new exception addresses a previous problem of transformer installations in semiconductor manufacturing.

WORKSHEET NO. 8
TRANSFORMERS

These questions are considered important to understanding the application of the *National Electrical Code®* to electrical wiring, and they are questions frequently asked by electricians and electrical inspectors. People working in the electrical trade must continue to study the Code to improve their understanding and ability to apply the Code properly.

DIRECTIONS: Answer the questions and provide the Code reference or references where the necessary information leading to the answer is found. Space is provided for brief notes and calculations. An electronic calculator will be helpful in making calculations. You will keep this worksheet; therefore, you must put the answer and Code reference on the answer sheet as well.

1.   A transformer for a Class 1 remote control circuit is installed adjacent to a motor starter and tapped from the motor branch circuit. Is the protection of this transformer required to be according to *Article 450?* (Yes or No)

   Code reference _____

2.   If the fuses shown in the primary wires of Figure 8.17 are properly sized according to *Section 450-3(b),* are they permitted to protect both the primary and the secondary conductors of the transformer circuit? (Yes or No)

   Code reference _____

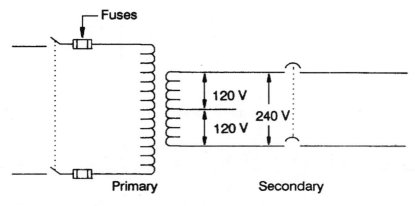

Figure 8.17   The primary and secondary windings for this transformer are protected from overcurrent by a set of fuses installed only on the primary.

3. A dry-type transformer is required to be provided with a nameplate that states the full-load ampere rating of the primary and the secondary windings. (True or False)

    Code reference _____

4. A 10-kVA dry transformer has an outer metal covering in direct contact with the core and no ventilation openings to the inside of the transformer. The transformer is installed in a room with walls constructed of a combustible material. The transformer operates with both primary and secondary windings at under 600 volts. The minimum permitted distance between the transformer and the wall or other combustible material is _____ inches.

    Code reference _____

5. A 500-kVA oil-insulated transformer is installed in a vault inside a building. The vault is ventilated to the outside of the building using natural circulation of air. After deducting the area for screens, gratings, and louvers, the minimum cross-sectional area of the ventilation opening shall be _____ square inches.

    Code reference _____

6. A single-phase load requires a 30-kVA transformer (under 600 V), but a 30-kVA transformer is not available. Two 15-kVA transformers are available and are installed with properly sized overcurrent protection on the primary and on the secondary, as shown in Figure 8.18. Secondary ties are sized to handle the load. Is it permitted to operate these transformers in parallel to obtain the desired 30-kVA capacity? (Yes or No)

    Code reference _____

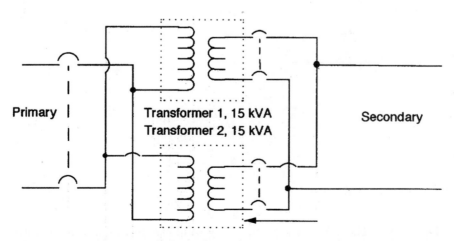

Figure 8.18   Two 15-kVA identical transformers are connected in parallel to supply a 30-kVA load.

7. A curb at least _____ inches high shall be provided at all doors entering a vault containing an oil-insulated transformer.

   Code reference _____

8. Replacement parts are permitted to be stored in a transformer vault if placed a distance of at least 4 feet from electrical equipment. (True or False)

   Code reference _____

9. A machine with a single-phase full-load current rating of 12 amperes operates at 120 volts. This machine is operated from a 480-volt supply using a 2-kVA transformer without internal thermal overload protection, as shown in Figure 8.19. The primary and secondary wires are AWG number 14 copper, type THHN, and both the primary and secondary overcurrent protective circuit-breakers are rated at 15 amperes. Is the primary overcurrent protection properly sized? (Yes or No)

   Code reference _____

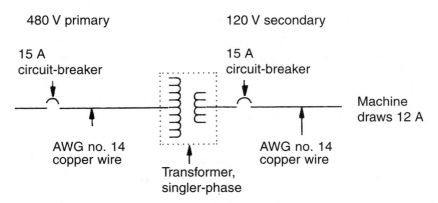

Figure 8.19   A 2-kVA transformer is installed to supply a 15-ampere, 2-wire, 120-volt circuit from a 480-volt supply with 15-ampere overcurrent protection on both the primary and the secondary.

10. A purpose for a zig-zag transformer is the conversion of a 3-phase, 3-wire ungrounded electrical system to a 4-wire grounded distribution system. (True or False)

    Code reference _____

11. Ventilation for a transformer shall be adequate to dispose of heat produced due to internal losses when the transformer is operating at _____.

    Code reference _____

12. Is a neon sign outline lighting transformer for a commercial building permitted to be installed in an attic? (Yes or No)

    Code reference _____

13. A 3-phase transformer primary is 480 volts, and the primary conductor is AWG number 6 copper, type THW and protected at 45 amperes. The transformer secondary is 208/120-volt, 4-wire supplying a panelboard with a 100-ampere main circuit-breaker. The secondary AWG number 3 copper, type THW conductors between the transformer and the secondary overcurrent protective device of Figure 8.20 are considered to be a tap. All terminations are rated 75°C. The secondary tap overcurrent protection is permitted to be located at a distance from the transformer secondary not to exceed _____ feet.

    Code reference _____

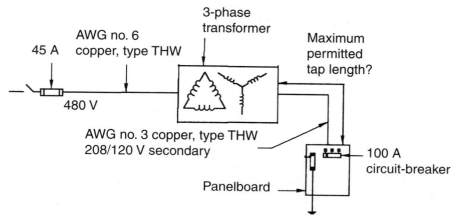

Figure 8.20   The overcurrent protection for the secondary windings of the transformer in this installation is located in the panelboard supplied by the transformer.

14. A building contains one large 480-volt, 3-phase pump motor and a general lighting circuit, as shown in Figure 8.21. A single-phase, 3-kVA dry transformer provides power for the one 20-ampere, 120-volt general lighting circuit. A single fusible disconnect containing time-delay fuses will be used to protect the transformer primary winding and circuit wires and the secondary winding and circuit wires. The maximum size fuse permitted on the primary to protect the transformer windings and the circuit wires is _____ amperes.

Code reference _____

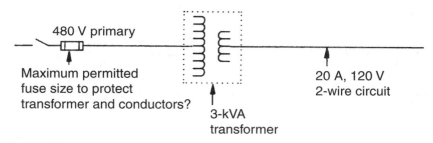

Figure 8.21   A 3-kVA transformer is used to provide a 20-ampere, 120-volt, single-phase circuit with overcurrent protection provided only on the 480-volt primary.

15. Is a fusible disconnect or circuit-breaker on the primary side of a transformer required to be within sight of the transformer when the secondary is protected at not more than 125 percent of secondary full-load current? (Yes or No)

Code reference _____

16. A 75-kVA, 3-phase transformer installation, shown in Figure 8.22, has the 480-volt primary connected delta, and the secondary is connected 120/208-volt, 4-wire wye. The secondary wires are AWG number 3/0, type THW copper terminating in a 200-ampere panelboard with a main circuit-breaker. The length of the secondary circuit to the panelboard is 5 feet. All wire terminations are 75°C rated. The primary wire is AWG number 3, type THW copper. All wires are in conduit. The primary is protected by time-delay fuses located at some distance away in another room. The maximum permitted rating of primary circuit fuses is _____ amperes.

    Code reference _____

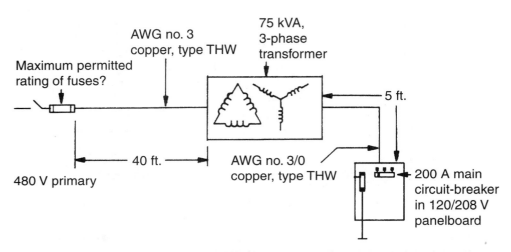

Figure 8.22   A 75-kVA, 3-phase transformer supplies a 200-ampere, 120/208-volt, 3-phase panelboard from a 480-volt supply.

17. A 500-VA control transformer is installed for a motor starter enclosure to provide 120 volts for control from a 480-volt supply. Fuseholders are provided with the transformer to protect the primary. The maximum primary fuse rating permitted is _____ amperes. (See *Section 240-6*, but smaller is permitted.)

    Code reference _____

18. A 230-volt, single-phase, 5-horsepower motor drawing 28 amperes will be operated from a 208-volt feeder with a boost transformer used to raise the voltage. The installation is shown in Figure 8.23. A feeder is tapped for the motor circuit with the tap ending at a transformer primary fusible disconnect. The transformer is rated 120/240-volt primary, 12/24-volt secondary, and 0.75 kVA. The input voltage boost is from 208 to 230 volts. Determine the maximum fuse rating permitted for the primary of the boost transformer for this application.

    Code reference _____

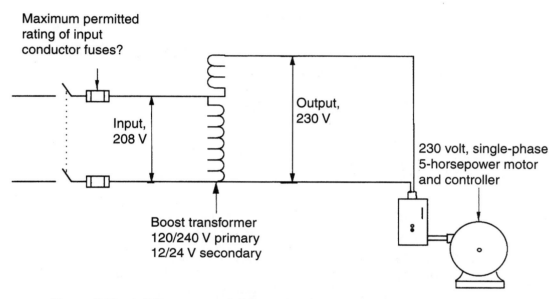

Maximum permitted
rating of input
conductor fuses?

Input,
208 V

Output,
230 V

230 volt, single-phase
5-horsepower motor
and controller

Boost transformer
120/240 V primary
12/24 V secondary

Figure 8.23   A 5-horsepower, 240-volt, single-phase motor is powered from a 208-volt supply through a boost and buck transformer.

19. A Class 2 transformer used for a particular furnace control circuit is designed to be an inherently limited power source with overcurrent protection not required, other than the branch circuit overcurrent device. If the transformer operates with a 120-volt primary and a 24-volt secondary, the maximum permitted rating of the transformer is _____ volt-amperes.

    Code reference _____

20. Transformers used to supply lighting fixtures in a swimming pool shall be listed for such applications, and they shall be of a two-winding type having a grounded metal barrier between the primary and secondary windings. (True or False)

    Code reference _____

21. A 0.5-kVA control transformer, rated 120-volt 2-wire secondary and 240-volt primary, has a secondary full-load current of 4 amperes. The wires are AWG number 14, type THWN copper in conduit. The installation is shown in Figure 8.24. Is the secondary considered adequately protected by 2-ampere, 250-volt fuses placed in the primary control circuit wires? (Yes or No)

Code reference _____

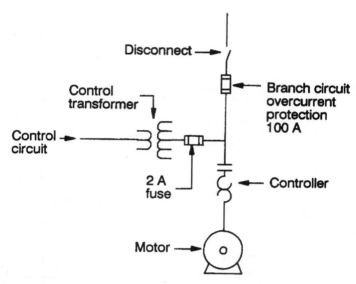

Figure 8.24 A control circuit for an electric motor is tapped to the motor branch circuit with overcurrent protection provided by fuses in the control transformer primary circuit.

22. A 37.5-kVA 3-phase transformer is used to supply a 100-ampere 3-phase lighting panelboard with 120/208-volt 4-wire power. The primary is 480 volts from an ungrounded electrical system. The secondary wires between the transformer and the panelboard are type THWN copper, AWG number 3 in conduit. If the structural metal of the building is effectively grounded, the secondary grounding electrode conductor from the grounded conductor shall be connected to the structural metal at the nearest available point. (True or False)

Code reference _____

23. The grounding electrode conductor of the secondary electrical system of the separately derived system in question 22 shall be bonded to the transformer secondary grounded conductor only at the transformer. (True or False)

    Code reference _____

24. The transformer of question 22 is considered to be a separately derived system. The secondary phase conductors are type THWN copper, AWG number 3. Assume the grounding electrode is the metal frame of the building. The minimum size copper grounding electrode conductor for the transformer installation is AWG number _____.

    Code reference _____

25. Transformers used with electric discharge tubing for outline lighting shall be identified for the use, and shall be limited in rating to _____ volt-amperes.

    Code reference _____

ANSWER SHEET   Name _____

No. 8   TRANSFORMERS

Answer	**Code reference**
1. _____	_____
2. _____	_____
3. _____	_____
4. _____	_____
5. _____	_____
6. _____	_____
7. _____	_____
8. _____	_____
9. _____	_____
10. _____	_____
11. _____	_____
12. _____	_____
13. _____	_____
14. _____	_____
15. _____	_____
16. _____	_____
17. _____	_____
18. _____	_____
19. _____	_____
20. _____	_____
21. _____	_____
22. _____	_____
23. _____	_____
24. _____	_____
25. _____	_____

UNIT 9

Hazardous Location Wiring

OBJECTIVES

Upon completion of this unit, the student will be able to:

- explain the difference between a Division 1 and a Division 2 (or a Zone 0, Zone 1, and Zone 2), Class I hazardous location.
- explain the difference between a Division 1 and a Division 2, Class II hazardous location.
- describe the conditions that constitute a Class III hazardous location.
- explain the function of an explosion-proof enclosure.
- select the proper atmospheric group when given the name of a common flammable vapor or dust.
- explain a wiring method permitted for use in a Class I hazardous location.
- explain a wiring method permitted for use in a Class II, Division 1 and Division 2 hazardous location.
- explain the special bonding requirements when double locknuts are used at an enclosure feeding a circuit in a hazardous location.
- state the minimum number of threads required on a piece of rigid metal conduit to be used in a Class I hazardous location.
- answer wiring installation questions relating to electrical installations in hazardous locations, as described in *Articles 500, 501, 502, 503, 504*, and *505*, including the special occupancies described in *Articles 510, 511, 513, 514, 515*, and *516*.
- state at least four significant changes that occurred from the 1996 to the 1999 Code for *Articles 500, 501, 502, 503, 504, 505, 510, 511, 513, 514, 515*, or *516*.

CODE DISCUSSION

An area is considered hazardous because of the highly flammable nature of a vapor, gas, dust, or solid material that may be easily ignited to cause fire or an explosion. Classification of the type of hazardous material, and the boundaries of the hazardous area, is covered in *Article 500*, with general requirements for wiring in these areas. *Articles 501, 502, 503, 504*, and *505* cover specific wiring requirements for the type of hazardous area involved. *Article 504* covers the installation of intrinsically safe wiring systems in any classified location. Intrinsically safe wiring systems are designed such that even if a fault or a short-circuit occurs, sufficient energy is not available to ignite the hazardous vapor, dust, or material in the surrounding area. *Article 505* describes an alternate method to *Article 501* of classifying Class I hazardous locations into zones, and describes the wiring methods permitted to be used for Zone 0, Zone 1, and Zone 2. *Articles 510* through *516* deal with specific types of occupancies where hazardous conditions exist. These articles bring about uniformity of installations. Many occupancies are unique, and therefore, it is necessary to depend on *Articles 500* through *505* to make decisions about the wiring. There are other National Fire Protection Association (NFPA) documents that may be necessary to determine the wiring requirements or to decide the boundaries of the hazardous area. It may be necessary to involve an expert, such as a registered professional engineer, with knowledge of the process and materials to decide the boundaries of the hazardous area, as well as to determine when a material constitutes a hazard.

Article 500 covers some explanation of the nature of the conditions in a hazardous location as far as electrical equipment is concerned. The types of hazardous vapors, gases, and dusts are categorized into Groups A through G. Flammable vapor and gas can also be categorized into Groups IIA, IIB, and IIC when installed under the zone system of *Article 505*. The markings required on electrical equipment for use in a hazardous location are covered. Hazardous locations are arranged into Class I, Class II, and Class III. Class I hazardous locations are described in *Section 500-7* and *Section 505-9*. Class II locations are described in *Section 500-8,* with Class III locations described in *Section 500-9*.

Article 501 covers the specific requirements and wiring methods permitted in a Class I hazardous location, which is one where flammable vapors or gas are or may be present in sufficient concentration to be ignited. The wiring methods permitted are covered in *Section 501-4,* and some common methods are illustrated in Figure 9.1. The wiring methods permitted in a Class I Division 1 location are threaded rigid metal conduit, threaded intermediate metal conduit, and type MI cable. Some additional cables and raceways are permitted in a Class I Division 2 location. These are covered in *Section 501-4(b)*. The requirements for providing seals in the wiring system are covered in *Section 501-5*. The types of motors permitted are covered in *Section 501-8,* and lighting fixture markings and installation are covered in *Section 501-9*. *Section 501-16* covers the requirements for grounding and bonding, which in some ways are different from the general requirements of *Article 250*. Refer to *Section 500-3(d)* for a caution about preventing sparking during fault conditions by making sure threaded fittings are wrench-tight. This can be prevented by adequate grounding and bonding of the equipment.

Article 502 gives the wiring requirements for a Class II hazardous location where combustible dust is present in the air or may become suspended in the air. The wiring methods permitted are given in *Section 502-4*. It should be noted there are differences in requirements for a Division 1 and a Division 2 hazardous area. Sealing the wiring system is covered in *Section*

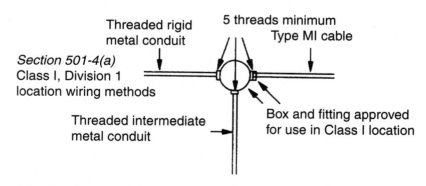

Figure 9.1 Common wiring methods permitted in a Class I Division 1 location are threaded rigid metal conduit, threaded intermediate metal conduit, and type MI cable with approved fittings.

502-5. The type of motors permitted for use in a Class II location and the ventilation of the motor are covered in *Sections 502-8* and *502-9.* Lighting fixture installation is covered in *Section 502-11.* The bonding and grounding requirements are similar to those of the Class I hazardous location and are covered in *Section 502-16.*

Article 503 covers wiring in areas where there will be flammable fibers and flyings. The key here is flammable. A woodworking area where wood flyings that are easily ignitable and collect on surfaces and equipment is a typical example. A sawmill, on the other hand, usually involves wood particles that are heavy in weight and have a high moisture content. This material is generally not considered to be easily ignitable, and thus would not be considered to be a Class III hazardous location. Wiring methods permitted for use in a Class III area are stated in *Section 503-3.* Motors are covered in *Section 503-6,* and lighting fixtures in *Section 503-9.* Grounding and bonding are covered in *Section 503-16.*

Article 504 covers the requirements for the installation of intrinsically safe wiring systems in hazardous classified locations. *Section 504-2* provides definitions important for the application of this article, such as the definition of intrinsically safe circuits. *Article 504-20* states that any wiring method suitable for similar conditions in unclassified locations is permitted to be used as a wiring method for intrinsically safe wiring systems in classified locations. *Section 504-30* specifies separation requirements for intrinsically safe wiring from power and light wiring systems. The minimum separation is 2 inches if the wiring of both systems is fixed in place. There are exceptions, stated in *Section 504-30.* It is necessary to maintain the 2-inch nominal separation everywhere in the building, not just in the classified area. This is illustrated in Figure 9.2. *Section 504-80* specifies that terminals shall be identified to prevent accidental interconnection of normal power circuits and intrinsically safe wiring systems. The intrinsically safe wiring system shall be identified with a permanent label affixed to the wiring at intervals with a minimum spacing of 25 feet. The label shall read "Intrinsic Safety Wiring."

Article 505 provides an alternative method of classifying Class I hazardous locations as well as specifying the wiring methods. This article separates the Class I area into three zones, Zone 0, Zone 1, and Zone 2. As described in *Section 505-9(a),* in a Zone 0 of a Class I location, an ignitable concentration of gas or vapor is present continuously or is normally present for extended periods of time. The rules for wiring in Zone 0 are provided in *Section 505-15(a),* but essentially the only wiring permitted is intrinsically safe wiring, nonincendive circuits, nonconductive optical fiber cable, or similar systems with an energy-limited electrical supply that has been approved for the purpose. *Section 505-9(b)* describes Zone 1 of a Class I location as one where an ignitable concentration of gas or vapor is likely to be present for limited periods of time during normal operation, system breakdown, or repair. In general, a Class I Division 1 location, as described in *Section 500-7(a),* is separated into Zone 0 and Zone 1 when the rules of *Article 505* are applied. The wiring methods permitted in Zone 0 are permitted in Zone 1 as well as the wiring methods permitted for Class I Division 1 locations as described in *Article 501.*

In Zone 2 of a Class I location, an ignitable concentration of gas or vapor is not likely to exist during normal operation, and if an ignitable mixture becomes present, it will only exist for a short period of time. Usually the short-duration ignitable gas or vapor concentration in Zone 2 is the result of an accident or an unexpected system component rupture or failure. For Zone 2, the wiring methods used in more hazardous Class I, Division 1 or Zone 0 or 1 are permitted, as well as the wiring methods described in *Article 501* for a Class I Division 2 location.

Typical equipment intended for use in classified hazardous locations is required to be marked to show the class, group, and operating temperature. But equipment to be installed in a Class I location in accordance with *Article 505* is required to be marked with the class, zone, and symbol indicating the equipment is built to American standards (AEx), the type of protection designation, gas classification group, and temperature classification. An example is shown in *Figure 505-10(b)(1)* of the Code. The types of protection system designations are listed in Code *Table 505-10(b)(1)* and are described later. Temperature classifications are described in *Table 505-10(b)(3)*. The gas groups as applied to *Article 505* are labeled differently than as applied to *Article 501*. As applied to *Article 505,* an ignitable concentration of gas or vapor is labeled Group II. But Group II is subdivided into Group IIA (same as Group D), Group IIB (same as Group C), and Group C (same as Groups A and B).

Different methods of preventing ignition of a flammable vapor in the area are used when wiring is installed according to *Article 505*. Intrinsically safe equipment cannot release enough energy to ignite a flammable vapor, and it would have a protection system designation of **ia** or **ib.** The term **explosion-proof** is not generally used with respect to wiring installed according to *Article 505*. The corresponding term is **flame-proof,** which carries a protection system designation **d.** For a complete discussion of protection system designations, refer to *Section 505-10* in the major code change portion of this unit.

Section 501-1 is the key for wiring in a Class I hazardous location. An area described as Class I, Division 1 or 2 according to *Section 500-7* is to be wired in accordance with the rules of the Code and as modified by *Article 501*. But according to *Section 501-1,* if the class I area is subdivided into zones rather than divisions, then the rules of *Article 505* shall apply. Therefore, wiring in a Class I location is permitted to be selected and installed in accordance with either *Article 501* or *Article 505*. There is an additional requirement if the class location is to be installed according to *Article 505*. A qualified registered professional engineer is required to supervise the definition of zone boundaries, equipment selection, and wiring system selection according to *Section 505-6(a)*.

Article 510 is a short article that states that *Articles 511* through *516* apply to special occupancies where hazardous areas exist. It also states that the general provisions of the Code apply, unless specific requirements are given in the article that apply to the type of occupancy.

Article 511 applies to commercial garages where repair is done to automobiles. The hazardous locations are defined, and the wiring methods permitted in those hazardous areas are

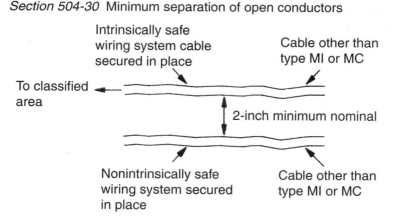

Section 504-30 Minimum separation of open conductors

Figure 9.2   The intrinsically safe wiring shall be separated a minimum distance of 2 inches from lighting, power, and Class I circuits everywhere in the building, not just in the classified location.

described in *Article 501*. Special requirements are placed on the installation of lighting fixtures above a hazardous area. When wiring is installed in these facilities, as much wiring as possible is located outside of the hazardous area to reduce the cost of the installation.

Article 513 covers the requirements for installing wiring in an aircraft hangar. There are similarities in the requirements for a commercial garage. The hazardous area includes a larger portion of room because of the height of aircraft and the location of fuel tanks in the wings. The wiring methods permitted for the hazardous areas are described in *Article 501*.

Article 514 covers the classification and specific wiring requirements for gasoline dispensing and service stations. It is important to be familiar with the type of equipment used in these areas. *Table 514-2* describes the boundaries of the hazardous areas. For example, in the case of a gasoline dispenser, the space within the dispenser is considered to be a Class I Division 1 hazardous location. The space in all directions horizontally from the dispenser and down to the grade level is considered to be a Class I Division 2 location. This Division 2 location extends outward from the dispenser 20 feet in all directions and up to a height of 18 inches above grade level. This is illustrated in Figure 9.3. For a dispensing unit, *Section 514-5* requires that each circuit conductor be wired so that the disconnecting means opens all of the wires, including a neutral conductor.

Article 515 covers wiring installed in hazardous areas around bulk storage plants. This is an area where gasoline or other volatile flammable liquids are stored in tanks where the tanks have a capacity of one carload or more. The boundaries of the hazardous area are described in *Table 515-2*.

Article 516 applies to the wiring installed in the area of spray application, dipping, and coating processes. A paint spray area is a typical application. The boundaries of the hazardous location are described in *Section 516-2*. All dimensions are described in this section and shown in *Figure 516-2(b)(1)*, *Figure 516-2(b)(2)*, and *Figure 516-2(b)(4)* of this *Article 516*. A typical dimension shown on the diagram is **5 ft. R,** which means a radius of five feet. In the case of a dipping process, it is important to make a determination of the point where the object that was dipped is no longer considered to be a vapor source. The definition of a vapor source is found in *Section 516-2(a)(4)*. The limits of the hazardous location for an open tank dipping process are shown in an illustration of an open tank dipping process in *Figure 516-2(b)(5)*. Frequently, it is possible to locate wiring, controls, motors, and lighting fixtures outside of the hazardous area.

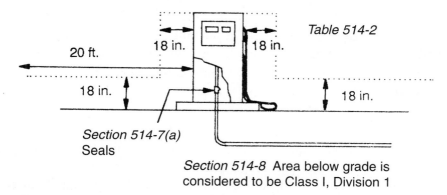

Figure 9.3   The limits of the hazardous locations for a gasoline dispensing unit are found in Code *Table 514-2*.

HAZARDOUS LOCATION WIRING FUNDAMENTALS

The first step for wiring in a hazardous location is to determine the type of material that makes the area hazardous. Next, determine if the type of facility is considered to be a special occupancy, which is described in one of the articles of the Code, for example, a paint spray booth or a service station. Then it will be possible to determine the areas classified as hazardous. Achieving a safe and low-cost wiring system is the result of locating as much electrical equipment as possible out of the hazardous area. In the case of an industrial process, chemicals may be involved that require the assistance of officials or engineers other than the electrical inspector to determine the type of hazard. The National Fire Protection Association has additional publications that deal with most materials that would result in an area being classified as a hazardous location.

Types of Hazardous Locations

The types of hazardous materials are described in *Article 500* or *Article 505* with specific wiring requirements for each class covered in *Articles 501, 502, and 503*.

- Class I   Groups A, B, C, and D, ignitable gas or vapor
- Class I   Groups IIC, IIB, and IIA, ignitable gas or vapor
- Class II   Groups E, F, and G, combustible dust
- Class III Easily ignitable fibers or flyings

Explosion characteristics of different air mixtures of flammable vapors are different; therefore, flammable vapors are separated into Groups A, B, C, and D when the wiring is installed according to *Article 501*, and the same vapors are separated into Groups IIA, IIB, and IIC when the wiring is installed according to *Article 505*. The Code does not provide a list of typical vapors that fall into each group. Instead, the Code gives one typical example of a vapor in each group. For Groups B, C, and D, vapors are placed according to ignition performance tests. The Code uses maximum experimental safe gap (MESG) and minimum ignition current (MIC) ratio to place vapors into the groups. Pressure of the vapor air mixture is an important factor with respect to flammability. One criteria is the maximum experimental safe gap under test conditions between adjacent metal parts. Another criteria is the minimum current to cause ignition. The ignition characteristics of some common materials such as gasoline are well known, and specific requirements can be established for the installation of wiring. For other materials, special training may be needed. A summary of the Groups A, B, C, and D is shown in Table 9.1, along with the MESG and MIC ratio for each group.

Once the hazardous material has been identified, the degree of hazard must be determined. A Division 1 area is one in which the hazardous material is likely to be present. A Division 2 area is one in which, under normal operating conditions, a hazardous material is present in dangerous quantities only under accidental spills or other unusual circumstances. The final condition is the nonhazardous area. The division line between hazardous and nonhazardous may be a physical barrier, or it may simply be a distance limit. Class I zones were discussed earlier in this unit.

Ratings of Equipment

Electrical equipment other than conduit, wire, and some fittings will be marked if suitable for hazardous locations. The equipment shall be marked with the class and the group of hazardous material. The National Electrical Manufacturers Association (NEMA) has established a numbering system for different types of enclosures. Typical motor and control enclosure designations for hazardous locations are listed as follows:

Table 9.1   Group designations for flammable vapor air mixtures that apply when wiring is installed according to *Article 501*.

Group	MESG	MIC ratio	1999 Code vapor list	1996 Code vapor list
A or IIC	NA	NA	Acetylene	Acetylene
B or IIC	Less than or equal to 0.45 mm	Less than or equal to 0.4	Hydrogen	Gases containing more than 30% hydrogen by volume: butadiene, ethylene oxide, propylene oxide, and acrolein
C or IIB	Greater than 0.45 mm, but less than or equal to 0.75 mm	Greater than 0.4, but less than or equal to 0.8	Ethylene	Ethyl, ether, ethylene
D or IIA	Greater than 0.75 mm	Greater than 0.8	Propane	Acetone, ammonia, benzene, butane cyclopropane, ethanol, gasoline, hexane, methanol, methane, natural gas, naphtha, and propane.

- Class I   Hazardous gas or vapor
              NEMA 7 explosion-proof enclosure
- Class II   Flammable dust
              NEMA 9 dust-ignition-proof enclosure
              Motors are permitted to be totally enclosed pipe-ventilated
- Class III   Flammable flyings and fibers
              NEMA 4 dusttight enclosure
              Motors are permitted to be rated totally enclosed, totally enclosed pipe-ventilated, or totally enclosed fan-cooled

Objectives of Explosion-Proof Enclosures

An explosion-proof wiring system is installed in Class I hazardous locations. The wiring system is installed with the assumption that it is impossible to prevent the entry of the hazardous gas or vapor into the wiring system. A flammable mixture of gas or vapor and oxygen may accumulate in a heat-producing or an arc-producing portion of the wiring system, such as a switch. Seals installed in conduit entries to enclosures limit the internal combustion to a small portion of the wiring system, as shown in Figure 9.4.

It is, therefore, assumed that an internal explosion cannot be prevented. The explosion produces extreme internal pressure. The enclosure, seals, conduit, and fittings shall have sufficient strength to withstand the pressure of the internal explosion. This is why rigid metal conduit or intermediate metal conduit (IMC) is required, and why joints must have a minimum of five threads fully engaged. This is also why explosion-proof enclosures are so massive. Therefore, it is absolutely necessary to install all bolts and screws on enclosure covers.

Heat is produced during an internal explosion, and this heat will raise the outside surface temperature of the enclosure. This surface temperature shall be kept below the ignition

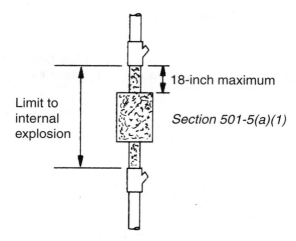

Figure 9.4   Seals confine combustion in the wiring system to a limited portion of the electrical system.

temperature of the gas or vapor on the outside, or an explosion is likely to occur. This is another reason why explosion-proof enclosures are so massive. There must be enough metal mass to absorb the heat of the internal explosion.

The high pressure developed during an internal explosion will cause the ignition gases to eventually escape through joints and threads. The covers and threads shall be tight enough to retard the leakage of products of combustion. If leakage is too fast, the escaping gas will be hot enough to ignite the vapor on the outside. Therefore, slowly escaping gas will cool before it gets to the outside, as shown in Figure 9.5. All threaded joints must be tight, with at least five threads engaged. Machined metal surfaces must be clean with no scratches. A grain of sand on the machined surface of a cover, or a scratch, will allow hot gas to escape, and an external explosion may occur.

Sealing Fittings

The thickness of the compound in a Class I sealing fitting is specified in *Section 501-5(c)(3)*. The minimum thickness of the sealing compound is not permitted to be less than the trade size of the sealing fitting. The minimum thickness under any circumstances is 5/8 inch. If a sealing fitting is made for trade size 3/4-inch conduit, then the minimum sealing

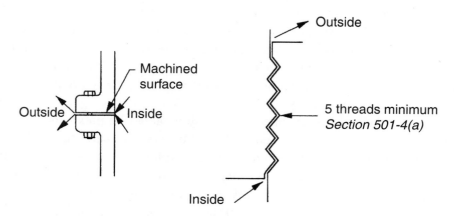

Figure 9.5   Explosion-proof equipment is designed to absorb the heat from escaping combustion vapors through the threads or a metal surface machined to a specified clearance.

compound thickness in the sealing fitting is 3/4 inch. This is illustrated in Figure 9.6. If a 1-inch trade size sealing fitting is used with 3/4-inch conduit, using a reducer at the fitting, the minimum thickness of sealing compound in the fitting is now 1 inch.

It is also important to make sure the conductors are separated so that the sealing compound will flow around each conductor, leaving no voids for vapor to pass. To help in the separation of conductors in the sealing fitting, standard fittings are listed for only 25 percent fill. This means that the total cross-sectional area of the conductors is not permitted to exceed 25 percent of the cross-sectional area of the sealing fitting based on the cross-sectional area of rigid meal conduit as listed in *Table 4* in *Chapter 9*. This would mean that the IMC or rigid metal conduit is permitted to be filled to only 25 percent of the conduit cross-sectional area. There are some ways around this problem. There are sealing fittings rated for 40 percent fill. Another solution is to use an oversized sealing fitting with reducers. The following example will show how to determine the minimum size of sealing fitting required for a particular installation.

Example 9.1   Four AWG number 6, type THWN conductors are run to an explosion-proof enclosure where a sealing fitting is required to be installed within 18 inches of the enclosure. Determine the minimum trade diameter sealing fitting required for this installation.

Answer: Look up the cross-sectional area of AWG number 6, type THWN conductors in *Table 5* in *Chapter 9* and find 0.0507 square inches. Next, multiply the individual conductor cross-sectional area by 4 to get the total cross-sectional area of the conductors (4×0.0507 in^2 = 0.2028 in^2). There is no 25 percent cross-sectional area column in *Table 4* in *Chapter 9*, so divide the cross-sectional area of the wires by 0.25 to get the minimum total cross-sectional area of the fitting required, which in this case is 0.8112 square inches

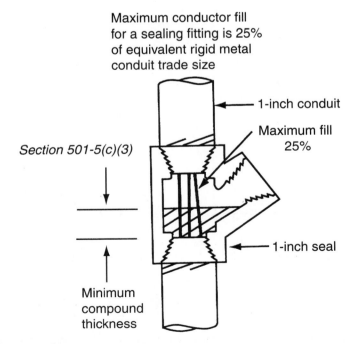

Section 501-5(c)(6)

Maximum conductor fill for a sealing fitting is 25% of equivalent rigid metal conduit trade size

1-inch conduit

Maximum fill 25%

Section 501-5(c)(3)

Minimum compound thickness

1-inch seal

Figure 9.6   Standard Class I sealing fittings are rated for 25 percent fill, and the thickness of the sealing compound is not permitted to be less than 5/8 inch or the trade size of the sealing fitting.

(0.2028 ÷ 0.25 = 0.8112 in²). Now go to the 100 percent cross-sectional area column of the rigid metal conduit section of *Table 4* in *Chapter 9* and find a size that is not smaller than 0.8112 square inches, which is trade size 1 inch.

The conduit is permitted to be sized according to a 40 percent fill. The total cross-sectional area of the conductors in this example is 0.2028 square inches. Look up the minimum trade size conduit from the 40 percent fill column of *Table 4* and find 3/4 inch. It is permitted to use 3/4-inch rigid metal conduit in this case with a 1-inch trade size sealing fitting and an explosion-proof reducer from 1 inch to 3/4 inch.

Equipment Grounding and Bonding

Grounding is extremely important in hazardous locations. Extra care must be taken to make sure the grounding path is of low impedance. An adequate grounding system will conduct current without allowing the enclosure of the equipment to develop a voltage above ground. This is important to prevent the case of the electrical equipment from arcing to ungrounded metal equipment and structural supports. A normally harmless static or small-fault induced arc can cause an explosion in a hazardous location.

It is also important that all threaded connections be made wrench-tight to prevent arcing at the joint in the event the conduit and metal enclosures are needed to conduct fault current. An arc at a loosely joined conduit connection can cause an explosion. Special bonding requirements are also required at certain areas of hazardous locations, as shown in Figure 9.7. Double locknuts are not permitted to serve as the equipment grounding for a circuit in a hazardous location. It is necessary to bond directly from conduit entries into an enclosure even if the enclosure is not in the hazardous area. Code *Sections 501-16, 502-16,* and *503-16* cover the special bonding requirements for wiring to hazardous locations.

Dust-Ignition-Proof Equipment

The principle of installing wiring in a Class II area is to prevent the entry of flammable dust. Dust is a solid material that is heavier than air. Gaskets will prevent the entry of dust

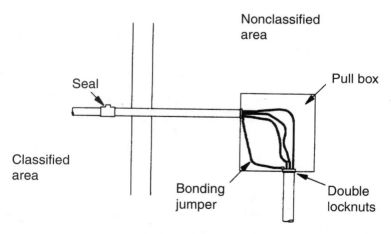

Figure 9.7 **Double locknuts are not considered adequate bonding for conduit systems supplying wiring in a hazardous location, and therefore special bonding is required even in the nonhazardous area.**

particles. In calm air, dust will settle. If dust is prevented from entering heat- and arc-producing parts of the wiring system, a fire or explosion will be prevented.

Enclosures for Class II areas are not required to be of equally heavy construction as Class I enclosures. Seals are not required to isolate arc- and heat-producing components. Sealing is required simply to prevent the entry of dust. The same extra care taken to ensure a good grounding system is required because arcs, due to poor bonding, can set off a general explosion if a fault occurs.

Flammable Fibers and Flyings

The primary problem in these types of facilities is to prevent the entry of heavier-than-air fibers and flyings such as cotton, textile fibers, and dry wood fibers. Enclosures generally have hinged doors with a latch and a gasket around the edges of openings. Enclosures and equipment must be selected to prevent the surface temperature from rising above the ignition temperature of the fibers or flyings that may collect on the enclosure. This temperature limitation must be achieved even when fibers and flyings have accumulated in layers on wiring and equipment.

Special Occupancies

The general wiring requirements for hazardous locations are provided in Code *Articles 501, 502, 503, 504,* and *505.* Later articles provide special wiring requirements and rules for determining the hazardous classified area.

The local electrical inspector, building code, or other organizations or agencies will provide additional requirements for types of hazardous facilities not specifically covered in the Code. Some manufacturers of electrical equipment and wiring materials for installation in hazardous locations publish excellent literature that is helpful when installing wiring in hazardous areas.

MAJOR CHANGES TO THE 1999 CODE

These are the changes to the 1999 *National Electrical Code®* for electrical systems that correspond to the Code sections studied in this unit. The following analysis explains the significance of only the changes from the 1996 to the 1999 Code, and this analysis is not intended to be used in place of the Code. Refer to the actual section of the 1999 Code for the exact wording and meaning of each section discussed. Changes are indicated in the Code with a vertical line in the margin. If material has been deleted or moved to another part of the Code, the location of the deletion is indicated with a dark dot in the margin.

Article 500   **Hazardous (Classified) Locations Class I, II, and III, Divisions 1 and 2**

Article 500 was reorganized to improve clarity and avoid confusion. Most references to Class I, Zone 0, 1, and 2 locations were moved to *Article 505.* The title of *Article 500* was changed to clarify that only Class I, II, and III, Division 1 and 2 hazardous locations are covered, not those areas that are classified as zone locations.

500-1: The scope for this article was changed in order to make *Article 505* stand alone and to avoid confusion. *Articles 500* though *504* cover electrical equipment and wiring in Class I, II, and III, Division 1 and 2 hazardous locations. A new fine print note was added to this section that refers to *Article 505* for the installation of electrical wiring and equipment in those areas that have been classified as Class I, Zone 0, 1, and 2 locations.

500-2: General provisions of the Code apply to classified hazardous locations unless modified in *Articles 501* through *504*. This is not new but was the *Exception* to *Section 500-2* in the previous edition of the Code.

500-3(b): This is a new requirement that all areas classified as a hazardous locations must be documented to indicate the class and division for those areas of a wiring installation to which they apply. This documentation is required to be made available to personnel who design, inspect, install, maintain, or operate electrical equipment in those areas. Clearly indicated directions on a set of plans, or detailed descriptions in a set of specifications, can be considered adequate documentation.

500-3(d): This section specifies that conduit shall be threaded with national pipe threads (NPT) using a **standard conduit die.** This section also requires that conduits shall be made wrench-tight. The fine print note in the previous edition of the Code that dealt with enclosures with metric conduit entries was rewritten and included as mandatary language. If equipment is provided with metric threads, listed NPT adapters are now required to be provided with the equipment or the equipment is required to be identified that it utilizes metric threads, as illustrated in Figure 9.8.

500-4(f)(2): Now there is a definition for nonincendive equipment that was not found in the previous edition of the Code. Under normal operating conditions, this type of equipment cannot cause ignition of any vapor or dust mixture by means of heating or arcing.

500-5(a): This section is used for the determination of groups of flammable vapors for Class I locations. These groups were reworded in order to establish groups that are defined by enforceable language. Vapors and gases in Groups B, C, and D are now placed into groups based on how the materials perform by testing for either the maximum experimental safe gap (MESG) or the minimum ignition current (MIC) ratio. The MESG is the

Sections 500-3(d) and 505-3(b)

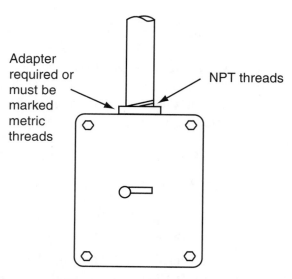

Class I hazardous location

Figure 9.8   If enclosures marked as suitable for installation in Class I, II, or III classified locations are supplied with metric threads, either an adapter converting the national pipe threads (NPT) shall be provided or the equipment shall be clearly labeled that it has metric threads.

maximum gap under explosive conditions that a mixture of hot gases and flames will not ignite the flammable atmosphere surrounding the equipment. A summary of Groups A, B, C, and D is shown in Table 9.1. Though no change of intent occurred in this section, the list of typical vapors found in each group was reduced to only one example for each group, thus making it difficult for personnel in the field to make determinations without referring to other documents. For example, propane is the only Group D vapor listed. Group D vapor commonly encountered by electricians is gasoline, which is not specifically mentioned.

Article 501 **Class I Locations**

501-1: A new paragraph was added to the general requirements for Class I locations that now permits the use of Class I, Zone 0, 1, or 2 equipment to be installed in Class I Division 2 locations. The Class I, Zone 0, 1, or 2 equipment is required to be listed and marked in accordance with *Section 505-10,* designed for the same type of vapor, and have a suitable temperature rating for it to be installed in a Division 2 location.

501-4(a)(1) Exception 3: This is a new exception that permits the use of type ITC instrument tray cable as a fixed wiring method in a Class I Division 1 location provided it is an industrial building with **restricted public access.** For these installations, the type ITC cable is required to have a vaportight continuous aluminum sheath, have an overall jacket of polymeric material, and be listed for use in a Class I Division 1 location. This is a special cable limited to circuits operating at not over 150 volts and with current not exceeding 5 amperes and sizes as small as AWG number 22.

501-5(a)(1): This section covers the seal requirement for installations in Class I Division 1 hazardous locations. As in the previous edition of the Code, conduit entries into enclosures that contain devices that produce arcs, sparks, or high temperatures are required to be sealed. The change in this section is that high temperature is now defined. High temperature is now defined as any temperature exceeding 80 percent of the autoignition temperature of the vapor for which the area is classified.

It is no longer permitted to consider a factory-sealed enclosure as the seal required for adjacent equipment requiring a seal, as illustrated in Figure 9.9. This is not a change of intent, but in the previous edition of the Code, it was not a requirement because it was simply a statement in the fine print note.

501-5(a)(1) Exception: It is not required to seal conduits entering enclosures that contain arc-, spark-, or high temperature–producing devices when the devices are hermetically sealed, immersed in oil, or enclosed within a factory-sealed explosion-proof chamber. This is the same basic provision that was in the previous edition of the Code except that it only applied to conduits of sizes 1 1/2 inches and smaller. This size restriction was removed; therefore, any conduit that enters these types of enclosures is not required to have additional seals, as illustrated in Figure 9.10.

501-5(a)(2): A new provision is placed in the seal requirements for installations in a Class I Division 1 location where the conduits enter pressurized enclosures. If the conduits are part of the pressurized system, seals are not required. However, if the raceway is not part of the pressurized system, seals are required within 18 inches of the enclosure. A new fine print note was also added for this section, indicating that the seal should be placed as close as practical to reduce purging dead air space.

501-5(b)(2): Placement of a conduit seal either inside or outside the boundary of a Class I Division 2 location is required. Now the location of the seal is to be within 10 feet of the

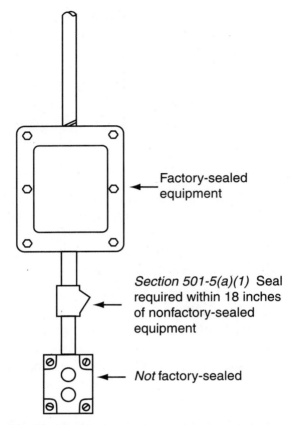

Factory-sealed
equipment

Section 501-5(a)(1) Seal
required within 18 inches
of nonfactory-sealed
equipment

Not factory-sealed

**Figure 9.9 If nonfactory-sealed equipment requiring a seal is installed
adjacent to a piece of factory-sealed equipment, it is necessary to install a
separate seal for the nonfactory-sealed equipment.**

boundary. In the previous edition of the Code, a distance from the boundary to the seal
was not specified except in the case of leaving a Class I Division 1 location.

501-5(b) Exception 3: This exception allows installation of conduits without seals when they
are run between a Class I Division 2 location and an enclosure or area that is unclassified
due to pressurization. In the previous edition of the Code, type Z pressurization was
required, and the exception dealt specifically with the seal requirement only at the boundary, not at an enclosure.

501-5(c)(3): To ensure proper seals in Class I, Division 1 and 2 locations, the Code now
requires the thickness of the sealing compound to be at least the trade size of the sealing
fitting, but not less than 5/8 inch. In the previous edition of the Code, the minimum sealing compound thickness was based on the trade size of the raceway entering the seal. This
is a change only where a larger seal is used with an explosion-proof reducer to a smaller
conduit. If a seal is not rated for 40 percent conductor fill, then an oversized seal is
needed if the conduit is filled to the 40 percent maximum fill.

501-5(c)(6): This section limits the cross-sectional area of conductors in a conduit seal. In the
previous edition of the Code, the maximum fill of a seal was 25 percent of the cross-
sectional area of the equivalent trade diameter conduit for the fitting. Now the 25 percent
fill is based on the cross-sectional area of rigid metal conduit. This would impact hazardous location installations only where IMC is threaded into an approved seal. For a

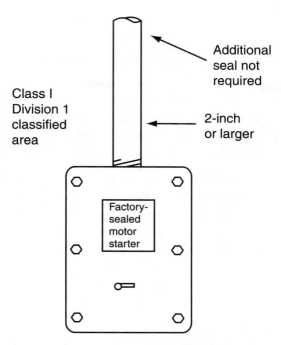

Figure 9.10 It is no longer required to provide an additional seal in the conduit trade size 2-inch and larger within 18 inches of an enclosure where the arc-producing or heat-producing components are in a factory-sealed chamber.

condition such as this, the area of the seal would be based on the equivalent size of rigid metal conduit rather than the actual IMC that is installed.

Question: Assume a 1 1/4-inch trade size sealing fitting is installed with 1 1/4-inch trade size IMC in a Class I Division 1 location. The sealing fitting is only listed for 25 percent fill. Determine the maximum number of AWG number 8, type THW conductors permitted in this conduit run.

Answer: The cross-sectional area of the sealing fitting is required to be taken at not more than the equivalent trade size of rigid metal conduit. From *Table 4, Chapter 9* the area of 1 1/4-inch rigid metal conduit is 1.526 square inches. The conductor cross section is not permitted to exceed 25 percent of this area, which is 0.3815 square inches (0.25 × 1.526 in² = 0.3815 in²). Next look up the cross-sectional area of an AWG number 8, type THW conductor in *Table 5, Chapter 9* and find 0.0556 square inches. Divide the cross-sectional area of the conductor into 25 percent of the area of the sealing fitting to get the maximum number of 6.86, which is rounded down to six conductors. If the cross-sectional area of IMC could have been used, then seven conductors would have been permitted in the sealing fitting.

501-5(d)(1) Exception: For cable seals in a Class I Division 1 location, a new exception was added to the Code. It no longer is necessary to remove the shielding material for cables or the separation of twisted pair cables provided the termination fitting is designed to minimize the entrance of hazardous vapors into the cable.

501-5(e)(1) Exception 1: For cables that enter areas or enclosures that are classified as a Class I Division 2 location, a new exception was added to the Code. Cables are not required to be sealed at the boundary of the unclassified location if the Class I area is under type Z pressurized.

Article 502   **Class II Locations**

502-4(b): This section provides a list of approved wiring materials that are permitted to be installed in a Class II Division 2 location. Type MI cable is now permitted to be installed in cable trays in these classified areas.

502-4(b) Exception 2: This is a new exception in the Code. For Class II Division 2 areas, there are no spacing requirements for type MC cable installed in ladders, ventilated cable trays, or troughs. In the previous edition of the Code, the cables would have to have been spaced by a distance of not less than the largest diameter of two adjacent cables.

502-6(a)(1): The types of enclosures necessary to house switches, motor controllers, and over-current devices that are installed in a Class II Division 1 location have changed. These types of devices that are intended to interrupt current shall be installed in an approved dust-ignition-proof enclosure. In the previous edition of the Code, it was necessary to have the enclosure and the enclosed equipment approved as a complete assembly for the Class II location.

Article 503   **Class III Locations**

No significant changes occurred in this article.

Article 504   **Intrinsically Safe Systems**

504-30: For open wiring, wiring within enclosures, and wiring in raceways, cable trays, or cables, the separation between nonintrinsically safe conductors and intrinsically safe conductors is (1.97 inches) or 50 millimeters. Dimensions are being converted to round metric numbers, and 50 millimeters is approximately 2 inches.

504-60(b): This section cross-references other sections in the Code for the requirements of bonding of metal raceways installed outside of the classified locations that are used for intrinsically safe wiring. *Article 505* is independent of the other articles dealing with classified hazardous location wiring, and specific references to installation techniques in those articles are necessary if they are to apply when the Class I wiring is performed according to *Article 505*.

504-70: When required to be installed, seals for conduits, tubing, and cable are now required to be explosion-proof or flame-proof for intrinsically safe systems. The word **flame-proof** was added to this section because it is common terminology for seals used in *Article 505*.

Article 505   **Class I, Zone 0, 1, and 2 Locations**

Article 505 was completely rewritten to stand separately from *Article 500* and *Article 501* except for specific requirements that are referenced in those articles. There are major differences in the way wiring is installed with respect to requirements in the previous edition of the Code, and as compared to wiring installed according to *Article 501*. In *Article 501*, equipment is required to be explosion-proof, whereas in *Article 505* the corresponding term is flame-proof. A much more elaborate system of marking equipment is used to indicate suitability for installation in an area classified into zones.

505-1: The new scope states that wiring installed according to this article is an **alternative** to wiring installed according to *Article 501*. This statement was in *Section 500-3*.

505-3(a): Each room, section, or area is to be classified individually. Criteria for classification are the properties of the flammable vapors or liquids present and the likelihood that a flammable concentration may be present. Even though rooms or areas may be evaluated separately, the likelihood of a source in one room resulting in a flammable concentration in an adjacent room must be considered. The term **pyrophoric material** is used to describe materials that may ignite spontaneously or emit sparks when struck, which are not to be classified. This is the same requirement found in *Section 500-2.*

505-3(b): Conduits are to be threaded, and this section specifies NPT with a 3/4-inch-per-foot taper like for wiring installed under *Article 501.* Connections are to be made wrench-tight to ensure explosion-proof or flame-proof integrity of the system, but there is no requirement that there be five threads fully engaged, as is the case in *Section 501-4(a).*

505-3(b)(1): The previous edition of the Code required threaded rigid metal conduit and threaded intermediate metal conduit as the wiring method when conduit was used with listed explosion-proof fittings and boxes. Now the only requirement is that the conduit and fittings be listed.

505-3(b)(2): Equipment provided with metric threads is required to be identified as having metric threads or a metric-to-NPT adapter is to be provided. This is the same rule as in *Section 500-3(d),* which applies to wiring installations under *Articles 501, 502, 503,* and *504.*

505-4: The protection techniques for electrical and electronic equipment installed in Class I locations that are classified by zones have changed greatly. Equipment is required to be marked with the type of protection system employed. These changes are new to *Article 505* and are meant to make it more in line with the current international standards. The acceptable protection techniques are flame-proof, purged and pressurized, intrinsic safety, oil immersion, increased safety, encapsulation, and powder filling. These are summarized in *Table 505-10(b)(1).*

Flame-proof protection is identified by the letter **d.** Type **d** equipment is permitted to be installed in Zone 1 areas and is designed to withstand an internal explosion that would cause the ignition of the vapors surrounding the equipment.

Equipment known as **p** is purged and pressurized. This type of equipment reduces the chance of the possible ignition of the surrounding flammable gases through the ventilation of air from an unclassified location. Purged and pressurized equipment is permitted to be installed in Zone 1 and 2 hazardous locations.

Intrinsic safety equipment is identified by two different designations. Type **ia** is intrinsically safe equipment that is intended for use in a Zone 0 location, while type **ib** is intrinsically safe equipment designed for Zone 1 locations. Equipment identified with the designation **ia** or **ib** has an associated apparatus such as a barrier that is connected to the intrinsically safe equipment but is generally not installed in the classified area.

Equipment that is provided with type **n** protection is permitted to be installed in Zone 2 locations. This type of equipment is designed in a manner such that it is not capable of igniting a surrounding hazardous vapor. Furthermore, type **n** equipment is broken down into three subcategories: **nA, nC,** and **nR.** The subcategories represent different design characteristics of the equipment or enclosures.

Protection provided by oil immersion is identified by the letter **o.** Through the use of this type of protection the electrical equipment or part of the equipment is immersed in a nonflammable liquid. Any spark or arc from this type of equipment cannot ignite the surrounding vapors or gases. This type of equipment is intended to be installed in a Zone 1 location.

Type **e** equipment is known as increased safety equipment. This type of equipment can be installed in Zone 1 hazardous locations. This type of equipment is characterized as not producing any arcs or sparks under normal operating conditions. Also, measures are taken to reduce the likelihood of surface temperatures of the equipment being of a level that would ignite the surrounding hazardous vapors or gases.

Encapsulation, type **m,** equipment is approved for Zone 1 locations. With this type of protection, arc-producing contacts are encased in a compound so that any resultant arc can in no way ignite gases or vapors surrounding the enclosure.

Powder-filled protection, type **q,** is very similar to encapsulation except that a filling powder surrounds the spark-producing contacts rather than a rigid compound. Type **q** equipment is for installation in Zone 1 locations.

505-6(a): This is a requirement similar to *Section 500-3* in the previous edition of the Code. A qualified registered professional engineer is required to supervise area classification, equipment selection, and selection of wiring methods. The previous edition of the Code required the wiring to be installed under the supervision of an engineer. Now only the selection of wiring methods is required to be under the supervision of an engineer.

505-6(b): A new precaution is now found in the Code for instances when a structure or location is classified by both the zone and division system. Class I, Zone 0 or Zone 1 classified areas may not border Class I, Division 1 or Division 2 locations. It is permitted to have Class I Zone 2 locations border, but not overlap, Class I Division 2 locations. This is illustrated in Figure 9.11.

505-6(c): Now the Code permits locations that were previously classified as divisions and wired according to *Article 501* to be reclassified by zones and wired according to *Article 505.* This reclassification is only permitted when there is only one vapor type or source present within the hazardous area.

505-7: This was *Section 505-5* in the previous edition of the Code, which provided the group designations for the different gases and vapors. Groups IIC, IIB, and IIA have been redefined utilizing the maximum experimental safe gap (MESG) and the minimum ignition current (MIC) ratio to catagorize the volatile vapors and gases; however, no change of intent occurred. The MESG and the MIC ratio are summarized for the different groups in Table 9.1. Group IIC is equivalent to Groups A and B in *Article 500.* Likewise, Group IIB is still equivalent to Group C, and Group IIA is equivalent to Group D.

505-7(d): A new provision in the Code permits enclosures and other equipment to be listed for a specific vapor or any combination of vapors. Equipment listed in such a manner will be marked accordingly. For example, equipment marked such as IIA + H_2, would be permitted to be installed in an area for any Group IIA vapor or hydrogen. Hydrogen happens to be in Group IIC. If equipment is marked as suitable for Group II, then it is permitted to be installed in an area with a Group IIA, IIB, or IIC vapor.

505-10(b)(1): Equipment that is marked as suitable for installation according the *Article 501,* such as Class I, Group C and D, temperature class T2, is permitted to be also marked according to the requirements in *Section 505-10(b)(2).*

505-10(b)(2): The previous edition of the Code required that equipment installed according to *Article 505* be marked with the class, zone, group, and temperature class. Now the marking must include the designation AEx to indicate it has been built to United States specifications. It also must be marked as to the type of protection system employed. For example, if the heat- or arc-producing components are encapsulated, then it will be marked as protection system **m.** If the equipment is rated as intrinsically safe, it will be marked with protection system **ia** or **ib,** depending upon whether it is suitable for

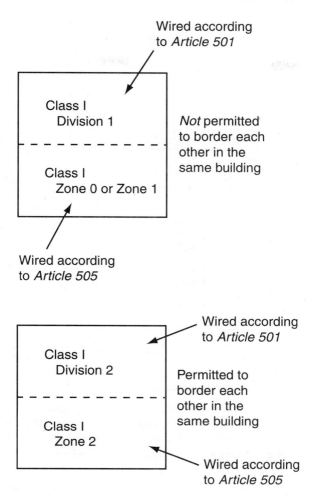

Figure 9.11 **When some Class I wiring in a facility is wired according to *Article 501* and other areas are wired to *Article 505,* Division 1 and Zone 0 and 1 locations are not permitted to border each other, but Division 2 and Zone 2 areas are permitted to border each other.**

installation in Zone 0. Typical markings for equipment suitable to be installed according to the zone system are shown in Figure 9.12.

505-10(b)(3): Temperature classification of zone-rated equipment is contained in this section. A new exception was added that exempts nonheat-producing equipment and heat-producing equipment that has a maximum operating temperature of 100°C from being required to be marked with a temperature rating.

505-10(c): Some form of documentation is now required for all areas that are classified as either Zone 0, 1, or 2 locations. This is a new requirement for this article. The documentation is required to be available to any individual whose is authorized to install, inspect, maintain, design, or operate electrical equipment at or in the hazardous location. Documentation can be on the plans, or it may be contained in the specifications for the installation. A similar requirement was added to *Section 500-3(b)*.

505-15(a): Intrinsically safe wiring is still the only wiring method permitted in a Class I Zone 0 hazardous location; however, the seal requirements for these locations have been changed. Cables installed in Zone 0 are required to be sealed at the first termination point after entering the area. Also, these seals are not required to be flame-proof or explosion-proof.

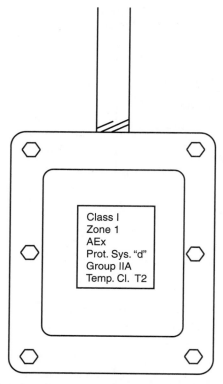

Figure 9.12 **Markings on equipment suitable to be installed in a Class I hazardous location under the zone system shall be marked with the class, zone, United States specifications symbol, protection system designation employed, vapor group, and temperature classification.**

505-15(d): A new requirement exists in the wiring methods of Class I zone-classified areas regarding the use of flame-proof equipment that utilizes flange joints. When such equipment is installed, clearances must be maintained between the flange opening and solid obstacles such as mounting brackets, pipes, other electrical enclosures, and similar solid objects. *Table 505-15(d)* provides the minimum clearances that are required based upon the vapor group present. Clearances less than those provided in *Table 505-15(d)* are permitted if the enclosure is listed for a smaller separation.

505-20: This section is used to select equipment that can be installed in a Class I, Zone 0, 1, or 2 location. In the previous edition of the Code, equipment was required to be **approved.** Now it is required to be **listed.**

505-20(c) Exception 3: Open motors not listed or labeled for being suitable for Zone 2 locations are permitted to be installed as long as they do not have any arc- or spark-producing devices such as brushes or switches.

505-21: This new section concerns the marking and the installation of motors and generators in a Zone 1 location that utilizes type **e,** increased safety, as a method of protection. It is required that these motors have special markings on their nameplates. One possible marking is a specific overload device that is required for the protection of the motor. When such a motor is used, the requirements of *Article 430,* for the setting of the overload device, cannot be used. Some of the other markings that are required on these types of motors include I_N/I_A and t_E. The time factor (t_E) is the time that the motor overload device must operate to prevent the motor from exceeding its limit temperature for the classified area. The current ratio (I_N/I_A is the starting current ratio (initial starting current/rated current).

Article 510   Hazardous (Classified) Locations—Specific

510-2: The general rules of the Code are in *Chapter 1* through *Chapter 4*. *Chapter 5* rules are not considered to be general rules, and now this section specifically states that the provisions of *Articles 500* through *504* also apply to the wiring installed in facilities covered in *Articles 511* through *516*.

Article 511   Commercial Garages, Repair and Storage

No significant changes were made to this article.

Article 513   Aircraft Hangars

513-1: A new scope was added in place of the definition that was in the previous edition of the Code. *Article 513* applies to those buildings in which aircraft containing volatile liquid fuels are stored or maintained. Because the previous edition of the Code did not have a scope, it was not exactly clear to what this article applied.

513-2: A new section that contains definitions was inserted. Definitions of portable and mobile equipment are now provided. Portable electrical equipment is defined as being able to be moved by a single person without any mechanical aids such as wheels. Electrical equipment that can be moved only with mechanical aids is considered to be mobile. These definitions are particularly important when it comes to selecting equipment and materials used to supply portable equipment.

513-4: This section requires that wiring and equipment placed in Class I locations of aircraft hangars must be done in accordance with *Article 501*. This electrical equipment is now permitted to be placed in vaults, ducts, and pits with other services such as water or waste disposal. In the previous edition of the Code, this electrical equipment was prohibited from being placed in these underfloor compartments with services other than compressed air.

513-8: The seal requirements for aircraft hangars are covered in this section. In the previous edition of the Code, it was not clear whether raceways buried in or under the floors of aircraft hangars were classified as Class I locations if the conduit systems did not actually penetrate the hazardous areas. All underfloor raceway systems in aircraft hangars are to be considered Class I locations, and seals are now required and are to be installed according to the rules in *Section 501-5*, as illustrated in Figure 9.13.

Article 514   Gasoline Dispensing and Service Stations

514-6: This new section was added to the Code and is intended to cover the disconnecting means and how it relates to safety of personnel performing maintenance and service on dispensing units. A means to disconnect all sources of power is required to be provided for each dispensing device. The requirement for disconnecting the sources of power to each dispensing device is already covered in *Section 514-5*. Also, this section requires that the sources of feedback from remote pump control wires and other devices are now required to be disconnected when maintenance is being performed.

Article 515   Bulk Storage Plants

515-7: This section is used for cases when gasoline and now any other flammable liquids or gases are dispensed at bulk stations. The requirements of *Article 514* apply to these locations at

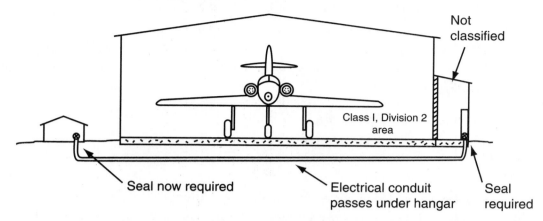

Section 513-8 Seals are required for raceways embedded in a floor or run under a floor where the area above is a Class I area

Not classified

Class I, Division 2 area

Seal now required

Electrical conduit passes under hangar

Seal required

Figure 9.13   Seals are required for a run of raceway embedded in the floor or run under a floor above which is a Class I area.

bulk stations. In the previous edition of the Code, it was not clear whether dispensed flammable materials other than gasoline were required to meet the requirements of *Article 514*.

Article 516   Spray Application, Dipping, and Coating Processes

No significant changes were made to this article.

WORKSHEET NO. 9
HAZARDOUS LOCATION WIRING

These questions are considered important to understanding the application of the *National Electrical Code®* to electrical wiring, and they are questions frequently asked by electricians and electrical inspectors. People working in the electrical trade must continue to study the Code to improve their understanding and ability to apply the Code properly.

DIRECTIONS: Answer the questions and provide the Code reference or references where the necessary information leading to the answer is found. Space is provided for brief notes and calculations. An electronic calculator will be helpful in making calculations. You will keep this worksheet; therefore, you must put the answer and Code reference on the answer sheet as well.

1. The Code states that it is important to exercise more than ordinary care with regard to installation of wiring in a hazardous location that is classified by divisions. (True or False)

    Code reference _____

2. In addition to the class of hazardous location, and sometimes the division and operating temperature, approved equipment shall be marked with the hazardous material _____ to indicate suitability of that equipment to be installed in an area where the particular hazardous vapor or dust is present.

    Code reference _____

3. Is type UF cable permitted to supply a gasoline dispensing pump in the yard of a dwelling in the manner shown in Figure 9.14? (Yes or No)

    Code reference _____

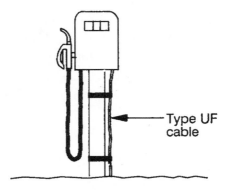

Figure 9.14   Type UF cable is used to supply a gasoline dispenser that is not subject to physical abuse.

4. Threaded conduit joints in a Class I Division 1 location shall be made with at least _____ threads fully engaged.

   Code reference _____

5. Areas within a grain elevator containing electrically operated grain handling equipment where combustible dust is frequently in the air are considered Class _____ Division _____ hazardous locations.

   Code reference _____

6. In the Class I Division 1 hazardous area shown in Figure 9.15, conduit seals shall be installed in each conduit within at least _____ inches of the motor starter where the current-interrupting contacts are not enclosed within a hermetically sealed chamber or immersed in oil.

   Code reference _____

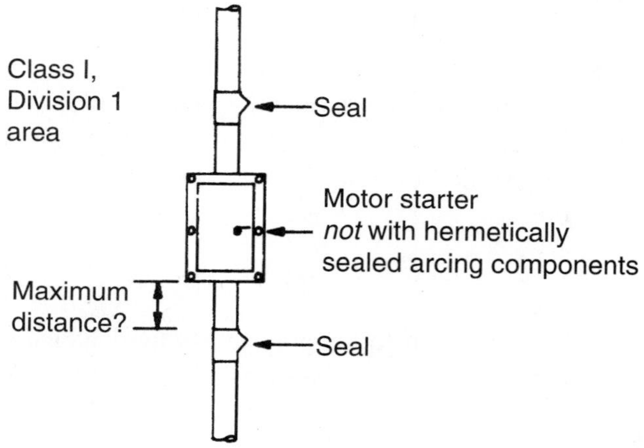

Figure 9.15   The motor starter is installed in a Class I hazardous location.

7. A 3/4-inch rigid metal conduit containing eight AWG number 10, type THWN copper conductors is installed in a Class I Division 1 location. The minimum permitted explosion-proof 25 percent fill seal that can be installed in this raceway is trade size _____-inch.

   Code reference _____

8. The local building code requires that the office area of a commercial garage be under positive ventilation pressure with respect to the outside and the service areas. The ventilation rate for the office area is four air changes per hour. Is type NM cable permitted to be used as a wiring method in the nonhazardous office areas of the commercial garage? (Yes or No)

   Code reference _____

9. Is it permitted to install a nonexplosion-proof, capacitor-start, single-phase motor in a Class I Zone 2 hazardous location of a chemical processing plant? (Yes or No)

   Code reference _____

10. The wiring method that shall be used in a Class I Division 1 hazardous location is type MI cable, threaded rigid metal conduit, or threaded _____.

    Code reference _____

11. A hard usage flexible cord with approved bushed fittings, such as type SJO, is permitted to supply portable electric utilization equipment in a Class I Division 2 hazardous location. (True or False)

    Code reference _____

12. If a gasoline dispenser is supplied with a 120-volt circuit, a disconnecting means such as a switch or a circuit-breaker shall be provided that will open only the ungrounded conductor of the circuit to the dispenser. (True or False)

    Code reference _____

13. Are the double locknuts permitted to serve as bonding at the pull box located in the nonhazardous area for the installation shown in Figure 9.16 where the wiring supplies a circuit in a Class I hazardous area? (Yes or No)

Code reference _____

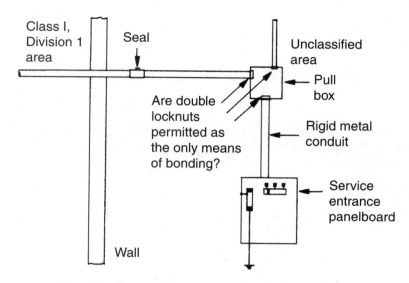

Figure 9.16   Wires for circuits in a hazardous location run through a pull box, which is located in a nonhazardous area.

14. Is a parking garage where no maintenance is performed on vehicles considered to be a Class I hazardous location from the standpoint of wiring? (Yes or No)

Code reference _____

15. Copper electrical wire with any moisture-resistant type of insulation is permitted for use inside conduit supplying a gasoline dispenser where condensed gasoline vapors may collect on the conductors. (True or False)

Code reference _____

16. Refer to the illustration of the wiring to the outdoor gasoline dispenser in Figure 9.17. The Class I Division 2 hazardous area extends outward from the edge of the gasoline dispenser in all directions a minimum distance of _____ feet.

   Code reference _____

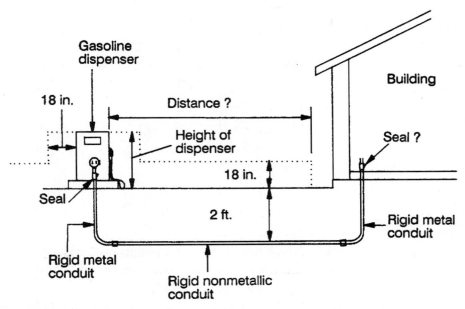

Figure 9.17   Wiring is run underground from the service station to a gasoline dispenser using rigid nonmetallic conduit in direct contact with the earth.

17. Refer to Figure 9.17 for the wiring supplying a gasoline dispenser. A seal is required in the conduit from the gasoline dispenser when it emerges up through the floor in the service station. (True or False)

   Code reference _____

18. Refer to Figure 9.17 for the wiring supplying the gasoline dispenser. Is rigid nonmetallic conduit permitted for the underground portion of the run to the gasoline dispenser where the entire length of rigid nonmetallic conduit is 2 feet below the surface of the ground and the conduit is in direct contact with the earth? (Yes or No)

   Code reference _____

19. A lighting fixture is installed in a Class II Division 1 area of a grain elevator, as shown in Figure 9.18, where a 12-inch stem is threaded into a cover of a box. The box is properly supported and listed for use in a Class II location. Is the installation as shown in Figure 9.18 permitted by the Code, provided the stem is threaded rigid metal conduit or threaded IMC? (Yes or No)

Code reference _____

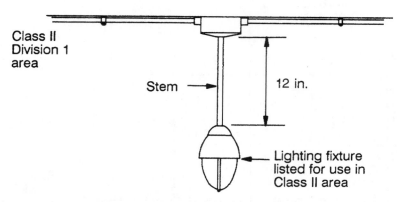

Class II
Division 1
area

Stem →

12 in.

Lighting fixture
listed for use in
Class II area

Figure 9.18   The Class II listed lighting fixture is installed in a Class II hazardous location with a 12-inch rigid conduit stem from the box to the lighting fixture.

20. The mechanical ventilation system provides less than four air changes per hour for the service area of a commercial garage. The service area is considered to be a Class I Division 2 hazardous location from the floor surface up to a distance of _____ inches above the floor.

Code reference _____

21. Standard fluorescent industrial fixtures with exposed lamps located over areas where vehicles are driven are permitted to be installed in a service area of a commercial garage, provided they are mounted at a height of at least 12 feet above the floor. (True or False)

Code reference _____

22. An area used for storage of packaged feeds and grain, containing no equipment for handling the material in the bulk, was considered a Class II Division 2 hazardous location by the authority having jurisdiction. Is type UF cable permitted to be used for the power wiring in this storage area? (Yes or No)

    Code reference _____

23. Is wiring of an intrinsically safe circuit run as AWG number 14 copper, type THWN conductors permitted to be run in the same metal conduit with type THWN power conductors, provided the conduit is outside of the classified hazardous area? (Yes or No)

    Code reference _____

24. A 12-foot horizontal length of rigid metal conduit with one coupling approved for use in a Class II location connects a NEMA 1 enclosure in a nonhazardous area to a dust-ignition-proof enclosure in a Class II hazardous area, as shown in Figure 9.19. Is a sealing fitting required to be installed in the conduit run between the two enclosures? (Yes or No)

    Code reference _____

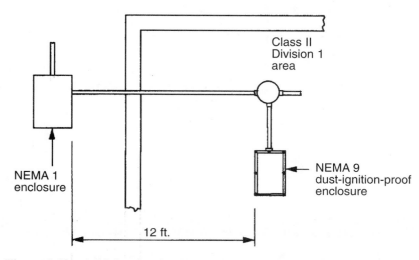

Figure 9.19   A 12-foot horizontal length of rigid metal conduit connects a NEMA 1 enclosure to a NEMA 9 enclosure in a Class II hazardous area.

25. A paint spray booth where the spray equipment cannot operate without the ventilation equipment operating is shown in Figure 9.20. Are the lights shown mounted outside of the spray booth required to be listed for use in a Class I Division 2 hazardous location? (Yes or No)

Code reference _____

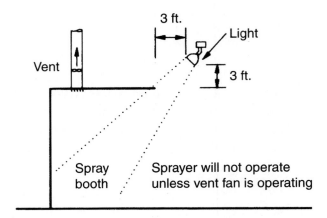

Figure 9.20   The paint sprayer cannot operate unless the ventilation system is operating for this paint spray booth.

ANSWER SHEET   Name _____

No. 9   HAZARDOUS LOCATION WIRING

Answer	**Code reference**
1. _____	_____
2. _____	_____
3. _____	_____
4. _____	_____
5. _____	_____
6. _____	_____
7. _____	_____
8. _____	_____
9. _____	_____
10. _____	_____
11. _____	_____
12. _____	_____
13. _____	_____
14. _____	_____
15. _____	_____
16. _____	_____
17. _____	_____
18. _____	_____
19. _____	_____
20. _____	_____
21. _____	_____
22. _____	_____
23. _____	_____
24. _____	_____
25. _____	_____

UNIT 10

Health Care Facilities

OBJECTIVES

Upon completion of this unit, the student will be able to:

- name two types of patient care areas of a health care facility.

- describe when an anesthetizing location is also considered to be a hazardous location.

- give examples of the types of electrical equipment to be supplied by the life safety branch of the essential electrical system of a health care facility.

- give an example of the type of electrical equipment to be supplied by the critical branch of the essential electrical system of a health care facility.

- explain the purpose of a reference grounding point for a patient care area of a health care facility.

- name a specific location of a health care facility where a receptacle outlet is required to be listed for hospital use.

- describe how equipment grounding is required to be installed from a receptacle outlet at a bed location of a critical care area to the service panel that supplies the receptacle.

- state the minimum number of receptacle outlets required at the bed location of a general care area of a hospital.

- state the minimum number of circuits required at a patient bed location of a critical care area to be supplied by the critical branch of the essential electrical system.

- answer wiring questions about installations in health care facilities from Code *Articles 517* or *660.*

- state at least three significant changes that occurred from the 1996 to the 1999 Code for *Articles 517* or *660.*

CODE DISCUSSION

An essential step to understanding the wiring requirements of a health care facility is knowledge of the definitions at the beginning of *Article 517.* This article addresses the following types of health care facilities: (1) clinics, medical offices, dental offices, and outpatient facilities; (2) nursing homes and residential custodial care facilities; and (3) hospitals. Other articles also are essential to the wiring installation for health care facilities. These are *Article 700,* which covers emergency systems and illumination for building egress, *Article 701,* covering standby electrical systems, and *Article 760* on fire-alarm systems.

Article 517 covers specific electrical wiring requirements for health care facilities that are not covered elsewhere in the Code. *Part A* of this article provides definitions necessary for the understanding and uniform application of the Code to health care facilities. In a multifunction building, the appropriate part of this article shall apply to an area with a specific function.

Part B specifies the wiring methods that can be used in health care facilities. *Section 517-13(b)* specifies that branch circuits supplying patient care areas shall be run in metal raceway such as rigid metal conduit, intermediate metal conduit, electrical metallic tubing, or cable assemblies such as type MI cable, or type MC where the outer metal jacket is an approved grounding means. Grounding and bonding requirements for receptacles and equipment in patient care areas are also covered in *Part B*. If a branch circuit originating from a panelboard of the essential electrical system and a branch circuit originating from the normal power system panelboard serve the same patient care area, the grounding bus of the two panelboards shall be bonded together, as specified in *Section 517-14*. Grounding and bonding in the patient vicinity of a critical care area are covered in *Section 517-19(c)*. All receptacles supplying power at over 100 volts to patient locations shall be grounded with an insulated copper equipment grounding conductor run with the circuit conductors in metal raceway, as stated in *Section 517-13(a)*.

The patient vicinity is defined in *Section 517-3* as the area within 6 feet of the patient bed. In a general care area, a minimum of four receptacles is required to serve the patient bed location. These may be single receptacles, or they may be multiple receptacles on the same yoke. For example, a duplex receptacle would count as two receptacles. The receptacles are required to be listed as hospital grade. A minimum of two branch circuits is required to serve a patient bed location in a general care area. At least one of the branch circuits is required to be supplied by the normal electrical system and one from the emergency system, as stated in *Section 517-18*, unless the room is supplied from two separate emergency power sources. These branch circuit and receptacle requirements are illustrated in Figure 10.1.

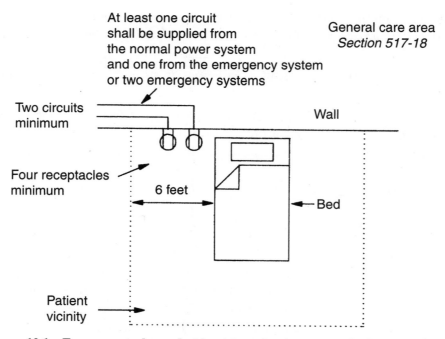

Figure 10.1　Four receptacles and at least two circuits are required to serve a patient bed location of a general care area.

A critical care area is required to have six receptacles at the patient bed location, as shown in Figure 10.2. At least two circuits are required, one of which is required to be supplied from the emergency electrical system, as stated in *Section 517-19*. The receptacles are required to be listed as hospital grade. At least one receptacle at the patient bed location is required to be supplied from the normal power system or other emergency system.

Part C provides the requirements for the types of electrical systems for a health care facility. There is a normal power system providing power for various circuits throughout the health care facility. An emergency electrical system is required in certain types of health care facilities, such as hospitals. This emergency electrical system is permitted to consist of a life safety branch and a critical branch. These systems are illustrated in Figure 10.3. In addition, an equipment system shall be provided in most types of health care facilities. All three of these systems make up the essential electrical systems. For a hospital, these systems are covered in *Section 517-30* through *Section 517-35*. For nursing homes and limited care facilities, these systems are covered in *Section 517-40* through *Section 517-44*.

Part D deals with wiring in inhalation anesthetizing locations. Where flammable anesthetics are administered, the area is considered a classified hazardous location. An isolated power system is required where flammable anesthetics are used. Wiring requirements for other than hazardous anesthetizing locations are found in *Section 517-61(c)*.

Part E deals with X-ray installations. Rating of supply conductors, disconnect, and over-current protection is covered by this part. Wiring of the control circuit and grounding is also covered.

Part F covers requirements for communications, signaling systems, data systems, fire-alarm systems, and systems operating at less than 120 volts. The key issue here is grounding. It is important that a patient not be exposed to hazard through the grounding of electrical equipment.

Isolated power systems are covered in *Part G*. Isolated power systems are required to supply circuits within areas where flammable anesthetizing agents are used. The requirements

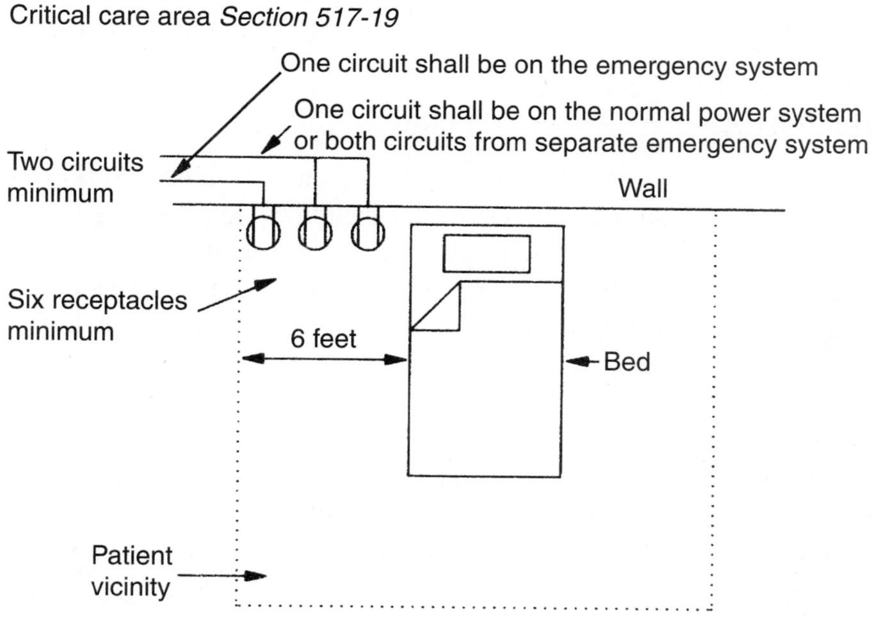

Figure 10.2 Six receptables are required to serve the patient bed location of a critical care area. These are served by a minimum of two branch circuits, one of which is supplied from the emergency electrical system.

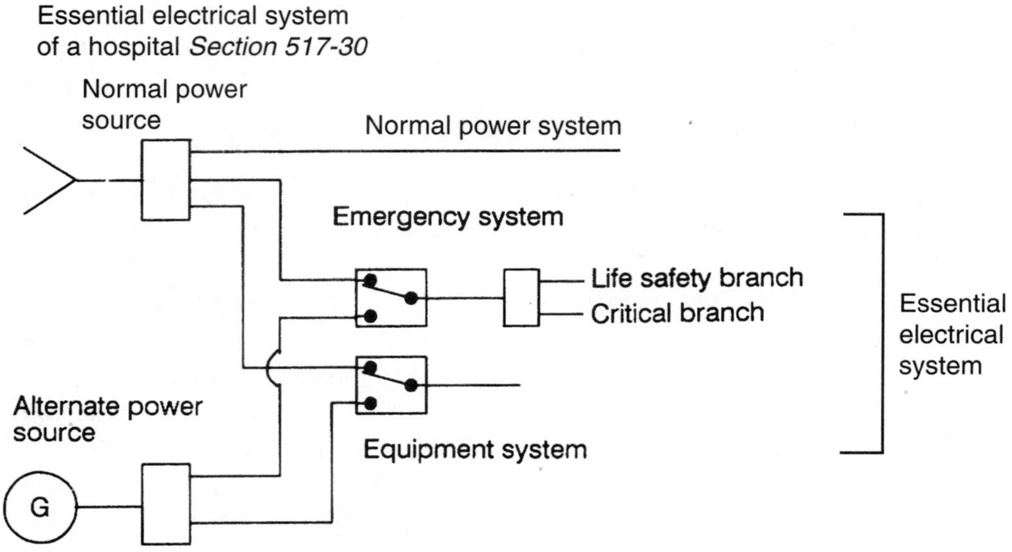

Essential electrical system
of a hospital *Section 517-30*

Figure 10.3   The essential electrical system consists of the emergency system and the equipment system.

which isolated systems are required are found in *Section 517-61(a)(1)*. The installation of the isolated power system is covered in *Section 517-160*.

Article 660 covers the installation of X-ray equipment for nonmedical and nondental use. The article provides specifications for minimum circuit rating and wire size, and the disconnect for the equipment. Wiring of the control system and grounding of the equipment are also covered.

Article 680, Part F covers the wiring to and in the area around therapeutic pools and tubs in health care facilities. Specifications are placed on receptacles and lighting fixtures in the area. Grounding and bonding of equipment are the main emphases of this part. Bonding and grounding of metal parts within 5 feet of the inside edge of the hydromassage unit are important for safety. The permitted methods of grounding and bonding are given in *Section 680-62*.

WIRING IN HOSPITALS

Several very important issues deal with wiring in health care facilities and hospitals in particular. Reliability of electrical supply is necessary for equipment and lighting needed for human life support. Lighting and equipment are needed for egress from the building in case of an emergency. Special measures must be taken to ensure that equipment in a patient area is grounded in such a way that differences in voltage will not be present between equipment such that a hazard to the patient will be created.

Definition of Areas Within a Hospital

Patient vicinity: The area within 6 feet horizontally from the perimeter of the bed, and within 7 1/2 feet above the floor. Figure 10.1 and Figure 10.2 illustrate the patient vicinity.

Patient bed location: The intended location of the patient bed, or the procedure table of a critical care area.

General care area: An area where the patient is not generally connected to electrical equipment. If so, the connection is basically external to the body, and there is no apparent hazard of electrical current affecting life-supporting organs of the body.

Critical care area: This is an area where a patient may be put in intimate contact with electrical equipment that may produce a real hazard of electrical shock to essential body organs, or where the reliability of the equipment is necessary for life support. See the Code for specific locations that come within this definition.

Wet location: An area that is made intentionally wet for some specific purpose. Wet locations are covered in *Section 517-20*. Refer to *Article 680, Part F* for installation requirements for therapeutic pools and tubs in health care facilities.

Anesthetizing location: An area intended for administration of flammable or nonflammable inhalation anesthetic agents in the course of examination or treatment.

Essential electrical system: This electrical system is required to have an alternate source of power. The emergency electrical system and the equipment electrical system are part of the essential electrical system. The building electrical system is made up of the normal power system and the essential electrical system. This system includes lighting circuits and equipment considered necessary for minimal operation and life safety. Power is provided for the life safety branch and the critical branch. See Figure 10.3. There is usually a delay before power is provided to the equipment systems branch.

Life safety branch: This system provides light and power for the emergency systems of *Article 700,* such as lighting for egress, exit signs, and alarm and communications systems. This is covered in *Section 517-32* and is illustrated in Figure 10.3.

Critical branch: This system supplies power in selected areas for illumination and receptacles considered essential for protection of life. For hospitals, specific locations are specified in *Section 517-33*. A patient bed location in a critical care area shall have at least one branch circuit supplied from the critical branch, as shown in Figure 10.3.

Hospital Receptacle Requirements

Hospital grade receptacles are required to be installed in patient care areas, above the hazardous portion of an anesthetizing location, and in certain other locations. These devices are usually identified by a visible green dot, as shown in Figure 10.4. Some devices are made of a clear material to allow employees to determine visually if an electrical malfunction occurs. Required hospital grade receptacles for the general care patient area are found in *Section 517-18(b),* for the critical care area in *Section 517-19(b),* and listed for hospital use above hazardous anesthetizing locations in *Section 517-61(b)(5).*

Receptacles with equipment grounding terminals insulated from the yoke are permitted to be used in hospitals for such devices as electronic equipment in which electrical noise may be a problem. These receptacle outlets with the insulated grounding terminals are required to be distinctively identified from the front. They have an orange triangle visible on the receptacle, as shown in Figure 10.4, *Sections 410-56(c), 230-146(d),* and *517-16*. They shall also be identified as hospital grade.

In the case of a critical care area, there shall be at least one branch circuit from the emergency system, and that receptacle or receptacles shall be identified. This is necessary to make sure essential equipment is attached to an emergency circuit if normal power is lost. It is also required that the location of the branch circuit overcurrent device panelboard be indicated so the device can be reset quickly if necessary to maintain operation of life-saving equipment. These requirements are found in *Section 517-19(a).*

Grounding in Hospitals

Proper installation of the equipment grounding system, particularly in hospitals, is extremely important. The goal, especially in critical care areas, is to prevent any two pieces of

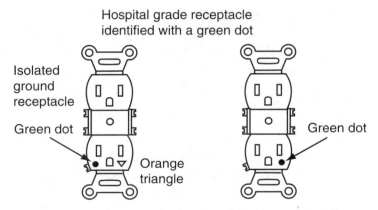

Figure 10.4   A receptacle identified as hospital grade may have the equipment grounding terminal insulated from the yoke of the receptacle and is identified by a green dot and an orange triangle.

metal equipment that may be contacted simultaneously by a person from developing a voltage difference harmful to a patient. Receptacles and fixed equipment in patient care areas shall be grounded with an insulated copper equipment grounding conductor. Several types of grounding points are discussed in the Code and are defined as follows:

- Patient equipment grounding point: This is a grounding bus with plug jacks available for the grounding of equipment to be operated in the patient bed location and listed for the purpose. The patient equipment grounding point is not required, as stated in *Section 517-19(c)*. A patient equipment grounding point is illustrated in Figure 10.5
- Reference grounding point: This is a terminal bus that is a convenient collecting point for equipment grounding conductors and bonding wires. A reference grounding point is required for circuits serving a critical care area. This requirement is found in *Section 517-19(b)*. A reference grounding point is illustrated in Figure 10.6. A reference grounding point is permitted to be separate from but is required to be bonded to the equipment grounding bus of the essential electrical system branch circuit panelboard.

The patient equipment grounding point and the reference grounding point are frequently the same point for critical care areas. When the room is large, these grounding and bonding points may be separated. A reference grounding point is required to be installed. It may be the

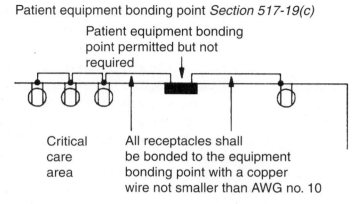

Figure 10.5   If a patient equipment grounding point is provided in a critical care area, the receptacle outlets are required to be bonded to that point.

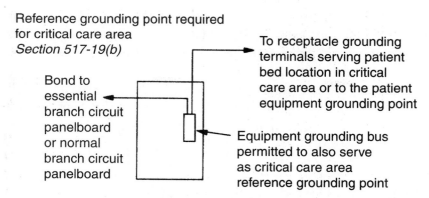

Figure 10.6   A reference grounding point is required to be provided for circuits supplying the patient vicinity of a critical care area.

equipment grounding bus of the electrical panel serving the area. A patient equipment grounding point is permitted but not required. If installed, the patient equipment grounding point shall be connected directly to the reference grounding point by means of a continuous length of insulated copper conductor. The reference grounding point of the room is bonded to the panelboard grounding bus supplying power to the room. Circuits in the critical care patient area are supplied from a panelboard of the normal power system and a panelboard of the essential electrical system. The equipment grounding buses of each of these panelboards serving the patient care area are required to be bonded together by means of a continuous length of insulated copper conductor, as required by *Section 517-14*. This is illustrated in Figure 10.7.

A patient bed location is permitted to have receptacles supplied from both the emergency and the normal power panelboards. The equipment grounding buses of these panelboards shall be bonded together to make sure a difference in voltage between equipment at the bed location cannot develop to a sufficient level to be harmful to the patient.

Hospital Required Electrical Systems

Essential electrical systems must be connected to the normal power system and the standby power source through a transfer switch. The essential electrical system may consist of a separate critical branch and a life safety branch. If there is a life safety branch, it is not permitted to occupy the same raceway or cable with other wiring. This requirement is found in *Section 517-30(c)(1)*. The critical branch and the life safety branch are not permitted to share

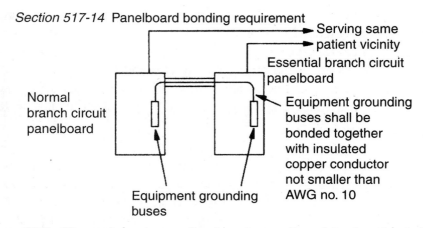

Figure 10.7   The equipment grounding bus of a panelboard serving circuits in a patient vicinity shall be bonded.

the same enclosures, raceways, or cables. The emergency wiring system in a hospital shall be in metal raceway. The types of loads permitted to be connected to the life safety branch of a hospital are limited to egress illumination, exit signs, alarm systems, emergency communication systems, and illumination at the generator location.

Power circuits within an anesthetizing location classified as a hazardous location shall be isolated from the normal distribution system. The lower 5 feet of an anesthetizing room where flammable agents are used shall be considered to be a Class I Division 1 hazardous location, as stated in *Section 517-60(a)(1)*.

MAJOR CHANGES TO THE 1999 CODE

These are the changes to the 1999 *National Electrical Code®* for electrical systems that correspond to the Code sections studied in this unit. The following analysis explains the significance of only the changes from the 1996 to the 1999 Code, and this analysis is not intended to be used in place of the Code. Refer to the actual section of the 1999 Code for the exact wording and meaning of each section discussed. Changes are indicated in the Code with a vertical line in the margin. If material has been deleted or moved to another part of the Code, the location of the deletion is indicated with a dark dot in the margin.

Article 517   **Health Care Facilities**

517-3: The definition of **emergency systems** was modified to make it clear that the emergency system requirements in *Article 700* can be superseded by the requirements in *Article 517*.

517-13(a) Exception 3: In the previous edition of the Code, this exception permitted lighting fixtures that were more than 7 1/2 feet above the floor to be grounded by means other than an insulated equipment grounding conductor. Now switches located outside the patient vicinity, which is defined as within 6 feet of the bed location, are not required to be grounded with an insulated equipment grounding conductor.

517-18(a) Exception 3: A general care area patient bed location is required to be supplied by at least one circuit from the normal electrical system and at least one circuit derived from the emergency electrical system. Now both circuits are permitted to be from the emergency system if the two emergency circuits originate from separate transfer switches, as illustrated in Figure 10.8.

Figure 517-30(b): This figure shows a line diagram of a large electrical system for a hospital. To clarify the intent of *Section 517-31,* the phrase **emergency system** was added to the figure after the critical branch and the life safety branch. Power to these branches of the emergency system is required to be restored within 10 seconds after the interruption of the normal source. *Section 517-30(a)(1)* describes the essential electrical system as being comprised of the emergency system and the equipment system. *Section 517-30(a)(2)* describes the emergency system as being comprised of the life safety branch and the critical branch. The term **emergency system** should also have been added to *Figure 517-30(a)*.

517-30(b)(5): Loads not specifically named in this section are permitted to be served by generating equipment provided the loads are supplied by a separate transfer switch not supplying essential electrical system loads. These loads are not permitted to be transferred to the generator if they will likely cause an overload. If an overload should occur, these loads are required to be the first loads disconnected. The life safety branch is now recognized in this section, correcting an oversight in the previous edition of the Code.

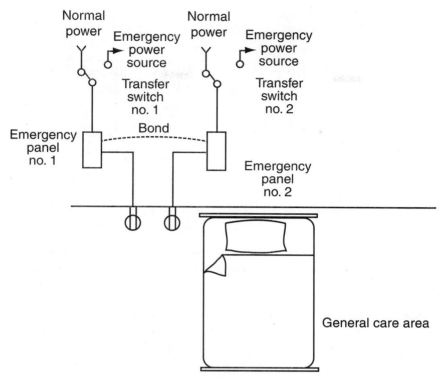

Section 517-18(a) Exception 3 Both required circuits
permitted from emergency system if supplied
through separate transfer switches

**Figure 10.8   The two required bed location circuits in a general care area are permitted to
be from the emergency system if the circuits are supplied by separate transfer switches.**

517-30(b)(6): This new provision makes it clear that if in the case of a hospital there are
contiguous facilities, all are permitted to be supplied from the same alternate power
source.

517-30(c)(3) Exception 5: This was *Exception 6* in the previous edition of the Code. Flexible
metal raceways and cable assemblies are permitted to be used in **listed** prefabricated
medical headwalls and now in **listed office furnishings.**

517-33(a): This section lists areas and equipment that must be supplied from the critical branch
of the hospital emergency electrical system. The only change in this list is in item *(9)*.
The critical branch of the emergency system is now permitted to supply power to all
single-phase fractional horsepower motors that are necessary for the operation of the facility. In the previous edition of the Code, only fractional horsepower exhaust fans that were
interlocked with 3-phase motors were permitted to be connected to the critical branch.

517-33(c): All receptacles or their faceplates that are supplied by the critical branch must have
distinctive color markings. This new requirement is designed to make these receptacles
easily recognizable from all other receptacles. Methods of color marking are shown in
Figure 10.9.

517-34(a): The previous edition of the Code required all compressed air systems serving medical functions to be connected to an alternate power source by delayed automatic connection. Now compressed air systems are permitted to be placed on the critical branch,
which would mean automatic restart in 10 seconds.

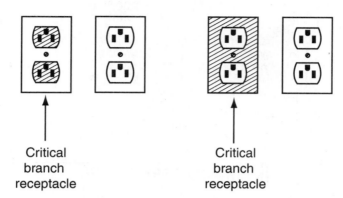

Section 517-33(c) Critical branch receptacle or
faceplate distinctive color marking required

Critical
branch
receptacle

Critical
branch
receptacle

**Figure 10.9   Receptacles supplied from the critical branch of the emergency system shall have a
distinctive color marking, or their faceplates shall have a distinctive color marking.**

517-40(c): Nursing homes and limited care facilities may have their essential systems supplied
by an essential system from a hospital. The change in this section is that it now covers
nursing homes and limited care facilities that are on the same site as well as those that
are contiguous.

517-63(a): It is now required that at least one battery-powered emergency lighting unit be
installed in all anesthetizing locations. These emergency lighting units are required to be
installed according to *Section 700-12(e)*. In the previous edition of the Code, the lighting
requirements for these locations were unclear.

517-160(a)(4): In the previous edition of the Code, an isolated branch circuit was only permitted to serve an anesthetizing locations and no other locations. Now an isolation transformer is permitted to serve only a single operating room. However, the Code does
consider a anesthetic induction room as part of the operating room. In cases where an
anesthetic induction room serves two or more operating rooms, an isolation transformer
for one of the operating rooms is permitted to serve this area.

517-160(a)(5): For health care facilities, isolated power systems utilize the colors orange,
brown, and yellow. For a 125-volt circuit, the colors are orange and brown. To maintain
consistency for all isolated power systems that supply 125-volt, 15- and 20-ampere
receptacles, the orange conductor is to be connected to the terminal identified for the
grounded conductor, as indicated in Figure 10.10.

Article 660   X-Ray Equipment

No significant changes were made to this article.

Section 517-160(a)(5) For 125-volt, 15- or 20-ampere
receptacles on isolated power systems,
connect the orange wire to the silver terminal

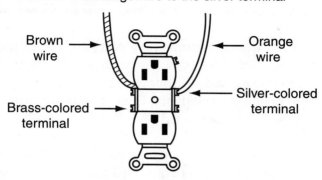

**Figure 10.10   The orange conductor of an isolated 125-volt circuit connected to a
15- or 20-ampere receptacle shall be connected to the silver-colored terminal
intended for connection of the grounded conductor in a normal power system.**

WORKSHEET NO. 10
HEALTH CARE FACILITIES

These questions are considered important to understanding the application of the *National Electrical Code®* to electrical wiring, and they are questions frequently asked by electricians and electrical inspectors. People working in the electrical trade must continue to study the Code to improve their understanding and ability to apply the Code properly.

DIRECTIONS: Answer the questions and provide the Code reference or references where the necessary information leading to the answer is found. Space is provided for brief notes and calculations. An electronic calculator will be helpful in making calculations. You will keep this worksheet; therefore, you must put the answer and Code reference on the answer sheet as well.

1. The essential electrical system of a hospital is made up of an equipment system and an emergency system. The critical branch and the _____ are supplied power by the emergency system.

   Code reference _____

2. Receptacles in patient care areas of hospitals shall be grounded with an insulated copper equipment grounding conductor even when run in metal raceway. (True or False)

   Code reference _____

3. One duplex receptacle at the patient bed location of a critical care area of a hospital is supplied from the emergency power system. The other receptacle outlets at the patient bed location are supplied from the normal power system, as illustrated in Figure 10.11. Is the emergency circuit supplying the identified emergency receptacle at the bed location permitted to serve an identified receptacle not at the bed location? (Yes or No)

   Code reference _____

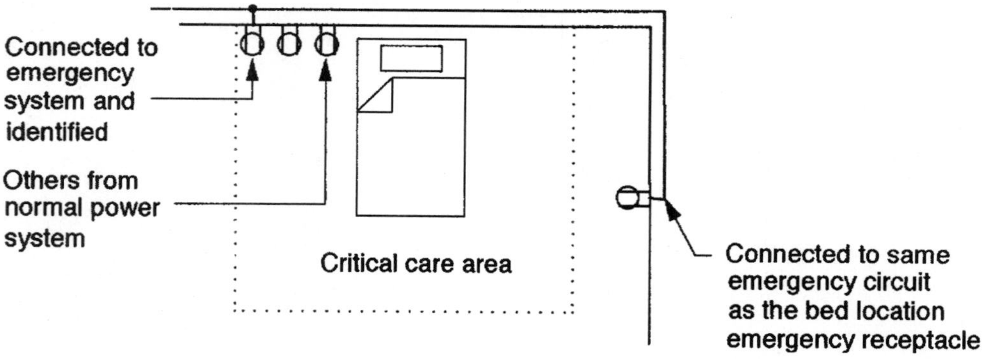

Connected to emergency system and identified

Others from normal power system

Critical care area

Connected to same emergency circuit as the bed location emergency receptacle

Figure 10.11   Is the emergency branch circuit that supplies the emergency receptacle at a critical care area patient bed location permitted to serve a receptacle not in the bed location?

4. A hospital with an essential electrical system is required to have at least two independent sources of power. (True or False)

   Code reference _____

5. Receptacle outlets with grounding terminals isolated from the metal yoke are permitted to be installed at the patient bed location in critical care areas of a hospital. (True or False)

   Code reference _____

6. For a general care area in a hospital, each patient bed location is required to be provided with two branch circuits, one of which shall originate from the emergency electrical system as illustrated in Figure 10.12. (True or False)

   Code reference _____

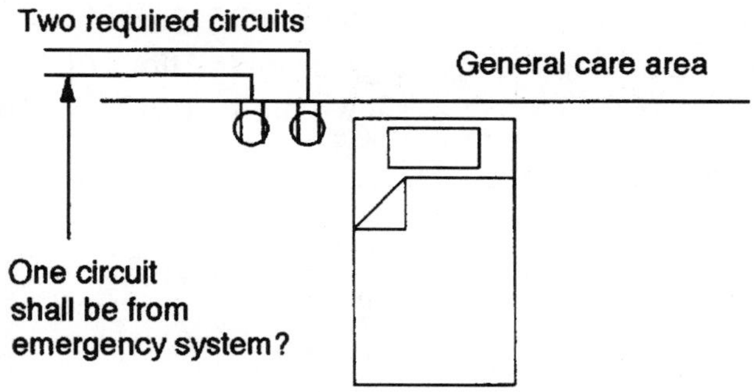

Figure 10.12   In a general care area, is one circuit at the patient bed location required to be supplied from the emergency electrical system?

7. Receptacles at the bed location of a critical care area supplied from the emergency system shall be provided with identification to indicate they are supplied from the emergency system. (True or False)

   Code reference _____

8. The six receptacle outlets required at the patient bed location in a critical care area of a hospital shall be permitted to be grounded to a patient equipment grounding point. (True or False)

   Code reference _____

9. Other than in bathrooms, or for therapeutic equipment, or near hydro-massage bathtubs, and where power interruption cannot be tolerated, areas of a hospital where ground-fault circuit-interrupters are required for electrical equipment supplied from 15- and 20-ampere, 125-volt receptacles are _____.

    Code reference _____

10. In an anesthetizing room where flammable inhalation anesthetics are employed, the portion of the room up to a level of _____ feet above the floor shall be considered a Class I Division 1 hazardous location.

    Code reference _____

11. A general care patient area of a hospital has the two bed location circuits served by a separate emergency system each through a separate transfer switch. Is it required to also supply an additional circuit to the bed location from the normal power system? (Yes or No)

    Code reference _____

12. An industrial fluorescent fixture with exposed lamps is permitted to be installed in the area of a therapeutic tub of a health care facility, provided the minimum mounting height from the floor to the bottom of the fixture is not less than 8 feet as shown in Figure 10.13. (True or False)

    Code reference _____

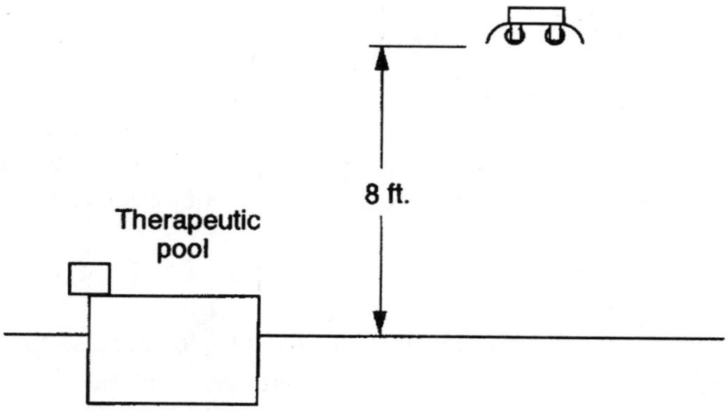

8 ft.

Therapeutic
pool

Figure 10.13   Is an industrial fluorescent fixture permitted to be installed in the area of a therapeutic pool if it is mounted 8 feet above the floor?

13. An X-ray machine is to be permanently installed in a veterinary hospital. The supply conductors and the X-ray equipment overcurrent protection rating is not permitted to be less than _____ percent of the long-term rating of the equipment or 50 percent of the momentary rating of the equipment, whichever is larger.

    Code reference _____

14. The lighted exit signs in a hospital are part of the life safety branch and are automatically transferred to the generator if normal power is lost. The exit signs are not powered from two power sources. The wiring supplying the exit signs is not permitted to share the same conduit with normal lighting and power circuits. (True or False)

    Code reference _____

15. Illumination of the means of egress in corridors, passageways, stairways, landings, and exit doors and all ways of approach to exits of hospitals shall have the source of electrical supply automatically restored within _____ seconds after interruption of the normal source.

    Code reference _____

16. The wiring in an other-than-hazardous inhalation anesthetizing location is permitted to be installed in electrical metallic tubing. (True or False)

    Code reference _____

17. A hospital has two separate operating rooms with one induction room. The minimum number of isolation transformers required to serve these operating facilities is _____.

    Code reference _____

18. A patient equipment grounding point, illustrated in Figure 10.14, shall be provided for each critical care patient vicinity. (True or False)

    Code reference _____

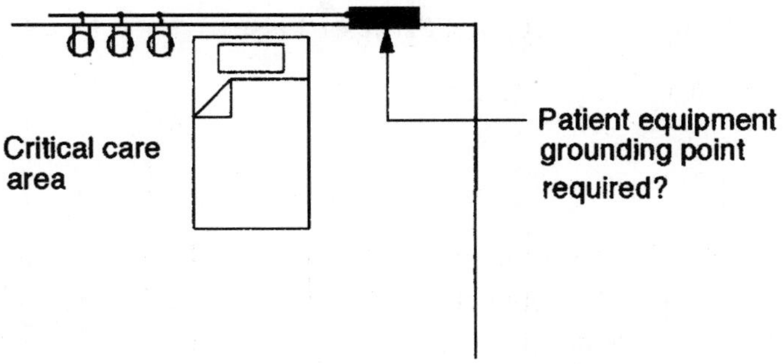

Figure 10.14   Is a patient equipment grounding point required to be installed in a critical care area?

19. Equipment systems of a hospital are permitted to be powered by the same standby generator as the essential electrical system, as well as being powered by the normal power supply. (True or False)

    Code reference _____

20. An intensive care unit of a hospital is considered to be a critical care area. (True or False)

    Code reference _____

21. The isolated power system within an inhalation anesthetizing area where flammable inhalation anesthetics are used is not permitted to be powered by a battery system. (True or False)

    Code reference _____

22. An extended care facility for the elderly has a life safety branch of an essential electrical system. Power shall be restored to the life safety branch in not more than _____ seconds if normal power is lost.

    Code reference _____

23. In a patient care area, type AC cable is permitted to be used to supply electrical outlets and equipment, provided it contains an insulated equipment grounding conductor and the external metal sheath of the cable is approved as an equipment grounding means. (True or False)

    Code reference _____

24. The essential electrical system of a hospital is permitted to be powered only from a standby generator, which is capable of supplying only part of the hospital load, driven by some form of prime mover and not connected to the utility power supply. (True or False)

    Code reference _____

25. The metal conduit for conductors of an isolated power system operating at more than 10 volts between conductors shall be grounded to the building equipment grounding system. (True or False)

    Code reference _____

ANSWER SHEET   Name _____

No. 10   HEALTH CARE FACILITIES

Answer   **Code reference**

1. _____   _____
2. _____   _____
3. _____   _____
4. _____   _____
5. _____   _____
6. _____   _____
7. _____   _____
8. _____   _____
9. _____   _____
10. _____   _____
11. _____   _____
12. _____   _____
13. _____   _____
14. _____   _____
15. _____   _____
16. _____   _____
17. _____   _____
18. _____   _____
19. _____   _____
20. _____   _____
21. _____   _____
22. _____   _____
23. _____   _____
24. _____   _____
25. _____   _____

UNIT 11

Emergency and Alternate Power Systems

OBJECTIVES

Upon completion of this unit, the student will be able to:

- determine the minimum size conductor permitted to connect a standby generator to a wiring system.
- explain the purpose of transfer equipment for a standby power system.
- explain ventilation and insulating needs for battery rooms.
- define terms used in the Code to describe solar photovoltaic systems.
- name the types of equipment and circuits that make up the emergency electrical system of a typical commercial building.
- explain how the wiring is to be run for emergency electrical circuits.
- explain the type of electrical equipment and circuits powered by a legally required standby electrical power system.
- describe a typical application where there is an optional standby electric power system.
- define an interconnected electric power production source.
- explain the difference between a supervised and an unsupervised fire-alarm system.
- explain how the wires are to be attached to a fire-alarm pull station.
- answer typical wiring installation questions from *Articles 445, 480, 690, 695, 700, 701, 702, 705,* and *760.*
- state at least three changes that occurred from the 1996 to the 1999 Code for *Articles 445, 480, 690, 695, 700, 701, 702, 705,* or *760.*

CODE DISCUSSION

Alternate electric power-producing systems are discussed in this unit, as well as the Code articles that cover their installation. These systems may be legally required for some buildings, such as a hospital, or they may be optional. Emergency electrical systems are discussed in this unit. The most simple emergency system consists of exit signs and lighting for building evacuation. Other buildings may have more extensive requirements. The electrical Code covers the installation of these systems, while other codes cover what systems and equipment are required in a particular building. The installation of fire-alarm systems is covered in this unit.

Article 445 covers the requirements for generators to be connected to wiring systems. The markings for generators are covered in *Section 445-3*. Overcurrent protection for the generator and the minimum ampacity of conductors running to the generator are also covered.

Article 480 deals with stationary storage battery installations. The main emphasis of this article is the rigid mounting of the batteries, ventilation and insulation of conductors and terminals, and clearances from the batteries to other live parts or grounded surfaces.

Article 690 applies to solar photovoltaic electric power systems including the dc-to-ac inverter, controllers, sun tracking system, and wiring of components in the system. Several important definitions are covered in *Section 690-2*. A solar photovoltaic cell is generally constructed from a silicon crystal, part of which has been treated with a material that gives it an excess of electrons. The other part is treated with a different material that gives the silicon an excess of positive charges known as holes. These two materials are joined to form a flat silicon sandwich. When sunlight strikes the photovoltaic cell, electrons are dislodged from the silicon atoms and migrate from one side of the sandwich to the other. The result is a negative charge on one-half of the photovoltaic cell and a positive charge on the other. If a wire is connected from the top to the bottom of a photovoltaic cell, current will flow. Two quantities used in *Article 690* of the Code are the open-circuit voltage of the photovoltaic cell and the short-circuit current of the cell. The cell output voltage will be at a maximum when the sun is shining on the cell and there is no current flow. As current flows, the output voltage will drop.

It is important to understand basic photovoltaic terminology in order to apply the rules of *Article 690*. A typical silicon photovoltaic cell produces approximately 0.5 volt open-circuit; therefore, many cells are connected in series to increase the voltage. Individual photovoltaic cells are grouped into an operating unit called a module. Modules may be connected in series and in parallel to form a solar panel. Solar panels are connected in parallel to form a photovoltaic array. The array is increased in size until the desired capacity in watts or kilowatts is obtained. A rough approximation of power output of a photovoltaic cell is about 1 watt for a cell that is 4 inches in diameter, although wattage varies depending upon a number of design factors. A photovoltaic system is shown in Figure 11.1 with some of the major components labeled.

Solar photovoltaic systems are sometimes interconnected with building wiring and the utility simultaneously to form what is known as an interconnected electric power production system, the wiring of which is covered by *Article 705* of the Code. There are several figures

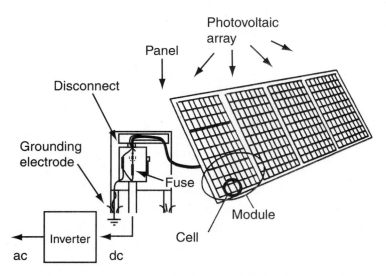

Figure 11.1   Photovoltaic cells are assembled into modules, which are then assembled into panels. A group of panels make up an array. Wiring requirements are covered in *Article 690*.

in the beginning of *Article 690* to help get an understanding of how different photovoltaic systems are interconnected to serve different purposes. *Part B* covers circuit conductor sizing and overcurrent protection. *Part C* specifies the disconnecting means requirements. It is important to remember that if light strikes the photovoltaic cell, a voltage is being produced. Also, when a system is interconnected with a utility system or run in parallel with another power production system, back-feeding may be a problem. This is why proper placement of disconnects is important. Wiring methods are covered in *Part D,* and grounding and grounded conductor sizing are covered in *Part G.*

Article 695 deals with the electrical power, wiring, and controls for fire pumps and associated equipment. This article assembles the electrical installation requirements related to fire-pumping installations together in one location in the Code. Much of the article is taken directly from *NFPA20-1996, Standard for the Installation of Centrifugal Fire Pumps.* The rules applying to fire-pumping systems may seem contrary to normal safety rules in the Code, but it is important to note that the purpose of the system is to reduce a fire emergency to allow the building to be evacuated and to assist fire-fighting personnel in dealing with the emergency. Hopefully, the operation of the pumping system will put out the fire or limit the extent of damage. In any case, the electrical system is to be installed for maximum reliability under heavy load conditions. The system is to be installed to minimize its exposure to damage. Overcurrent protection is set high enough to prevent premature shutdown of the pumping system. With a building on fire, and human life in danger, damage or burnout of pumping system motors and associated equipment is of little concern. *Section 695-3* directs that the power supply be reliable. In the case where the pumping system is supplied power by a tap to the normal power system, the tap ahead of the service disconnect is not permitted to be made in the service disconnecting means enclosure, *Section 695-3(a)(1).*

The supply conductors for the pumping system are not permitted to be protected from overload, but only protected from short-circuits as stated in *Sections 695-6(a)* and *240-3(a).* When the system is supplied from a dedicated transformer, overcurrent protection is not permitted to be installed on the secondary of the transformer. The minimum sitting of the transformer primary overcurrent protection shall be sufficient to continuously supply the sum of the locked-rotor currents of all motors of the system and the full-load current of the other loads of the system.

A separate service is required for a fire-pump installation. The power source is required to be reliable and capable of supplying the locked-rotor current of all pump motors and pressure maintenance pump motors, and the full-load current of any associated equipment. Power source requirements are discussed in *Section 695-3.* A separate service is permitted as shown in Figure 11.2. It is also permitted to tap ahead of the service disconnect. An on-site generator is also permitted to supply the fire pump and associated equipment. Rules for determination of the minimum size of generator are found in *Section 695-3(b)(1).* Obviously it is important to make sure the fire-pumping system is always ready for operation. Therefore, the system is required to be supervised to ensure that power is supplied to all system components, and a signal will indicate if any part of the system is not ready for operation. Disconnects must be capable of being locked in the "closed" position to make sure they are not inadvertently opened.

Article 700 deals with the installation of emergency electrical systems. The fine print notes to *Section 700-1* provide information that helps understand the type of electrical circuits that may be included as part of an emergency electrical system. The wiring of these systems is covered in *Part B,* and the sources of power for these systems are covered in *Part C.* Emergency illumination is the most common type of emergency system installed. The requirements for the circuits, equipment, and control of equipment are covered in *Parts D* and *E.*

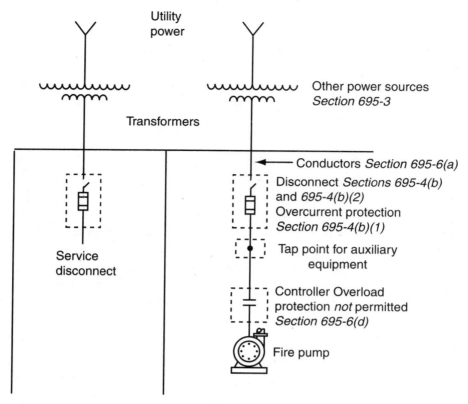

Figure 11.2 A fire-pump system is required to have a separate service capable of reliably supplying the locked-rotor current of the fire pump and pressure maintenance pump and the full-load current of the other associated equipment. The conductors and equipment are to be protected from short-circuits and ground-faults only.

Article 701 deals with permanently installed standby power systems legally required by municipal, state, federal, or other codes. These systems are not intended to supply power to emergency electrical systems. *Section 701-2* and the fine print note provide information that helps understand the type of circuits and equipment to be powered by this system. Wiring and overcurrent protection are covered. The legally required standby power source is not necessarily required to be a generator. Sources of power are covered in *Part C*.

Article 702 covers the installation of permanently installed optional standby power systems. A generator covered by this article is one that is permanently installed, including the prime mover for the generator. Portable generators, or generators powered by a nonstationary prime mover, are not covered by this article.

Article 705 deals with electrical power production systems connected to loads simultaneously with the normal power system. When the loads are not requiring the power produced by the on-site power production system, the power flows into the normal power supply network. This arrangement is often called cogeneration. Solar power, wind power, water power, and generation plants are typical examples of power sources that can be operated as interconnected electric power production sources. The requirements for the installation of these systems are covered in this article. It is important to check with the electric power supplier serving normal power to the property before connecting such a system. The electric power supplier also may have requirements for the installation of the system.

Article 760 covers the installation of fire-alarm systems. Electrical circuits are required to be supervised; that means there is an indication when there is an open circuit or a ground-fault. It is important these systems be installed with the greatest amount of reliability possible.

Much of the article deals with installation requirements that will ensure the greatest amount of reliability possible. Cable types, wiring methods, and wiring installation requirements for nonpower-limited fire-alarm circuits are covered in *Part B.* Power-limited fire-alarm circuits are covered in *Part C.* When cable is run for a power-limited circuit, several cables are permitted to be used as substitutions. These substitutions are covered in *Figure 760-61.*

EMERGENCY AND ALTERNATE POWER SYSTEMS

Understanding the fundamental purpose of alternate power source, emergency, and fire-alarm systems is important to understanding the Code requirements for their installation. A high level of reliability is necessary for these systems to perform their functions effectively. Wiring steps that may seem excessive at first may be necessary to ensure system reliability or safety. Some key points for system reliability and safety are covered in the following discussion.

Alternate Power Production Equipment

An alternate power production source as defined in this text is a source other than the normal power from the electric power supplier serving the area. There are many types of alternate power sources, such as wind-powered generators, water-powered generators, and generators powered by steam or internal combustion engines. Solar photovoltaic panels also can be used to produce electrical power, and there are others.

These alternate power production sources are connected to supply loads in various ways. The source may be connected to equipment or circuits that are not connected in any way to the normal power system. Another way is to connect the alternate source to equipment or circuits through a double-pole transfer switch, as shown in Figure 11.3. It is common for legally required and optional standby power production systems to be connected in this manner. The alternate power production source may be connected directly to the normal power source by following the rules of *Article 705.* Most electric power suppliers will install a kilowatt-hour meter that will measure the amount of energy the interconnected electric power production source supplies to the normal power system for use by other customers.

Emergency Electrical Systems

The primary function of the emergency electrical system, covered in *Article 700,* is to provide automatic illumination for safe exiting of a building and panic control during an emergency. This would require illumination in places where people assemble in the event that power to the normal illumination is interrupted. Illumination is required to safely exit the building, including power for the marking of exits from the building.

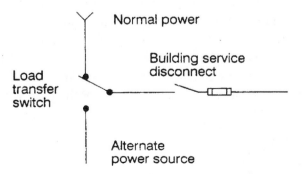

Figure 11.3   A load transfer switch prevents interconnection of the alternate power source to the normal power source.

The emergency electrical system also may include fire-alarm systems, power for fire pumps, ventilation where necessary to provide life safety, or other functions necessary to prevent other life-threatening hazards.

A self-contained, battery-operated exit sign may consist of a standard universal exit sign that is internally equipped with a battery, charger, and a means of automatically transferring to battery power on loss of normal operating power. Exit signs of this type, which are listed by an independent testing laboratory, are readily available with operational time ratings in excess of 4 hours. The following criteria are used in determining whether self-contained, battery-operated exit signs are acceptable in lieu of conventionally wired exit signs. Self-contained, battery-operated signs are permitted to be installed provided the following conditions are met:

- The power supply for the exit signs and illumination units shall be from the same branch circuit as that serving the normal lighting in the area, and shall be connected ahead of any local switches, *Section 700-12(e)*.
- The operational time rating of the units shall not be less than that required by the rules applicable to the facility in which they are installed.

Fire-Alarm Systems

A fire-alarm system, covered in *Article 760,* is an assemblage of components, acceptable to the authority having jurisdiction, that will indicate a fire emergency requiring immediate action. The system shall alert all occupants of a building in which it is installed when a fire emergency is present.

The authority having jurisdiction is the governmental, legally employed agency that can require the installation of a fire-alarm system with specified features, characteristics, functions, and capabilities. The authority having jurisdiction may be a person, firm, or corporation with financial or other interest in the protected property and whose authority lies in contractual arrangements between the affected parties. The electrical inspection authority has jurisdiction over installation methods, materials, and some operational characteristics of all systems.

Fire-alarm systems required to be installed by governmental agencies are designed and installed to save lives by alerting occupants to a fire emergency. Life safety systems may provide some property protection. These systems are often termed local protective signaling systems. Fire-alarm systems installed to protect or limit property loss are not required by governmental agencies. These optional systems are installed to protect high-value properties or to reduce insurance premiums.

Fire alarm systems required for life safety shall be the supervised type. This requirement is not in the *National Electrical Code®*, but is required by other codes. A supervised circuit will indicate a ground-fault or an open circuit with a distinctive audible signal at the control panel location and remote locations as required. Class A supervised circuits will continue to operate with a single open-circuit conductor. Class B supervised circuits will not operate with an open circuit. A Class A fire-alarm initiating circuit is shown in Figure 11.4. Supervision is provided by the design of the control panel and correct installation of the field wiring. Circuits required to be supervised are:

- The main operating power
- The initiating circuits
- The sounding device circuits

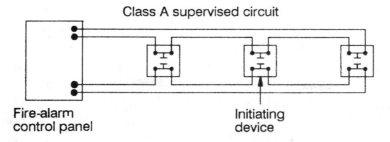

Figure 11.4   A Class A supervised fire-alarm initiating circuit.

The initiation circuit senses the fire emergency condition and permits initiation current to flow in the circuit. Initiating devices are of several common types: pull stations, products of combustion detectors, heat detectors, and water flow switches. Rules not a part of the electrical Code are provided for the proper location of these initiating devices.

The signaling circuit alerts building occupants to the fire emergency. Sounding devices installed on the signaling circuit are of several common types. Typical sounding devices are horns, buzzers, bells, chimes, electronic signals, and sirens. The type of sounding device used in a building must have a distinctive sound not used for any other purpose in the building. Sounding devices must be audible in all areas of the protected facility. Visual signals are now being provided for hearing-impaired persons by some building codes for barrier-free design. Visual signals, when installed, are required to be on a supervised circuit.

One type of initiating device is a heat detector. Several types of heat detectors are used for fire-alarm systems. One type operates at a predetermined fixed temperature. Another operates at a predetermined rate of rise in temperature. Both types may be combined in one unit.

There are several types of products of combustion detectors. The photoelectric type is designed to detect visible particles in smoke by obscuring the light or by light scattering. Another is the ionization type, which is designed to detect invisible products of combustion by using the electrical conducting nature of air in the presence of low-level radioactivity. The conductivity of air is changed by the presence of combustion products or other gas or vapor.

Duct detectors are installed to prevent toxic vapors from being moved from one part of the building to another. There are basically two types of duct detectors. The heat type is used to detect abnormal temperatures in air handling systems. They are permitted in systems under 15,000 cubic feet per minute air capacity by some codes. Some are similar to the fixed temperature detectors described earlier. This type is no longer acceptable for new construction by some state and local codes. Smoke-type duct detectors are adaptations of photoelectric or ionization-type detectors. They are installed to sample the air in a duct system to detect abnormal smoke in air handling systems. All air handling systems are not required to have automatic detectors. Refer to state and local codes for specific requirements.

Flame detectors are a type activated by visible light, light invisible to human eye, or flame flicker. These units are for special occupancies or hazards and are not required by codes, but they may be provided for an additional level of protection.

Wiring Fire-Alarm Systems

It is important to understand how supervised electrical circuits operate to understand how the wiring is to be installed. The wiring for the initiator circuit or the sounding circuit shall be installed such that the system will function when an initiation action has been taken. If the initiation circuit is not properly installed, it is possible for an initiation station to develop an open circuit at a later time without activating the trouble indicator. Specific requirements for wiring the fire-alarm system are found in *Article 760*. This discussion is intended to show the proper wiring of the fire-alarm circuits to make sure they will function properly.

The Class A initiating circuit requires four wires. A resistor is not required at the last initiating station. The current limiting resistor is built into the alarm control unit. One wire on each side of the circuit can open, which sounds the trouble signal, and the initiating circuit will still function. Figure 11.4 shows a Class A initiating circuit properly installed.

The Class B initiating circuit has only two wires. An end-of-line resistor is installed to complete the supervisory circuit. A trickle current flows through the wires at all times. A short-circuit, ground-fault, or an open wire will prevent proper operation, and a trouble signal will sound. A control unit with four terminals may be connected for a 2-wire Class B circuit. Figure 11.5 shows a properly wired Class B initiating circuit. An end-of-line resistor is used for a Class B supervised circuit.

The sounding circuit is also required to be supervised. One method is the series circuit, in which a trickle current passes through each sounding device connected in series. The fire-alarm control panel is adjusted for the number and type of sounding devices used. Figure 11.6 shows a series type sounding circuit.

A polarized sounding circuit is shown in Figure 11.7, and an end-of-line resistor is used. This system is supervised with a current flowing along one wire, then through the end-of-line resistor, and back to the alarm control panel on the other wire. Proper polarity must be observed for connecting to the sounding devices.

Wire splices and terminations are a source of problems in any wiring system. It is not uncommon to have a splice or termination fail and cause an open circuit. For this reason, it is not permitted to make a connection to an initiation station that depends on one splice or termination. If an open circuit should occur, it is possible for the supervisory circuit to function, but an initiation station may be inoperative. If a connection to an initiation station is made with two separate terminations or with a loop termination, an open circuit to an initiation station will most likely open the supervisory circuit. Proper connections to an initiating station are shown in Figure 11.8.

Improper connections of an initiating circuit are shown in Figure 11.9. The wires are connected to only one terminal of the initiating circuit. This is not acceptable because the failure of a connection can render an initiating station not operable without sounding a trouble signal. Devices may have pigtail wires rather than terminal screws. The 2-wire devices are not acceptable. Each initiating device must have four pigtail wires. For the single-wire connection, it is possible for the initiating station wire to come loose without indicating a trouble signal.

In the process of wiring in an initiating circuit, it may be necessary to make a tee tap. This procedure is a frequent source of error. If this is not done properly, the tee tap circuit will not be supervised. It is possible for an open to occur, rendering one or more initiating devices not operable without sounding a trouble signal. The proper method of making a tee tap for a Class

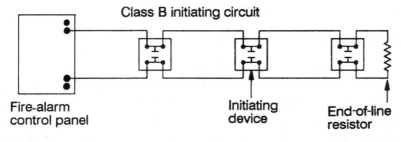

Figure 11.5   A Class B supervised fire-alarm initiating circuit.

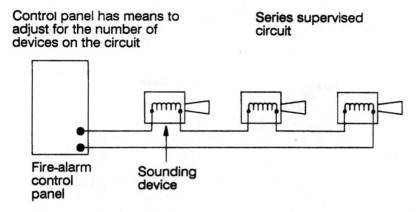

Figure 11.6   A supervised series type fire-alarm sounding device circuit.

A initiating circuit and a Class B initiating circuit is shown in Figure 11.10. Note that the initiating stations are all connected in a series circuit. The improper connections of the Class A and Class B circuits are shown in Figure 11.11. Note that with the improper tee tap, it is possible to have an open circuit in the tap portion of the circuit without interrupting the supervisory circuit current.

Computerized fire-alarm systems are frequently used on large buildings. A remote terminal unit is placed in an area or on a floor. Each remote terminal unit has an initiating circuit and a sounding circuit. The central computer interrogates each remote terminal unit for a status check. If an alarm is initiated, the sounding signal is given in a predetermined strategy. For large buildings, evacuation of the areas of greater danger is initiated first. The rules for supervision and wiring discussed earlier must be followed for the initiating and sounding circuits of each remote terminal unit.

MAJOR CHANGES TO THE 1999 CODE

These are the changes to the 1999 *National Electrical Code®* for electrical systems that correspond to the Code sections studied in this unit. The following analysis explains the significance of only the changes from the 1996 to the 1999 Code, and this analysis is not intended to be used in place of the Code. Refer to the actual section of the 1999 Code for the exact wording and meaning of each section discussed. Changes are indicated in the Code with a vertical

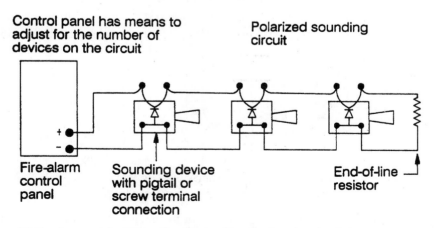

Figure 11.7   A supervised fire-alarm sounding device circuit of the polarized type.

Correct wiring methods

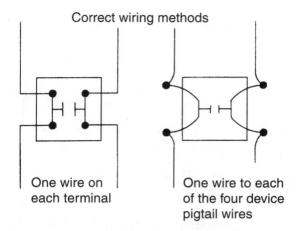

One wire on
each terminal

One wire to each
of the four device
pigtail wires

Figure 11.8 **Several acceptable methods of making connection to supervised fire-alarm circuit components.**

line in the margin. If material has been deleted or moved to another part of the Code, the location of the deletion is indicated with a dark dot in the margin.

Article 445 **Generators**

445-9: This is a new section in the Code, which provides minimum sizes for generator terminal housings. The requirements of *Section 430-12* are to be used for establishing the minimum dimensions of these terminal housings. Areas used solely for the purpose of factory connections are only bound by the requirements in *Section 430-12(d)*.

445-10: A disconnecting means is now required for generators. The disconnecting means is required to completely remove the circuits supplied by the generator from their power source as illustrated in Figure 11.12. This disconnecting function is accomplished with a transfer switch, but if a generator is connected to a wiring system other than through a transfer switch, a disconnecting means may be required. If the motor or engine driving the generator can be readily shut down, and the generator is not arranged to operate in parallel with another power source, a disconnecting means is not required. This new section to the Code provides no location or placement requirements for the disconnect.

Incorrect wiring methods

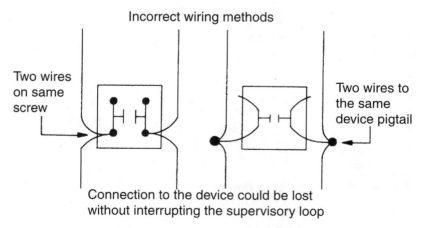

Two wires
on same
screw

Two wires to
the same
device pigtail

Connection to the device could be lost
without interrupting the supervisory loop

Figure 11.9 **These methods of connection to supervised fire-alarm circuit components are *not* acceptable.**

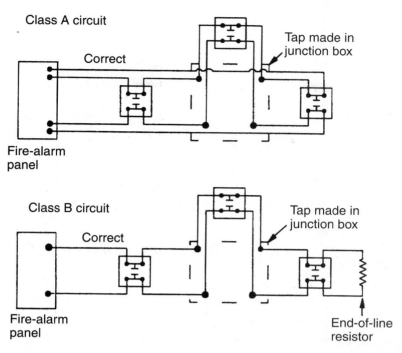

Figure 11.10   The proper method of making a tee tap in a Class A and a Class B fire-alarm initiating circuit.

Article 480   Storage Batteries

480-8(c): A new reference for the minimum working space requirements for battery systems is now in the Code. The requirements of *Section 110-26* apply to battery systems and allow space for battery maintenance and replacement. The clearance requirements in *Section 110-26* are to be measured from the edge of the battery rack. If batteries are

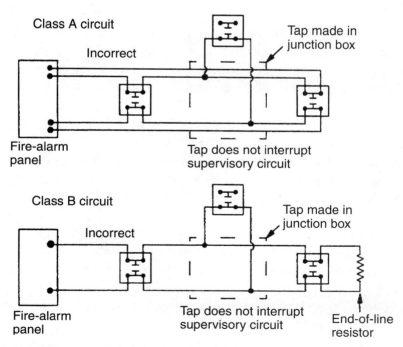

Figure 11.11   These methods of making a tee tap in a fire-alarm initiating circuit are *not* acceptable.

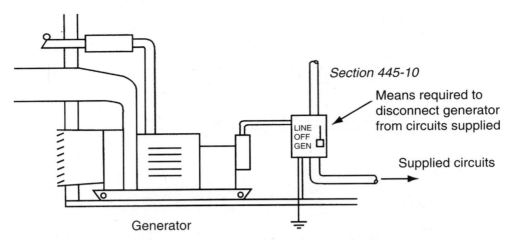

Figure 11.12 **A means is required to be provided to disconnect a generator from the circuits supplied.**

located in a room with a concrete block wall and the batteries operate at not over 150 volts to ground, the minimum distance from the edge of the battery rack to the wall is 3 feet.

Article 690 **Solar Photovoltaic Systems**

Figure 690-1(a): This figure shows the basic components of a photovoltaic array. The new figure is less specific than the figure it replaced in the previous edition of the Code. The figure does include a diagram of an ac photovoltaic module.

Figure 690-1(b): This is a new figure that has three parts to show how an interactive system is connected to a power production and distribution network, how the components of a hybrid system are typically interconnected, and how a stand-alone photovoltaic system including a battery is assembled and connected to the wiring in a building.

690-2: There is a new definition for an alternating current photovoltaic module. It is a self-contained unit that has ac power as the output. It contains the solar cells that produce dc power and an inverter that converts dc power to ac power. Components not necessary for the collection of solar energy and its conversion to ac power are not considered to be a part of this module.

Charge controller is a new term that is defined as a device that controls dc voltage, dc current, or both for the purpose of charging a battery.

The term **electrical production and distribution network** is new; it is an electrical system external to a photovoltaic power system. The utility ac power system or the ac power wiring in a building is an example of an electrical production and distribution network.

Hybrid system is a new term defined as a power system with multiple power sources other than the electrical production and distribution network. An example may be a solar photovoltaic system coupled to a small hydro-generator and a backup engine-driven generator.

System voltage for a photovoltaic system is defined as the maximum dc output voltage of a 2-wire or 3-wire system.

690-4(c): This section requires the wiring for photovoltaic source circuits to be such that the removal of a module or a panel will not interrupt the grounded conductor from another photovoltaic source circuit. When the sun shines on a solar cell, it produces a voltage even if it is not connected to a circuit. The issue is to prevent electrical shock to personnel working on the equipment. Systems operating at 50 volts or less are not considered to be a shock hazard. For these systems operating as a single source, removal of a module or panel is permitted to interrupt the grounded conductor to some parts of the array. Typical systems to which this exemption would apply utilize modules that operate at 12 volts dc frequently with two modules connected in series to provide a 24-volt dc source circuit as illustrated in Figure 11.13.

690-5: The previous edition of the Code required a ground-fault detection and disconnection system for a photovoltaic source mounted on a dwelling. Now there must be some kind of indicator in the dwelling to warn that a ground-fault occurred. *Section 690-5(c)* now requires a label at that indicator warning that because a fault was detected and disconnected, array conductors that normally were grounded may no longer be grounded. *Section 690-5(b)* requires that source conductors be disconnected automatically to interrupt a flow of fault current. If a grounded conductor is interrupted, the ungrounded conductor must also be interrupted at the same time.

690-6(a): This is a new section dealing with modules that have a built-in inverter and have an alternating current output. This section requires that the ac module be treated as a unit, and the internal wiring and components are not within the scope of the Code.

690-6(b): The output of the ac module is alternating current, and the output is to be considered the same as inverter output circuits.

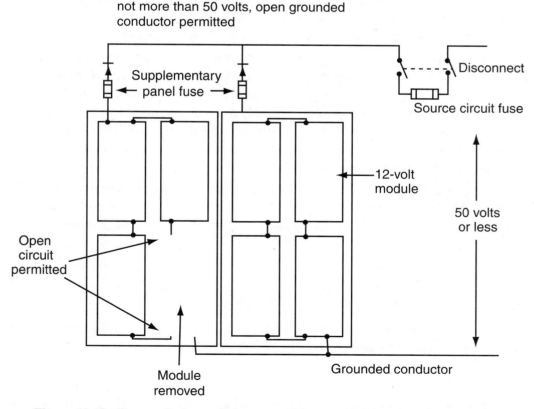

Figure 11.13 Removal of a module or panel is permitted to open the grounded conductor of a single source operating at not over 50 volts.

690-6(c): A disconnect is **permitted** to be provided to disconnect all ungrounded conductors from the photovoltaic source circuit as required in *Section 690-15* and *Section 690-17*. If multiple modules make up a source circuit, each module is **required** to be capable of being disconnected, but the disconnecting means is permitted to be some type of bolted, terminal, or similar means of disconnecting the conductors to the module.

690-6(e): The ac output of the module is **permitted** to have overcurrent protection.

690-7(a): The voltage produced by a photovoltaic system is higher when the source circuit conductors are open with no current flow compared to when the circuit is supplying a load with current flowing. Manufacturers rate photovoltaic units as a dc open circuit output voltage at 77°F. As temperature decreases, the open-circuit output voltage of photovoltaic modules increases. Open-circuit output voltage will be higher in winter than in summer. A new *Table 690-7* was added to the Code giving multiplier values to use to determine the maximum expected open-circuit output voltage of a photovoltaic system. When the sun shines on the photovoltaic cells, voltage is produced even when a load is not being supplied. The conductors and all other components subjected to this open-circuit voltage must have a minimum voltage rating sufficient for this maximum output voltage. To determine the maximum open-circuit voltage for a photovoltaic unit, multiply the rated open-circuit voltage given on the nameplate by the correction factor from *Table 690-7* for the lowest expected temperature in the area. For example, assume the daytime temperature in winter is not expected to go below minus 20°F. The correction factor from *Table 690-7* is 1.25.

690-8(b): The rated current of a photovoltaic system is required to be considered to be continuous current.

690-9 Exception: Where the short-circuit currents from the sources do not exceed the ampacity of conductors, and there are no parallel sources or back feed from an inverter that can put current into the conductors, then an overcurrent device is permitted to be omitted.

690-10: A stand-alone photovoltaic system is permitted to supply ac power to the service of a building at less than the rating of the service disconnect. Overcurrent protection for the inverter output conductors is required to be installed at the output of the inverter. A 2-wire, 120-volt system is permitted to supply a 3-wire premises wiring system, but multiwire circuits are not permitted to be supplied by the system. This type of system is shown in Figure 11.14. A warning sign is required to be located at the service panel stating that multiwire circuits are not permitted.

690-15: When a group of inverters or ac photovoltaic modules are connected to operate in parallel and are connected to be interactive with an electric production and distribution network, it is permitted to have one disconnect for all sources located on the common inverter output feeder.

690-17: Exception 2: All components of a photovoltaic system are required to be provided with a means of disconnection for maintenance purposes. This exception permits conductor connectors to be the means of disconnection for components.

690-18: This section now mentions three means to disable an array for installation or servicing. They are open-circuiting, short-circuiting, and opaque covering of the photovoltaic cells.

690-54: In the case of an interactive photovoltaic power system, the maximum ac operating output current and maximum operating ac voltage shall be marked at the disconnect location to the interactive connection.

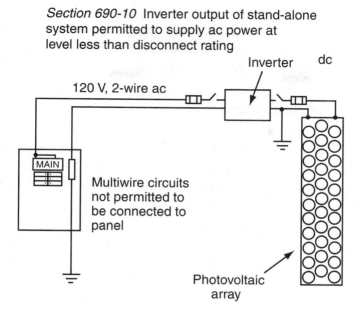

Section 690-10 Inverter output of stand-alone system permitted to supply ac power at level less than disconnect rating

Figure 11.14 A stand-alone, 120-volt, 2-wire photovoltaic system is permitted to supply power to a 3-wire electrical system at less than the rating of the service disconnect.

690-60: Inverters and ac modules intended to be connected as an interactive system shall be listed for interactive systems.

690-72: Specifications are now included in this section for the case where a photovoltaic system is permitted to be directly connected to batteries without means of charge control.

Part I: This is a new part specifying that in the case where systems operate with a maximum dc voltage over 600 volts, the requirements of *Article 490* for high-voltage installations shall apply. There is also a definition of maximum battery circuit voltage, which is the maximum voltage during charging.

Article 695 **Fire Pumps**

695-3: The source of power for a fire-pump installation is now required to be considered **reliable.**

695-3(a): There is now a requirement placed on the capacity of the power source for a fire pump when the supply is a single source. The single power source is required to be able to indefinitely deliver the sum of the locked-rotor currents of all fire-pump motors and the pressure maintenance pump motors, and the full-load currents for all of the associated equipment. This provision was required for overcurrent protection in the previous edition of the Code, but not for the power source.

695-3(b): There is a requirement on the minimum permitted capacity of a standby generator when the separate utility service or tap ahead of the main is not considered to be a reliable source of power. An on-site standby generator is required to have a capacity to allow the starting and running of all fire-pump motors and associated equipment. If the generator supplies other loads, automatic load shedding of optional loads is now permitted to ensure that the generator has the capability to supply the fire-pump load. Also, the supply for fire-pump equipment is not required to be taped ahead of the emergency generator disconnecting means, which would be between the disconnecting means and the generator.

690-3(b)(2): A new provision allows the use of two or more feeders as a power source for multibuilding complexes. There may be more than one fire-pump installation for a complex of buildings. In this case, the separate fire-pump installations are permitted to be supplied by separate feeders. The authority having jurisdiction must approve such an installation.

695-4(b)(2): The disconnecting means for a fire pump is required to be located remote from all other disconnecting means.

695-5(a): The size of the transformer supplying fire pumps is required to be rated for at least 125 percent of the fire-pump motor and pressure maintenance pump motors and 100 percent of the fire-pump accessory equipment. In the previous edition of the Code, the accessory equipment was taken at 125 percent.

695-5(b): As in the previous edition of the Code, overcurrent protection for a transformer is only permitted on the primary side of the transformer; this protection was limited to a maximum rating of 600 percent of the full-load current of the transformer, but this limit was removed. The minimum transformer rating is specified in *Section 695-5(a)*, and the minimum overcurrent protective device rating is specified in *Section 695-5(b)*.

695-5(c): New provisions are now found in the Code when two or more feeders are used as the power source for multibuilding complexes. The transformer supplying the fire-pump system is permitted to supply other loads. These loads are to be calculated in accordance with *Article 220*. The sizing of the transformer primary overcurrent device must be adequate to supply the other loads plus the locked-rotor currents of the fire-pump and maintenance pump motors and the full-load currents of all associated equipment. Yet the primary overcurrent protective device is required to meet the transformer overcurrent protection rules of *Section 450-3*.

695-6(b): Fire-pump conductors shall be kept completely independent from all other wiring beginning at the point where the conductor leaves the final disconnecting means and overcurrent device.

695-6(c): This section now has minimum requirements for the sizes of conductors supplying fire pumps and their associated equipment. The conductors are required to have a minimum ampacity of at least 125 percent of the sum of all full-load current of the fire-pump motor and pressure maintenance pump motor plus 100 percent of full-load current of other associated equipment. In the previous edition of the Code, the only requirement for conductor sizing was on the basis of voltage drop. The following example and Figure 11.15 will show how the minimum conductor size is determined for a fire-pump installation and the overcurrent protection located at the disconnecting means.

Example 11.1   A fire-pump installation in a building consists of a 75-horsepower, 3-phase, 460-volt design B motor and a pressure maintenance pump rated 1.5-horsepower, 3-phase, 460-volt, design B. Determine the minimum size copper type THWN conductors required to supply this load and the minimum rating of short-circuit and ground-fault protection for the installation. All conductor terminations are rated 75°C.

Answer: Look up the full-load currents of the 75-horsepower and 1.5-horsepower motors in *Table 430-150*, add them together, and multiply by 1.25. Then look up the minimum copper conductor size in *Table 310-16*, which is AWG number 1.

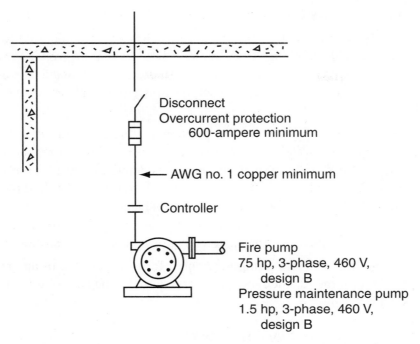

Disconnect
Overcurrent protection
        600-ampere minimum

◄── AWG no. 1 copper minimum

Controller

Fire pump
75 hp, 3-phase, 460 V,
        design B
Pressure maintenance pump
1.5 hp, 3-phase, 460 V,
        design B

Figure 11.15   The minimum conductor size for a fire-pump installation is required to be not less than 1.25 times the full-load current of the fire-pump and pressure maintenance pump motors plus the full-load current of other associated loads.

75 hp, 3-phase, 460 V, design B	96 A full-load
1.5 hp, 3-phase, 460 V, design B	3 A full-load
Total motor load	99 A × 1.25 = 124 A

The method for determining the minimum rating of short-circuit and ground-fault protection is found in *Section 695-4(b)(1)*. Look up the locked-rotor currents in *Table 430-151B* for each motor. The overcurrent device must be capable of carrying the locked-rotor current of each motor plus all other associated load indefinitely. Look up the next larger standard rating of overcurrent device in *Section 240-6* and find 600 amperes.

75 hp, 3-phase, 460 V, design B	543 A locked-rotor
1.5 hp, 3-phase, 460 V, design B	20 A locked-rotor
Total locked-rotor current	563 A

695-6(d) Exception: Now a disconnecting means is not required for the conductors that are run between storage batteries and the engine for a fire pump. In the previous edition of the Code, an exception only excluded overcurrent protection devices for these conductors.

695-6(e): Liquidtight flexible nonmetallic conduit, type B, is now permitted to be run between the controller and a fire pump.

695-10: This new section requires that equipment such as controllers, motors, and transfer switches that make up the switching and control equipment for fire pumps be listed for fire-pump service.

Article 700   Emergency Systems

700-6(c): Electrically operated automatic transfer switches are now required to be mechanically held closed. This would apply to transfer switches that utilize an electric-magnetic coil to maintain the position of the switch. For these types of switches, some form of a mechanical latch is now required to prevent the transfer switch from inadvertently changing position.

700-9(d): There are new requirements for protecting emergency system feeders. These requirements apply only to occupancies that have a capacity for more than 1,000 persons, or to buildings used for educational, business, residential, mercantile, or detention that are more than 75 feet in height. Feeder circuit wiring must be protected by embedding it in a minimum of 2 inches of concrete, using cables listed for a one-hour fire rating, using a listed thermal barrier system, or by installing it in a one-hour fire rated assembly. The fire-protection requirements for the emergency power sources are still in *Section 700-12*.

700-12(b)(2): When a fuel delivery pump is required for the operation of an emergency generator, the circuit for the fuel delivery pump is required to be connected to the emergency power system.

Article 701   Legally Required Standby Systems

701-7(c): This section covers the requirements of transfer switches used for legally required standby systems. The requirements in this section were reorganized with one significant change. Electrically operated automatic transfer switches are now required to be mechanically held closed. This would apply to transfer switches that utilize an electric-magnetic coil to maintain the position of the switch. For this type of switch, some form of a mechanical latch is now required to prevent the transfer switch from inadvertently changing position.

Article 702   Optional Standby Systems

No significant changes were made to this article.

Article 705   Interconnected Electric Power Production Sources

705-3: Exception: This section requires the application of *Article 705* and other certain articles of the Code to interconnected power systems. When an interconnected power system utilizes solar photovoltaic as a power source, the system is now only required to be installed in accordance with *Article 690*. In the previous edition of the Code, a solar photovoltaic system had to comply with *Articles 690* and *705*. The intent of this change is to reduce the amount of confusion in applying the requirements of this article to requirements for a solar-powered system.

Article 760   Fire-Alarm Systems

760-2: A new definition was added for **fire-alarm circuit integrity (CI) cable.** This type of cable will operate under fire conditions for a specified amount of time.

760-24 Exception 3: This new provision deals with listed power supplies for nonpower-limited fire-alarm circuits. Overcurrent protection for electronic power supplies was not covered in the previous edition of the Code. In the case of a single-voltage, 2-wire output, an overcurrent device on the input circuit is permitted to protect the output if the following rating is not exceeded. The overcurrent device rating on the input conductors is not to be greater than the fire-alarm circuit conductor ampacity times the ratio of the fire-alarm circuit voltage divided by the input circuit voltage.

760-30(a)(1): A new provision was added to the nonpower-limited fire-alarm wiring methods. Listed fittings, boxes, and enclosures are required when multiconductor nonpower-limited cables are spliced or terminated.

760-31: This section provides the listing and marking requirements for nonpower-limited fire-alarm cables. Paragraph *(c)* in the previous edition of the Code was deleted, which permitted the use of nonpower-limited circuit conductors that had only 300-volt insulation. It is now clearly stated in paragraph *(b)* that 600-volt insulation is required for nonpower-limited fire-alarm cable.

760-31(f): Nonpower-limited fire-alarm circuit integrity cables are considered to be resistant to the spread of fire and have a suffix **CI,** such as NPLFP-CI.

760-54: In the previous edition of the Code, this section provided the separation requirements for power-limited fire-alarm circuits from electric light, power, and Class 1 or nonpower-limited circuits. Now power-limited fire-alarm circuits must also be separated from medium- and high-power broadband communication circuits. Low-power broadband communication circuits, on the other hand, are permitted to occupy the same raceway or enclosure with power-limited fire-alarm cables.

760-54(c): This section was revised to make it clear that power-limited fire-alarm circuit conductors may not be strapped, taped, tied, or attached in any fashion to the exterior of any raceway unless permitted by *Section 300-11(b)*.

760-71(g): Power-limited fire-alarm circuit integrity cables considered to be resistant to the spread of fire are marked with a suffix **CI,** such as PLFP-CI.

WORKSHEET NO. 11
EMERGENCY AND
ALTERNATE POWER SYSTEMS

These questions are considered important to understanding the application of the *National Electrical Code*® to electrical wiring, and they are questions frequently asked by electricians and electrical inspectors. People working in the electrical trade must continue to study the Code to improve their understanding and ability to apply the Code properly.

DIRECTIONS: Answer the questions and provide the Code reference or references where the necessary information leading to the answer is found. Space is provided for brief notes and calculations. An electronic calculator will be helpful in making calculations. You will keep this worksheet; therefore, you must put the answer and Code reference on the answer sheet as well.

1.   The fire-pump installation shown in Figure 11.16 receives the source of power directly from a utility transformer. There is no central station monitoring, local signaling, or sealing program with weekly inspections to supervise the disconnect to make sure it is maintained in the closed position. Is it required to lock the switch in the on position as shown? (True or False)

Code reference _____

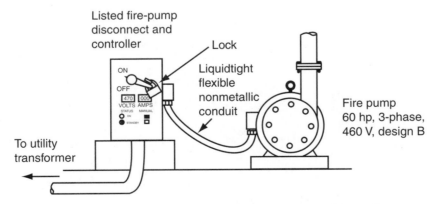

Figure 11.16   The fire-pump system is powered with a single 60-horsepower, 3-phase, 460-volt design B motor with no other loads.

2. A fire-pump installation consists of only one motor that is rated 60-horsepower, 3-phase, 460-volt, design B as shown in Figure 11.16. Conductor terminations are rated 75°C. The minimum size copper type THWN conductor permitted to be run from the controller to the fire-pump motor is AWG number _____.

    Code reference _____

3. Fire-alarm power-limited signaling circuit conductors are permitted to be run in the same conduit with normal power wires, provided the normal power wires have overcurrent protection not exceeding 20 amperes. (True or False)

    Code reference _____

4. Twenty lead-acid 3-cell batteries are connected in series, as shown in Figure 11.17. The Code-rated output of this battery set is _____ volts.

    Code reference _____

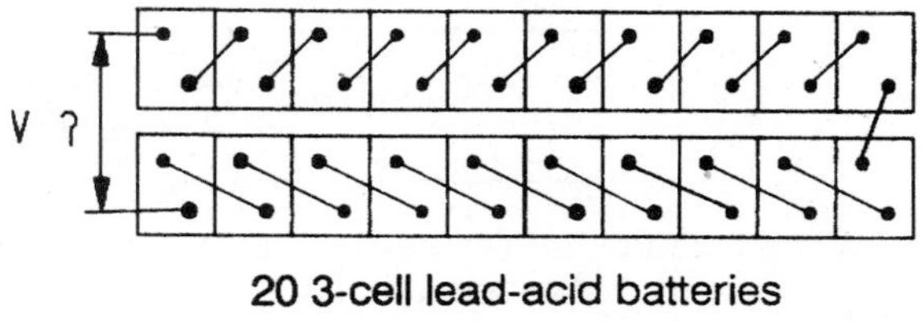

20 3-cell lead-acid batteries

Figure 11.17   Twenty lead-acid batteries, each with 3 cells, are connected in series.

5. Emergency systems are those systems legally required and classed as "emergency" by municipal, state, federal, or other codes, or by any governmental agency having jurisdiction. (True or False)

    Code reference _____

6. An emergency electrical system is permitted to be powered by a generator set that is manually started. (True or False)

    Code reference _____

7. An emergency illumination unit containing a rechargeable battery, charging system, lamp, and power failure relay device is permitted to be powered from any circuit serving the area it is required to illuminate. (True or False)

    Code reference _____

8. An area of a building is required to have emergency illumination by local codes. The only means of normal illumination in the area is high-pressure sodium lighting. Battery-powered emergency lighting units, which turn off when circuit power is restored, are permitted to be installed in the area. (True or False)

   Code reference _____

9. An area is required to be provided with emergency illumination by local codes. The area is likely to be in total darkness if the power to the lighting circuit fails. A single-lamp, battery-powered emergency lighting unit will provide adequate illumination for egress from the area. Is more than one of these single-lamp emergency illumination units, as shown in Figure 11.18, required to be installed in the area? (Yes or No)

   Code reference _____

Battery-powered
emergency
illumination
unit

Single lamp

Rechargeable battery
and charging system

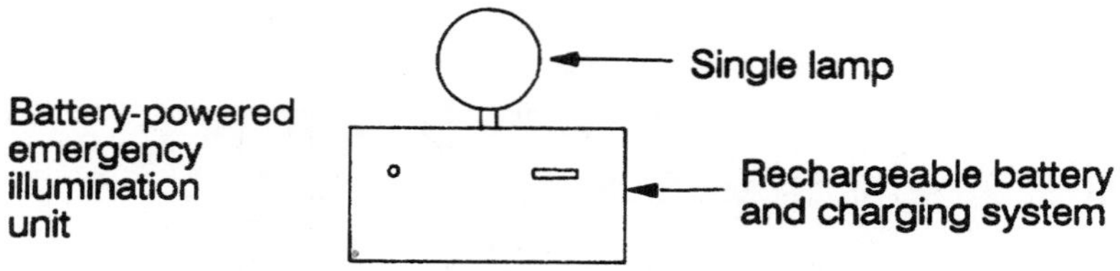

Figure 11.18   A single-lamp, battery-powered emergency lighting unit.

10. Nonpower-limited fire-alarm cable with circuit integrity that is resistant to the spread of fire will have the cable jacket marked with the type letters _____.

    Code reference _____

11. A panelboard for emergency circuits in a building is located remote from the main service equipment. Is it permitted to tap the service entrance conductors ahead of the service disconnecting means for the emergency service panel as shown in Figure 11.19? (Yes or No)

    Code reference _____

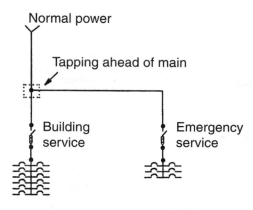

Normal power

Tapping ahead of main

Building service

Emergency service

Figure 11.19   The panelboard supplying emergency circuits in a building receives power by tapping the building service entrance conductors ahead of the main disconnect.

12. A solar photovoltaic source supplies power for loads that will use direct current. The power photovoltaic source is used to charge a set of storage batteries arranged so power can be drawn at 12 volts dc and at 24 volts dc. Three conductors are run from the battery compartment to a power distribution panel. One conductor operates at 24 volts and is protected by a 40-ampere fuse. Another conductor operates at 12 volts and is protected by a 40-ampere fuse. A common conductor serves as the return as shown in Figure 11.20. The 24-volt circuits and 12-volt circuits are permitted to operate at the same time. The minimum size copper common direct current return conductor permitted based on 75°C terminations is AWG number _____.

   Code reference _____

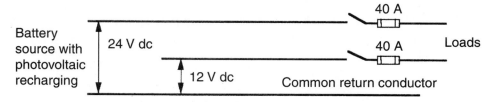

Figure 11.20   A common conductor serves as the return for 12-volt and 24-volt dc circuits each protected at 40 amperes.

13. A switch used to disconnect a dc photovoltaic array from the inverter and other circuit components is required to be rated as suitable for service equipment. (True or False)

   Code reference _____

14. A solar photovoltaic system operates interactively with a normal power system through a power conditioning unit. If voltage is lost on the normal power system, the solar photovoltaic system is permitted to remain connected to the normal power system ungrounded conductors to feed power into that system. (True or False)

    Code reference _____

15. The nameplate continuous current rating of a 208/120-volt, 3-phase generator permanently installed in a building is 200 amperes. The installation is shown in Figure 11.21. Conductor terminations are rated 75°C. The minimum type THWN copper conductor size permitted to be installed between the generator and the first overcurrent device for the generator is AWG number _____.

    Code reference _____

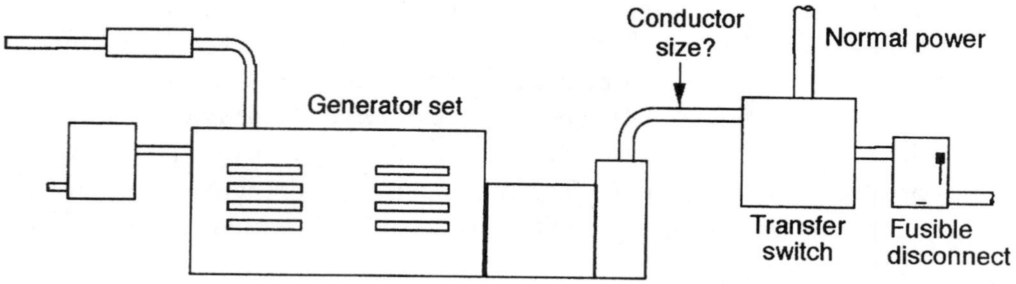

Figure 11.21   Determine the minimum size conductor located between the generator and the disconnect switch that is fused at 200 amperes.

16. A legally required standby power system is one intended to supply power automatically to selected loads if there is a failure of the normal power system. (True or False)

    Code reference _____

17. It is not permitted for wiring of the legally required standby power system to occupy the same conduit as wiring for circuits of the normal wiring system. (True or False)

    Code reference _____

18. A legally required standby power system is only permitted to supply power for the emergency system. (True or False)

    Code reference _____

19. If a business has an optional standby power system installed, a sign of the type shown in Figure 11.22 shall be installed at the building service equipment to indicate the type and location of the optional standby power source. (True or False)

Code reference _____

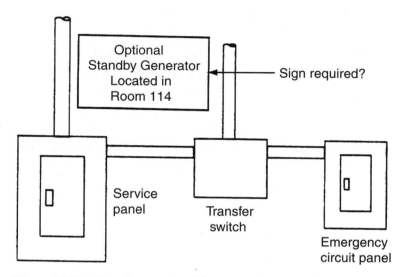

Figure 11.22   **Is a sign required at the service equipment location stating the location of an optional standby generator?**

20. A transfer switch for the connection of a standby generator is installed at the service entrance of a small commercial building. A portable engine-driven standby generator is connected to the building wiring at the transfer switch by means of a cord-and-plug to supply power for a sump pump and refrigeration equipment in the event of a power failure. This installation is considered an optional standby power system and shall be installed according to the rules of *Article 702*. (True or False)

Code reference _____

21. Cable type FPL is permitted for use with a power-limited fire-alarm system of a commercial building as open cable and not permitted to be used as open cable in air handling spaces or as an open cable riser between floors in buildings other than dwellings. (True or False)

Code reference _____

22. A conduit of a nonpower-limited fire-alarm system contains eight circuit wires supplying continuous loads where the current in six of the wires is less than 10 percent of the allowable ampacity of the conductors, and the current for the other two conductors is more than 10 percent of the conductor ampacity. Are the conductors required to have their ampacities derated for more than three current-carrying conductors in a conduit? (Yes or No)

Code reference _____

23. Copper wire, not tinned, size AWG number 14, type THWN is permitted to be installed in conduit for a nonpower-limited fire-alarm system initiation circuit regardless of the number of conductor strands. (True or False)

    Code reference _____

24. The emergency circuits in a building are connected to the normal power system through an automatic transfer switch. An on-site generator provides an alternate source of power in case of normal power loss. The system is illustrated in Figure 11.23. Is the on-site generator that supplies the emergency circuits also permitted to supply other nonessential loads provided the generator is sized adequately for the load and provided the nonessential load is supplied through a separate transfer switch? (Yes or No)

    Code reference _____

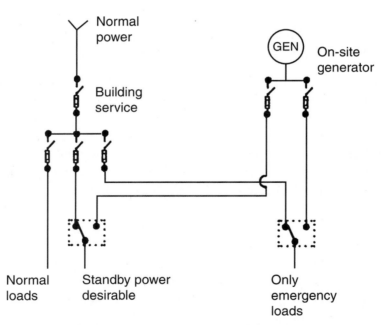

Figure 11.23   An on-site generator adequately sized for the load serves emergency circuits and nonessential loads through separate transfer switches.

25. A customer has a small hydroelectric plant producing single-phase, 120/240-volt power interactively connected to the normal power system such that when the customer does not need the power, it is sold to the electric power supplier. The interconnection of the hydroelectric plant to the normal power system shall be done in accordance with *Article* _____ .

    Code reference _____

ANSWER SHEET

Name _____

No. 11   EMERGENCY AND ALTERNATE POWER SYSTEMS

Answer	Code reference
1. _____	_____
2. _____	_____
3. _____	_____
4. _____	_____
5. _____	_____
6. _____	_____
7. _____	_____
8. _____	_____
9. _____	_____
10. _____	_____
11. _____	_____
12. _____	_____
13. _____	_____
14. _____	_____
15. _____	_____
16. _____	_____
17. _____	_____
18. _____	_____
19. _____	_____
20. _____	_____
21. _____	_____
22. _____	_____
23. _____	_____
24. _____	_____
25. _____	_____

UNIT 12

Industrial Electrical Applications

OBJECTIVES

Upon completion of this unit, the student will be able to:

- determine the minimum dimensions of cable tray permitted for given sizes and numbers of wires and cables.
- describe the conductor insulating method for integrated gas spacer cable.
- determine the maximum ampacity of wires and cables to be installed in cable tray.
- determine the maximum ampacity for a particular application to type TC cable or type ITC cable.
- determine the maximum ampacity for wires to be installed in cablebus.
- define the different methods of electrically heating pipelines and vessels.
- determine the maximum ampacity of branch circuit conductors for a non-continuous motor on a crane, hoist, or monorail hoist.
- explain the application of *Article 665* on induction heating and dielectric heating to the types of facilities.
- explain how dielectric heating works in comparison with induction heating.
- describe facilities to which *Article 668*, dealing with electrolytic cells, and *Article 669*, dealing with electroplating, shall apply.
- determine the minimum size supply conductor permitted for an industrial machine if the nameplate information is given.
- describe the different types of optical fiber cables and their applications.
- answer wiring installation questions from *Articles 318, 325, 340, 365, 427, 610, 630, 665, 668, 669, 670, 727,* and *770.*
- state at least three changes that occurred from the 1996 to the 1999 Code for *Articles 318, 325, 340, 365, 427, 610, 630, 665, 668, 669, 670, 727,* or *770.*

CODE DISCUSSION

This unit deals with wiring methods and materials commonly used in industrial installations but not necessarily limited to industrial use. Several wiring methods are discussed from *Chapter 3* of the Code that are used most frequently in industrial and commercial wiring. Installation of electric heating equipment for pipelines and vessels is also described.

Chapter 6 of the Code deals with the wiring of special equipment. Several types of special equipment for industrial applications are covered in this unit.

Article 318 deals with cable trays, which are defined in *Section 318-2*. A cable tray has a rectangular cross section and is a form of support system for conductors, cables, conduit, and tubing. A typical ladder-type cable tray is illustrated in Figure 12.1. Single conductor cable installed in cable tray is not permitted to be smaller than AWG number 1/0, as stated in *Section 318-3(b)(1)*. All single conductor cable installed in cable tray is required to be marked for use in cable tray on the exterior of the cable. Generally single conductor cables AWG number 1/0 and larger of types such as XHHW, THHN, THWN, THW, RHH, and RHW are listed for use in cable trays. Multiconductor cables type TC and type MC with three or four type XHHW insulated conductors plus an equipment grounding conductor are in common use. Individual conductor sizes range from AWG number 8 copper to 1,000 kcmil. Cables rated over 600 volts are usually type MV or type MC with conductors appropriate for the circuit voltage.

Standard widths of ladder-type and trough-type cable trays are 6, 9, 12, 18, 24, 30, and 36 inches. Typical lengths are 10 feet and 12 feet. There is no specific support spacing required in the Code. For a particular support spacing, cable tray is rated by the manufacturer for a maximum load on a pounds-per-linear-foot basis for cables and raceways supported by the cable tray and any other environmental loads. Standard support spacings for cable trays are 8, 12, 16, and 20 feet. A manufacturer will provide a load class designation for a particular cable tray consisting of a number and a letter. The number is the recommended maximum support spacing. The letter is the working load category, which is A for a 50-pound-per-linear-foot load, B for 75 pounds per linear foot, and C for 100 pounds per linear foot. Assume an aluminum ladder-type cable tray has a load classification 8B. The recommended maximum support spacing is every 8 feet, and with that spacing the load is not permitted to exceed 75 pounds per linear foot. When installed outside, it may be necessary to calculate the side pressure on a linear-foot basis for wind, or a vertical loading expected from ice or snow. In northern climates, ice and snow load is added to the conductor loading. Support spacing may be determined on the availability of points within a building from which to anchor supports. When spanning open areas, the cable is usually supported in a trapeze style. When run adjacent to a wall, a cantilever support to the wall is frequently used. These supports are illustrated in Figure 12.2. If the desired support spacing is not practical to achieve, cable trays are available with stronger side rails to support the weight for longer spans. Cable trays are not required to be mechanically continuous, but bonding is required. Thermal expansion and contraction may also need to be considered in some locations where the cable tray will be installed in an area where the temperature will change.

Section 318-7 covers the situation in which the cable tray is permitted to serve as the equipment grounding conductor. *Table 318-7(b)(2)* gives the maximum circuit rating for a given cable tray cross-sectional area for which the cable tray may act as the equipment grounding conductor. Much of the remainder of the article deals with the number of wires and cables

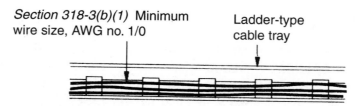

Section 318-3(b)(1) Minimum wire size, AWG no. 1/0   Ladder-type cable tray

Figure 12.1 **A ladder-type cable tray is permitted to support single conductor cables, marked for use in cable trays, and multiconductor cables as listed in *Section 318-3(a)(1)*.**

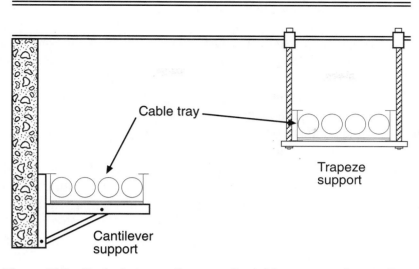

Figure 12.2   Typical means of support for cable trays are the cantilever support when attaching to side walls and the trapeze support when attaching to structural members above the cable tray.

permitted to be installed in cable trays. Examples of conductor installation and determination of the minimum permitted width of cable tray are explained later in this unit.

Methods for determining the width of cable tray for the types and sizes of cables to be installed are explained in *Section 318-9* for multiconductor cables in the cable tray, and in *Section 318-10* when single conductor cables are installed. How the cables are installed in the cable tray influences the method to be used to determine the conductor ampacity. If multiconductor cables are installed so that the cables have a space between them of at least one cable diameter, then conductor ampacity can be determined using the method of *Section 318-11(a)(3)*. If the minimum spacing between conductors is not maintained, the ampacity is determined according to *Section 318-11(a)(1)*. A similar situation exists for single conductor cables supported by cable tray. Two different methods are used to determine the ampacity of the same conductors. If the minimum spacing is maintained between conductors, the ampacity will be higher than if the spacing is not maintained. However, when the minimum spacing is maintained, a wider cable tray is frequently required.

Article 325 deals with integrated gas spacer cable where pressurized sulfur hexafluoride gas and dry kraft paper tapes are used as the conductor insulation. The conductor or conductors are contained within a flexible nonmetallic conduit. The pressurized gas-filled cable assembly helps prevent contaminants from entering the assembly and causing cable failure. As stated in *Section 325-20,* the minimum conductor size permitted is 250-kcmil solid aluminum. A single solid aluminum conductor not smaller than 250-kcmil is permitted in the cable, or the conductor is permitted to be formed using up to nineteen 250-kcmil solid aluminum rods laid parallel.

Article 340 covers the installation of type TC power and control cable. The uses permitted and not permitted are the main emphasis of this article. The ampacity of type TC cable is specified in *Section 340-7*. If the cable is smaller than AWG number 14, the ampacity is determined using the rules of *Section 402-5* for fixture wires. If the wire is AWG number 14 and larger, the ampacity may be determined using the rules of *Section 318-11*.

Article 365 covers the installation of cablebus. Cablebus is a factory-made cable and ventilated metal framework system that is usually field-assembled. *Section 365-2* states that cablebus is permitted to be used for branch circuits, feeders, and services. The minimum permitted size of insulated conductors is AWG number 1/0, which is covered in *Section 365-3(c)*. The conductors are supported periodically on insulating blocks made for the cablebus. The maximum permitted support spacing for horizontal and vertical runs of cablebus is covered in *Section 365-6*. Figure 12.3 shows a cross section of cablebus. The metal framework of cablebus is permitted to be used as an equipment grounding conductor, as stated in *Sections 365-2* and *250-118*. The grounding requirements are covered in *Section 365-9*.

The ampacity of the conductors is permitted to be determined using *Tables 310-17* and *310-19*, which are for single insulated conductors in free air. For circuits operating at more than 600 volts, *Table 310-69* and *Table 310-70* shall be used. Also, it should be noted in *Section 365-5* that the next standard rating overcurrent device is permitted to be used when conductor ampacity does not correspond to a standard rating overcurrent device provided the rating does not exceed 800 amperes. For example, a type THWN copper wire size 700 kcmil with a rating of 755 amperes, found in *Table 310-17*, is permitted to be protected with an 800-ampere overcurrent device.

Article 427 covers the installation of fixed electric heating equipment for pipelines and vessels. Several types of heating are defined in *Section 427-2*. One type is a resistance heating element attached to or inserted into a pipeline or vessel. An impedance heating system is one where electrical current flows through the wall of the pipeline or vessel, and the heat is produced by the impedance of the wall and the current flow. Skin-effect heating uses a ferromagnetic envelope attached to a pipeline or vessel, and electrical current flow through the envelope produces the heat. The impedance of the envelope and the current flow produce the heat. Induction heating is accomplished by using an external induction coil. Electric current is caused to flow in a pipeline or vessel wall by electromagnetic induction similar to a transformer. Induced wall current flow and the impedance of the wall produce the heat.

The minimum permitted ampacity of circuit conductors is required to be 125 percent of the total load of the heaters. This requirement is covered in *Section 427-4*, which points out that it is permitted to round up the conductor ampacity to the next standard size of overcurrent device as permitted by *Section 240-3(b)*. The type of disconnecting means permitted for the heating is covered in *Section 427-55*.

Special consideration is required for equipment grounding depending on the type of heating involved. In the case of resistance heating, the grounding requirements of *Article 250* shall apply. Grounding for impedance heating is covered in *Section 427-29*. *Section 427-48* covers

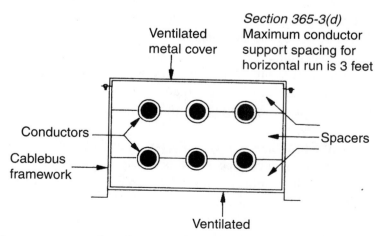

Figure 12.3   Cross-sectional view of a cablebus, which is permitted to be used for branch circuits, feeders, and services.

grounding for skin-effect heating. It is pointed out that the grounding requirement of *Section 250-30* for a separately derived system does not apply in the case of skin-effect heating.

Article 610 deals with the wiring methods and wiring requirements for cranes, hoists, and monorail hoists. Motors powering this equipment may have a duty cycle rating. That is, the motor may only be permitted to operate for 15 minutes, 30 minutes, or 60 minutes until it is required to cool before being operated again. If the motor has a duty cycle, then the conductor to the motor is permitted to carry a higher level of current than would be typical for a general circuit or a continuous load. *Table 610-14(a)* is used to determine the minimum wire size for time-rated motors. The article deals with the installation, support, and grounding of open and insulated conductors on hoists, cranes, and monorail hoists. All exposed noncurrent-carrying metal parts are required to be grounded.

Article 630 provides the requirements for wiring electrically powered welders. A nameplate or rating plate is required to be provided on the welder. Electric welders are of the arc type and the resistance type. Arc welders are of the ac transformer type and the motor-generator type. With the arc type, the metal parts to be joined are brought together and an arc is struck between the metal parts and a metal electrode. The metal electrode is melted away by the heat of the arc to provide extra metal for joining the parts. The actual load on a transformer-type arc welder is intermittent, which is taken into consideration when sizing the circuit components. Even during a continuous weld, current flows in pulses. Current flow is on for several cycles and off for several cycles. The duty cycle for a welder is determined based upon the number of cycles the welder is operating in a one-hour period of time. At the end of *Section 630-31* is a fine print note called explanation of terms. Duty cycle is explained in the third paragraph of that fine print note.

The rules for determining the size of conductors for an individual arc welder or a feeder supplying a group of arc welders are found in *Part B*. Overcurrent protection is required for the welder and for the conductors supplying the welder. One overcurrent device may provide protection for both the conductors and the welder, in which case the minimum value is determined according to *Section 630-12*.

Rules for sizing components for a resistance-type electric welder are found in *Part C*. With a resistance type, the metal pieces to be joined are clamped between two electrodes and current flows from one electrode to the other through the metal. Because the area of contact is small and the current level is high, the area is heated to the welding point due to the resistance of the metal. The weld is actually accomplished with only a few cycles but at very high current. Voltage drop is an important consideration when sizing conductors for a resistance welder. The number of cycles the resistance welder delivers depends upon the particular weld. Duty cycle for a resistance welder is determined by multiplying the number of cycles required for a weld times the number of welds per hour and dividing by 216,000, which is the total number a 60-hertz cycles per hour.

Information on the welder rating plate or nameplate needed for determining conductor size, overcurrent rating, and disconnect rating is based on the effective rated primary current, or the rated primary current times a multiplying factor obtained from a table in either *Section 630-11* or *Section 630-31*. The welder duty cycle is also needed to find the multiplier from the table. Duty cycle is permitted to be calculated based upon the weld to be performed, and the fine print note to *Section 630-31(b)* explains how duty cycle can be determined. According to *Section 630-11(a)*, if the primary rated current of a transformer-type arc welder is 40 amperes and the duty cycle is 70 percent, the minimum current rating for the supply conductors is found by multiplying the primary rated current by the multiplying factor found in *Section 630-31(a)*, which is 0.84. The minimum permitted ampacity for the supply conductors for this welder is 34 amperes (40 A × 0.84 = 33.6 A).

Welders draw some current when running idle (not welding) and a higher current when welding. Effective rated primary current (I_{1eff}) combines the conductor heating due to these two levels of current. The effective rated primary current (I_{1eff}) may be given on the welder rating plate. The fine print note to *Section 630-12(b)* gives a formula for determining the effective rated primary current. It is based on the rated supply current for the welder, the no-load supply current, and the duty cycle. Duty cycle for a welder may be fixed, or it may be adjustable. If it is adjustable, then effective primary current can be calculated. Because heat in a conductor is proportional to the square of the current (Equation 1.16), the rated supply current with load is squared and multiplied by the duty cycle. That value is added to the square of the no-load current times the percentage of time remaining, which is one minus the duty cycle. The square root of the sum of those two values is taken to get the effective primary current. The heating of the supply conductor is considered to be proportional to this value. The rated supply current under load and the rated no-load supply current can be measured, but proper instrumentation is needed because during welding the current is flowing in pulses not continuously.

Determining the minimum size of conductor for a welder and the maximum rating of overcurrent protection permitted is based upon the maximum rated supply current and the effective rated supply current (rated primary current and duty cycle may also be used). The welder is required to be protected from overcurrent at a level not exceeding 200 percent of either the maximum rated supply current or the rated primary current, whichever is given on the nameplate. The conductor is not permitted to be protected at a level greater than the ampacity of the supply conductor. It is permitted to round up to the next standard rating of conductor if this value does not correspond to a standard rating overcurrent device as listed in *Section 240-6*. The minimum permitted rating of supply conductor is determined from *Table 310-16* based upon the nameplate value of the effective rated primary current or the rated primary current and a multiplier found in *Section 630-11(a)*. The supply conductors are likely to have an ampere rating less than the maximum rated supply current; therefore, the overcurrent device sized to protect the conductors will frequently have a rating less than the maximum required to protect the welder. If a single overcurrent device is used to protect both the conductors and the welder as shown in Figure 12.4, the minimum rating is used unless the device opens under

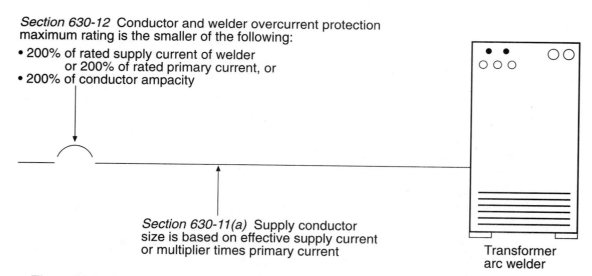

Section 630-12 Conductor and welder overcurrent protection maximum rating is the smaller of the following:

- 200% of rated supply current of welder
      or 200% of rated primary current, or
- 200% of conductor ampacity

Section 630-11(a) Supply conductor size is based on effective supply current or multiplier times primary current

Transformer arc welder

Figure 12.4 **One overcurrent device is permitted to protect both the supply conductor and the arc welder if the overcurrent device rating is not more than 200 percent of the ampacity of the conductor or 200 percent of the maximum rated supply current of the welder.**

normal operation, in which case the next standard rating is permitted. If a feeder supplies a group of arc welders, the method for determining the minimum size of feeder conductor is found in *Section 630-11(b)*. An alternate method of determining the minimum size of feeder conductor is described in the fine print note.

Article 665 deals with equipment for induction heating and dielectric heating of materials in industrial processes and scientific applications. An electrical conducting material is heated by the induction heating process. The material is placed in an electromagnetic field operating at a frequency of a few kilohertz to several hundred kilohertz. Electrical current is induced into the material and with the impedance of the material results in heating.

The dielectric process is used when the material to be heated is not an electrical conductor. The material to be heated is placed between two electrical plates to which is applied a varying electric field in the range of a few megahertz to over 100 megahertz. The varying electric field vibrates the molecules of the material, thus producing heat. The high-frequency electromagnetic field or the electric field may be produced by a motor-generator, or it may be produced by some other type of field-producing equipment.

With the use of electromagnetic and electric fields of an output circuit at frequencies in the kilohertz and megahertz range to produce heat, there are specific requirements of the various systems to protect personnel from exposure to these fields or from electrical shock. Also, it is important that requirements be followed to prevent unintended heating of components and wiring of the system. Requirements for guarding and grounding are covered in *Part B*.

Article 668 deals with the wiring to electrolytic cells for the production of a particular metal, gas, or chemical compound. Definitions important for the application of the article are found in *Section 668-2*. *Section 668-1* states that this article does not apply to the production of hydrogen, electroplating, or electrical energy. This article provides some general wiring requirements for electrolytic cells, but each process is unique and requires engineering design for the specific material and process. This article allows for individual process design. A fine print note in *Section 668-1* refers to the IEEE standard *463-1993* for *Electrical Safety Practices in Electrolytic Cell Line Working Zones*. Electrolytic cell line is defined in *Section 668-2*, and the cell line conductors shall meet the provisions of *Section 668-3(c)* and *Section 668-12*.

Article 669 deals with the installation of wiring and equipment for electroplating processes. *Section 669-9* requires overcurrent protection to be provided for the direct current conductors. Bare conductors are permitted to be run to the electroplating cells. *Section 669-6(b)* requires that when bare cell conductors operate at more than 50 volts, the conductors shall be guarded against accidental contact. When there is more than one power supply to the electroplating tanks, there shall be a means of disconnecting the direct current output of the power supply from the conductor to the electroplating tanks, as required by *Section 669-8*.

Article 670 covers the wiring requirements for industrial machinery, the definition of industrial machinery, and required nameplate information. Industrial machinery is defined in *Section 670-2*. *Section 670-3(a)* states the nameplate information to be provided on industrial machinery. The full-load current of the machine is required to be provided on the nameplate, as well as the full-load current of the largest motor or load. There is a separate NFPA standard for electrical wiring of the actual industrial machinery (*NFPA 79-1994*). If the industrial machine is provided with overcurrent protection, as permitted in *Sections 670-3(b)* and *670-4(b)* at the supply terminals, the supply conductor to the machine is permitted to be considered a feeder or tap from a feeder. When overcurrent protection is provided at the supply terminals, *Section 670-3(b)* requires that a label be placed on the machine stating that overcurrent protection is provided at the machine supply terminals, as shown in Figure 12.5. *Section 670-4(a)* specifies the minimum permitted ampacity of supply conductors to

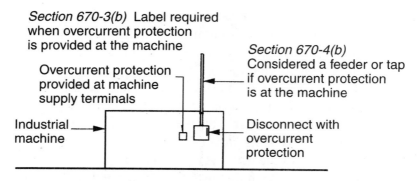

Figure 12.5   When overcurrent protection is provided at the supply conductor terminals of the industrial machine, the supply conductors are permitted to be a feeder or a tap from a feeder.

industrial machinery. The supply conductor ampacity shall not be less than the sum of 125 percent of the full-load current of resistance heating loads, 125 percent of the full-load current of the largest motor, and the full-load current of all other loads.

Article 727 provides specifications for the use and installation of instrument tray cable. It is permitted to be used in industrial facilities per *Section 727-2*. Instrument tray cable is not permitted to be installed with power, lighting, and nonpower-limited circuits, as stated in *Section 727-5*. There is an exception where the type ITC cable is permitted with other types of circuits when the type ITC cable has an approved outer metal sheath. Spacing requirements and other installation specifications are not provided to define what the Code means by type ITC cable not being installed with power, lighting, and nonpower-limited circuits.

Type ITC cable is not permitted to be used for circuits operating at more than 150 volts or more than 5 amperes, as stated in *Section 727-5*. The 150 volts is between conductors as well as to ground. The conductor material for type ITC cable is only permitted to be copper or a thermocouple alloy. A thermocouple is a junction of two dissimilar metals that produce a small voltage that changes as the temperature changes. In industry and research, thermocouples are used to measure the temperature of materials and processes. *Section 727-7* requires that the cable be marked as type ITC on the outer nonmetallic sheath. If the cable has a metal sheath, the marking is permitted to be on the nonmetallic sheath beneath the outer metal sheath.

Article 770 covers the markings and installation of optical fiber cable. The types of cable are described in *Section 770-5*. Conductive optical fiber cable contains a noncurrent-carrying material, which may be a metallic vapor barrier, or a metal member may be present to add mechanical strength. Composite cables contain current-carrying electrical conductors. Nonconductive optical fiber cable is permitted to occupy the same raceway or cable tray with conductors for light, power, and heating circuits. Conductive optical fiber cable is not permitted to occupy the same raceway cable tray or raceway with conductors of electric light, power, or Class 1 circuits, as stated in *Section 770-52(a)*.

The markings on optical fiber cable run in buildings shall be as stated in *Section 770-50*. The listing requirements, which are essentially the uses permitted of the various types, are summarized in *Table 770-50* and described in *Section 770-51*. Conductive optical fiber cable for general-purpose use is marked as type OFC and OFCG, and nonconductive optical fiber cable for general-purpose use is marked as type OFN and OFNG. When the letter **P** is included at the end of the type marking, such as type OFCP, the cable is suitable for installation in air handling spaces, as described in *Section 770-53(a)*. The letter **R** at the end of the type marking, such as type OFNR, designates that the cable is permitted to be run as cable from one floor to another of a building. Cable types OFC, OFCG, OFN, and OFNG are permitted to be run from one floor to another of a building if contained in metal conduit or in a shaft with fire stops

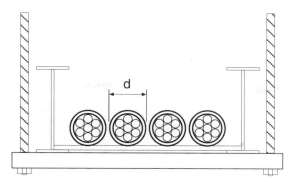

Figure 12.6   Four type XHHW copper 500-kcmil conductors are supported by aluminum ladder-type cable tray.

at each floor. In the case of a one- or two-family dwelling, as stated in *Section 770-53(b) Exception 2,* types OFNG, OFCG, OFC, and OFN are permitted to be installed between floors without metal raceway protection. Cable substitutions are permitted, as stated in *Table 770-53* and as illustrated in Code *Figure 770-53.*

SAMPLE CALCULATIONS

When cables are supported by cable tray, the allowable ampacity of the conductors depends upon whether the installation allows for cooling of the conductors when they are carrying current. If a solid cover for a length more than 6 feet is placed over a ventilated-channel or ladder-type cable tray, the conductor ampacity will be lower than if the top of the cable tray is open. Spacing of the conductors in the cable tray is also important. If a minimum width of space between the cables is maintained as described in *Sections 318-11(a)(3), 318-11(b)(3), and 318-11(b)(4),* the conductor ampacity will be higher. In all cases where the conductor spacing is maintained, the minimum width of the cable tray depends upon the diameters of the cables laid out in a single layer. In some cases, a single layer of cables is required but space is not required to be provided between the cables. In other cases, the cables are permitted to be placed in multiple layers. These rules are found in *Section 318-9* for multiconductor cables, and in *Section 318-10* for single conductor cables. The following several examples will illustrate the relationship between ampacity of conductors and width of cable tray.

Example 12.1   A feeder consisting of four single conductor cables size 500-kcmil copper with type XHHW insulation and 75°C terminations is run in aluminum ladder-type cable tray as shown in Figure 12.6. The cables are placed in the cable tray without maintaining a space between the cables. Determine the ampacity of the conductors and the minimum permitted width of cable tray.

Answer: The minimum width of cable tray is determined by the method in *Section 318-10(a)(2)* for single conductor cables. The cross-sectional area of the conductors is not permitted to exceed the area given in column 1 of *Table 318-10* for each width of cable tray. Look up the cross-sectional area of a 500-kcmil type XHHW conductor in *Table 5, Chapter 9* and find 0.6984 square inches. The total area of the conductors is 2.7936 square inches. According to column 1 of *Table 318-10,* a cable tray with a 6-inch width is permitted to have a single conductor cable fill of 6.5 square inches. The minimum width, then, is 6 inches.

$$4 \text{ conductors} \times 0.6984 \text{ in.}^2 = 2.7936 \text{ in.}^2$$

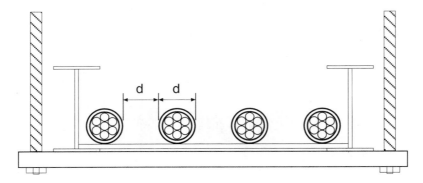

Figure 12.7   Four type XHHW copper 500-kcmil conductors are run in aluminum ladder-type cable tray with a space maintained between the conductors equal to the cable diameter.

Next determine the ampacity of the conductors according to *Section 318-11(b)(2)*. The ampacity is permitted to be 0.65 times the ampacity found in *Table 310-17,* which is 403 amperes.

$$0.65 \times 620 \text{ A} = 403 \text{ A}$$

Example 12.2   Determine the minimum width of cable tray and ampacity of the conductors when the conductors are installed in a single layer with a space between the conductors not less than the diameter (d) of the conductors as prescribed by *Section 318-11(b)(3)* and shown in Figure 12.7. Four single conductor cables size 500-kcmil copper with type XHHW insulation and 75°C terminations are run in aluminum ladder-type cable tray.

Answer: *Section 318-11(b)(3)* specifies that this method of determining conductor ampacity requires a maintained space between the conductors placed in a single layer. First look up the diameter of the conductor in *Table 5, Chapter 9* and find 0.943 inches. The minimum width of cable tray is then seven times the cable diameter, which is 6.60 inches. Choose a 9-inch minimum cable tray width.

$$7 \times 0.943 \text{ in.} = 6.60 \text{ in.}$$

Section 318-11(b)(3) simply specifies the minimum conductor ampacity can be found in *Table 310-17,* which is 620 amperes. Note the big difference in conductor ampacity between the previous example and this example, when a space is maintained between the conductors.

Example 12.3   Three type TC cables with three 250-kcmil XHHW copper conductors and an equipment grounding conductor are to be supported by aluminum ladder-type cable tray where a space is not maintained between the individual cables, as illustrated in Figure 12.8. All conductor terminations are rated 75°C, and the cable diameter is 1.76 inches. Determine the minimum width of cable tray permitted and the ampacity of the insulated conductors in the cable.

Answer: The minimum width of the cable tray is determined according to *Section 318-9(a)(1)*. The cables are required to be in a single layer, but no space is required between the cables. The minimum width of cable tray is not permitted to be less than the sum of the diameters of the cables in the cable tray, which is 5.28 inches. Therefore, a 6-inch cable tray is adequate.

$$3 \times 1.76 \text{ in.} = 5.28 \text{ in.}$$

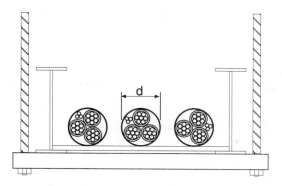

Figure 12.8   Three type TC cables with 250-kcmil copper type XHHW insulated conductors and a bare equipment grounding are supported by an aluminum ladder-type cable tray.

Next determine the ampacity of the 250-kcmil copper conductors in the cables according to *Section 318-11(a)*. Note in *Section 318-11(a)(1)* that the derating factors do not apply unless there are more than three current-carrying conductors in any one cable, and then they apply only to that particular cable. In this case, conductor ampacity is determined according to *Table 310-16*, and it is 255 amperes.

Example 12.4   For the cable tray installation of Figure 12.9, the three type TC 3-conductor cables are installed with a space maintained between them of one cable diameter. Each cable contains three XHHW insulated conductors size 250-kcmil copper and has a diameter of 1.76 inches. All conductor terminations are rated 75°C. Determine the minimum permitted width of aluminum ladder-type cable tray and the ampacity of the insulated conductors in the cable.

Answer: There is no rule for determining the minimum width of cable tray. It is simply a matter of adding the cable diameters. There are three cables of the same diameter, and there are two spaces between the cables of the same diameter. The minimum width of cable tray is not permitted to be less than 8.8 inches; therefore, the minimum standard width is 9 inches.

$$5 \times 1.76 \text{ in.} = 8.8 \text{ in.}$$

The conductor ampacity is determined according to *Section 318-11(a)(3)*, which specifies the method of *Section 310-15(c)*. That section is one that requires engineering

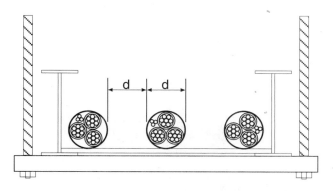

Figure 12.9   Three type TC cables each with three 250-kcmil copper XHHW insulated conductors and one equipment grounding conductor are installed in ladder-type cable tray with a space maintained between the cables equal to the diameter of the cable.

supervision, but the ampacity is usually determined from *Table B-310-3* in *Appendix B*, and in this case is 274 amperes.

Example 12.5   Three sets of triplexed conductors are installed in an aluminum ladder-type cable tray as shown in Figure 12.10. The individual conductors are type THWN, AWG number 4/0 copper, and all terminations are rated 75°C. *Section 318-8(e)* permits these conductors to be bundled rather than installed in a single layer. Determine the minimum width of cable tray permitted and the ampacity of the conductors.

Answer: The minimum width of cable tray is specified in *Section 318-10(a)(4)*. The width is not permitted to be less than the sum of the diameters of all of the conductors in the cable tray. Look up the diameter of an AWG number 4/0 type THWN conductor in *Table 5, Chapter 9*; it is 0.642 inch. Then multiply by nine conductors to get the minimum width of 5.78 inches. A 6-inch wide cable tray is permitted.

$$9 \text{ conductors} \times 0.642 \text{ in.} = 5.78 \text{ in.}$$

The ampacity of the individual conductors is determined according to *Section 318-11(b)(2)*. The ampacity found in *Table 310-17* is multiplied by 0.65 to get 234 amperes.

$$0.65 \times 360 \text{ A} = 234 \text{ A}$$

Example 12.6   Three triplexed bundles of type THWN copper conductors are installed in aluminum ladder-type cable tray with a space maintained between each bundle equal to 2.15 times the individual cable diameter, as shown in Figure 12.11. All conductor terminations are rated 75°C. Determine the minimum permitted width of cable tray and the ampacity of the conductors.

Answer: The width of the cable tray is simply the physical dimension necessary to accommodate the conductors when installed as required by *Section 318-11(b)(4)*. There are three bundles of cables; therefore, there will be two spaces with a minimum dimension of 2.15 times the individual conductor diameter. The bundles will lay in the cable tray with two conductors touching the cable tray; therefore, space must be allowed for a minimum of six conductors, as shown in Figure 12.11. Look up the diameter of an AWG number 4/0 type THWN conductor in *Table 5, Chapter 9;* it is 0.642 inch. The calculated minimum width permitted is 6.612 inches, which requires at least a 9-inch-wide cable tray.

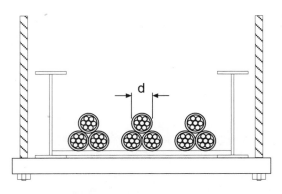

Figure 12.10   Three triplexed single conductor cable bundles size AWG number 4/0 with type THWN insulation are installed in aluminum ladder-type cable tray with no maintained space between the conductors.

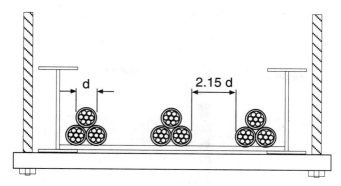

Figure 12.11   Three triplexed bundles of AWG number 4/0 copper type THWN conductors are installed in cable tray with a space not less than 2.15 times the individual cable diameter maintained between the cable bundles.

Width = 2 × (2.15 × d) + 6 × d
   = 2 × (2.15 × 0.642 in.) + 6 × 0.642 in.
   = 2 × 1.380 in. + 3.852 in.
   = 2.760 in. + 3.852 in. = 6.612 in.

Because a space between the triplexed bundles is maintained for cooling, the ampacity is permitted to be looked up in *Table 310-20,* and in this case is 287 amperes.

   When multiconductor cables of different sizes are placed in a cable tray together, the method of determining the width of cable tray required depends upon the size of conductors in the cables. There is one simple method when all cables contain conductors AWG number 4/0 and larger. *Section 318-9(a)(1)* requires that the cables be arranged in a single layer, and the width is equal to the cable diameters. There is another method when all conductors are smaller than AWG number 4/0. *Section 318-9(a)(2)* does not require cables with conductors AWG numbers 1/0 through 3/0 to be in a single layer. They can be placed in multiple layers as long as they are not stacked on top of cables with conductors AWG number 4/0 and larger. Cables AWG number 3/0 and smaller are not permitted to have a cross-sectional area greater than approximately 40 percent of the cross-sectional area of a standard usable depth cable tray, which is 3 inches. The areas given in column 1 of the table in *Section 318-9* are approximately 40 percent of the usable cross-sectional area of a standard cable tray.

   When cables with conductors larger and smaller than AWG number 4/0 are placed in the same cable tray, both of these methods of determining cable tray fill are combined in column 2 of the table in *Section 318-9.* The diameters of all cables with conductors AWG number 4/0 and larger are added together, then multiplied by 1.2, then subtracted from the cable tray width. If the standard depth of a cable tray is 3 inches, and the allowable fill is 40 percent, then the area of the cable tray that is permitted to be filled with conductors is the width of the cable tray times 3 inches times 0.4. Note that 3 inches times 0.4 is 1.2, which is the number in column 2 of the table in *Section 318-9.* Therefore, this is a calculation to determine 40 percent of the cross-sectional area of the remaining cable tray that is not occupied by conductors AWG number 4/0 and larger. An easier method may be simply to calculate 40 percent of the cross-sectional area of the cable tray after the sum of the diameters of the cables with conductors AWG number 4/0 and larger have been subtracted. For example, assume that four type MC cables with three 350-kcmil conductors with a diameter of 1.96 inches are installed in an 18-inch-wide ladder-type cable tray. Multiply the 1.96 inches by four cables to get 7.84 inches. Subtract the 7.84 inches from 18 inches to get a remaining dimension of 10.16 inches. Refer

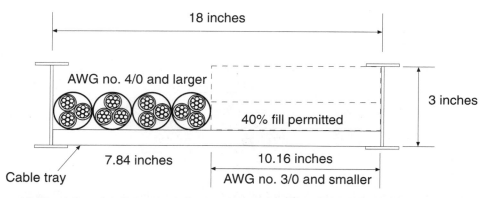

Figure 12.12   After the diameters of the cables with conductors AWG number 4/0 and larger are subtracted from the width of the cable tray, the cables with conductors AWG number 3/0 and smaller are permitted to fill only 40 percent of the remaining portion of the cable tray.

to Figure 12.12. Multiply the 10.16 inches by a 3-inch height, then multiply by 0.4 to get the cross-sectional area of the remaining portion of the cable tray that is permitted to be filled with cables with conductors AWG number 3/0 and smaller. The cross-sectional area permitted to be filled with these cables with conductors AWG number 3/0 and smaller is 12.19 square inches (10.16 in. × 3 in. × 0.4 = 12.19 in.²). The following example will illustrate how this method works. Column 4 of the table in *Section 318-9* uses a multiplier of 1.0 times the sum of the cable diameters because the allowable fill is only about 31 percent rather than 40 percent.

Example 12.7   A cable tray contains two type TC cables with three 350-kcmil conductors with a diameter of 1.98 inches and six type TC cables with three AWG number 2/0 conductors with a cable cross-sectional area of 1.35 square inches, as shown in Figure 12.13. Determine the minimum permitted width of ventilated-trough cable tray for this installation.

Answer: The method is described in *Section 318-9(a)(3)*. The cross-sectional area and diameter of multiconductor cables are not given in the Code. The best source for this information is to contact the cable manufacturer. Some manufacturers provide cable technical data on their web sites. The symbol **Sd** in column 2 of the table in *Section 318-9* is the sum of the diameters of all cables with conductors AWG number 4/0 and larger, which is multiplied by 1.2 to convert to square inches. In this example, there are two cables, so multiply the diameter by two to get 3.96 inches, then multiply that number by 1.2 to get 4.75 square inches.

$$2 \text{ cables} \times 1.98 \text{ in. diameter} \times 1.2 = 4.75 \text{ in.}^2$$

Next determine the cross-sectional area of all conductors AWG number 3/0 and smaller. There are six that are AWG number 2/0, so multiply 1.35 square inches by six to get 8.10 square inches.

$$6 \text{ cables} \times 1.35 \text{ in.}^2 = 8.10 \text{ in.}^2$$

The method is described in the text as trial and error. Choose one of the standard cable tray widths, and do the calculation. Start with a 12-inch width. The formula in column 2 is as follows:

$$14 - (1.2 \times Sd) = \text{cross-sectional area of cables with conductors 3/0 and smaller}$$

$$14 \text{ in.}^2 - (1.2 \times 3.96 \text{ in.}) = 14 \text{ in.}^2 - 4.75 \text{ in.}^2 = 9.25 \text{ in.}^2$$

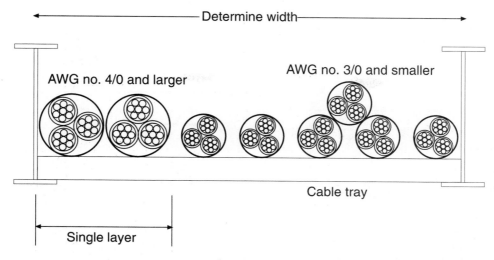

Figure 12.13 The cable tray contains two type TC cables with 350-kcmil conductors and six type TC cables with AWG number 2/0 conductors.

The cables with conductor size AWG number 3/0 are only 8.10 square inches; therefore, the 12-inch wide cable tray is adequate for these cables. An easier way to determine the minimum cable tray width may be to add the 4.75 to the 8.10 to get 12.85 square inches. Then move down column 1 until a cable tray width is found that has an area that is equal to or greater than 12.85.

MAJOR CHANGES TO THE 1999 CODE

These are the changes to the 1999 *National Electrical Code®* for electrical systems that correspond to the Code sections studied in this unit. The following analysis explains the significance of only the changes from the 1996 to the 1999 Code, and this analysis is not intended to be used in place of the Code. Refer to the actual section of the 1999 Code for the exact wording and meaning of each section discussed. Changes are indicated in the Code with a vertical line in the margin. If material has been deleted or moved to another part of the Code, the location of the deletion is indicated with a dark dot in the margin.

Article 318 **Cable Trays**

318-5(c): This section refers to corrosion protection of metal cable trays. Now it specifies that additional protection from corrosion applies only in the case of cable trays made of ferrous (generally steel consisting primarily of iron) materials.

318-6(a): This new 6-foot maximum support spacing requirement was added for single conductor cables passing between cable tray sections, or from a cable tray to a raceway, or from a cable tray to equipment, as illustrated in Figure 12.14. It is also required to bond between the cable tray sections or from the cable tray to the equipment or raceway.

318-6(j): Outlet boxes are now permitted to be supported by cable trays.

Article 325 **Integrated Gas Spacer Cable**

No significant changes were made to this article.

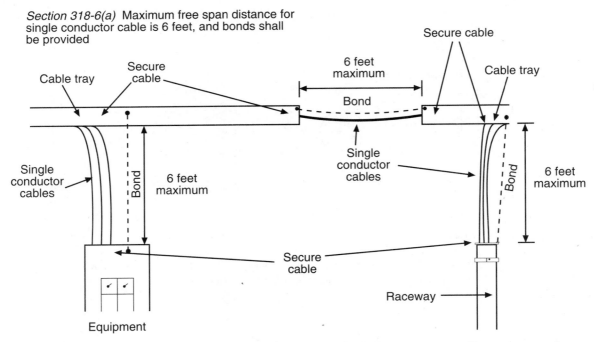

Section 318-6(a) Maximum free span distance for single conductor cable is 6 feet, and bonds shall be provided

Figure 12.14 **Single conductor cables are to be protected from physical damage if necessary and supported at intervals not to exceed 6 feet when passing between sections of cable tray, or passing from cable tray to raceway or equipment.**

Article 340   **Power and Control Tray Cable**

340-4: Type TC tray cable is permitted to be used in Class I hazardous locations where the wiring is installed according to *Article 505*.

Article 365   **Cablebus**

365-3(b): The tables to be used to determine the ampacity of conductors are listed. Even though cablebus was approved for use with conductors operating over 600 volts, only the low-voltage tables were listed for determining ampacity of conductors. Now *Table 310-69* and *Table 310-70* are specified to be used for medium-voltage cable installed in cablebus.

365-5: With respect to overcurrent protection for conductors, the last sentence of the previous edition of the Code was deleted that permitted rounding up to the next standard rating of overcurrent device only when the overcurrent device rating did not exceed 800 amperes. The rule did not change. Overcurrent device rating for a cablebus system is determined in accordance with the provisions of *Section 240-3*, and that same limitation is in that section.

Article 427   **Fixed Electric Heating Equipment for Pipelines and Vessels**

427-22: Ground-fault protection of equipment is no longer required on all resistance-type electric heating equipment. It is required only on heat tracing and heating panels.

427-23: A metal covering is required for resistance-type electric heating equipment. The change is that now this section requires any metal covering to act as an **effective** ground path.

Article 610   **Cranes and Hoists**

610-13(d): This section requires that all electrical conductors meet the requirements of conductors listed in *Table 310-13*. Alternate cable construction is covered in this section.

Now this section specifically excludes Class 1, Class 2, and Class 3 conductors if they meet the requirements of *Article 725*.

610-31: This section requires a disconnecting means to be installed for the runway conductors of a crane or hoist. Item 4 in the list was modified to no longer require the disconnect for the runway conductors to be placed within sight of the crane or hoist.

Article 630 **Electric Welders**

630-11(a): This section is used to determine the minimum ampacity of the supply conductors to individual arc welders. If on the rating plate of an arc welder a value for I_{1eff} (maximum effective supply current) is given, the ampacity of the supply conductors is required to be equal to or greater than this ampacity.

The maximum effective supply current (I_{1eff}) is considered to be a more accurate method of allowing for the heating effects on the supply conductors to an arc welder. Effective supply current takes into account the current drawn while welding and at idle times. The welder is drawing current when energized and not welding. A new fine print note to this section defines effective supply current.

If I_{1eff} is not given on the welder's nameplate, then the ampacity of the supply conductors is based on the same method used in the previous edition of the Code. This method utilized the duty cycle of the welder to determine an appropriate multiplier for the rated primary current.

630-12: The minimum permitted size of overcurrent protective device for an arc welder is covered in this section, which was modified to show the true intent of the Code. When the calculations for the overcurrent protection size do not correspond to those in *Section 240-6*, the next higher overcurrent device shall be permitted. In the previous edition of the Code, technically the size of the overcurrent device would have been the **nearest** standard size available.

630-12(a): A new term I_{1max} (maximum rated supply current at maximum rated output) will be given on welder rating plates. The term I_{1max} is explained in a new fine print note.

630-14: In the welding industry, the markings provided on welders are placed on **rating plates** as opposed to **nameplates.** This section was modified to use this language. Also, on these rating plates, it is now required to show the maximum rated supply current (I_{1max}) and the effective supply current (I_{1eff}) of arc welders.

Article 665 **Induction and Dielectric Heating Equipment**

665-63: The disconnect requirements for equipment other than motor-generators were changed by the addition of a new sentence. The disconnect is now required to be located within sight from the controllers for such equipment.

Article 668 **Electrolytic Cells**

668-2: This section provides definitions that are unique to electrolytic cells. A definition for **electrically connected** is now included in this section. However, this definition is not new to this article. In the previous edition of the Code, it was in a fine print note to *Section 668-3(b)*.

Article 669 **Electroplating**

There were no significant changes made to this article.

Article 670   **Industrial Machinery**

670-5: This section provides clearance requirements for industrial machinery in cases where only qualified persons will provide necessary maintenance. To clarify the intent of the voltage limitation given in this section, the phrase **line-to-line or line-to-ground** was added to this section.

Article 727   **Instrumentation Tray Cable**

727-1: A scope was added to the beginning of this article. This article is to be used to determine the usage and construction of type ITC cable when it is used for instrumentation and control circuits that operate at not more than 150 volts. The load placed on these cables is limited to 5 amperes.

727-4: This section provides the permitted uses of type ITC cables. This type of cable is now permitted to be installed in hazardous locations that fall under *Article 505*. In the previous edition of the Code, only *Articles 501, 502, 503*, and *504* were listed.

727-4(4): In the previous edition of the Code, it was permitted to install type ITC cable as open wiring provided it was protected by either a smooth metallic sheath, a continuous corrugated metal sheath, or an interlocking armor applied over the nonmetallic sheath. This provision was actually an exception to the uses-not-permitted section. This requirement was moved to this section to be in a more logical location. The significant change here is that the cable is now required to be supported at intervals not exceeding 6 feet. In the previous edition of the Code, there was no such requirement for these cables.

727-4(4) Exception 1: This new exception permits type ITC cable, without a metallic covering, to be run as open wiring between cable trays and other equipment in lengths not exceeding 50 feet provided the cable is supported at intervals not exceeding 6 feet. This practice is permitted when the cable is protected from physical damage.

727-4(4) Exception 2: If type ITC cable meets the crush requirements of type MC cable, it may be installed as open wiring between cable trays and equipment. The length limitation for this type of application is 50 feet, and the cable must be secured at intervals not exceeding 6 feet. These cables that can withstand the same level of abuse as type MC cables do not have to be protected against physical damage.

727-9: This new section limits the overcurrent protective device on type ITC cables to 3 amperes for AWG number 22 conductors and 5 amperes for larger conductors.

Article 770   **Optical Fiber Cables and Raceways**

770-2: This section contains definitions used in *Article 770*. Now definitions for **exposed, point of entrance,** and **optical fiber raceway** are included. An optical fiber raceway is a raceway that is designed to contain nonconductive optical fiber cables.

770-5(b): Optical fiber cables that utilize a metallic armor or sheath are now considered to be a conductive type of fiber-optic cable. The previous edition of the Code did not specifically recognize cable with this type of protection as a conductive cable.

770-6: This section places requirements on a raceway if it contains fiber-optic cables. In the previous edition of the Code, these requirements were in *Section 700-5*. This section applies to all types of fiber-optic cables. In the past, this section only pertained to nonconductive types of cables.

770-6 Exception: This exception permits the installation of a listed nonmetallic optical fiber raceway. It can now be listed for installation in a plenum, used as a riser, or used for general-purpose installations. No fill requirements are found in the exception. Now the fill requirements of *Chapter 3* and *Chapter 9* apply only to those installations where the fiber-optic cable is installed in the raceway with other power conductors.

770-50 Exception 1: Provided that the length of fiber-optic cable installed in a building is not greater than 50 feet, no listing or marking requirements apply to the cable. The change in this section was to clarify how the 50 feet are to be measured. The measurement is made from the point of entry to the termination point at an enclosure.

770-51(g): In the previous edition of the Code, this section provided the listing requirements for fiber-optic cables and raceways. This section now recognizes and provides listing requirements for general-purpose, riser, and plenum optical fiber raceways.

770-52: In the previous edition of the Code, this section provided the separation requirements for fiber-optic cables and electric light, power, and Class 1 circuits. Separation requirements were added to this section for nonpower-limited fire-alarm circuits and medium-power broadband communication circuits. Now it is not permitted to install conductive type of fiber-optic cables with nonpower-limited fire-alarm circuits and medium-power broadband communication circuits. Nonconductive and composite types of fiber-optic cable are permitted to be installed in the same raceways with nonpower-limited fire-alarm circuits and medium-power broadband communication circuits. Low-power broadband communication circuits, on the other hand, are permitted to occupy the same raceway or enclosures with any type of optical fiber cable.

770-53(b): The word **only** was added to this section to make it clear that only types OFNR and OFNP cables are permitted to be installed in riser optical fiber raceways.

WORKSHEET NO. 12
INDUSTRIAL ELECTRICAL
APPLICATIONS

These questions are considered important to understanding the application of the *National Electrical Code®* to electrical wiring, and they are questions frequently asked by electricians and electrical inspectors. People working in the electrical trade must continue to study the Code to improve their understanding and ability to apply the Code properly.

DIRECTIONS: Answer the questions and provide the Code reference or references where the necessary information leading to the answer is found. Space is provided for brief notes and calculations. An electronic calculator will be helpful in making calculations. You will keep this worksheet; therefore, you must put the answer and Code reference on the answer sheet as well.

1.   An aluminum ladder-type cable tray 18 inches wide in an industrial facility, with 3-inch side rails, has a load classification of 12B and is supported at 12-foot intervals. A 2-inch rigid metal conduit passes under the cable tray at a right angle approximately at the midpoint of the cable tray span, as illustrated in Figure 12.15. The conduit is supported about 10 feet from the point where it passes under the cable tray. Assuming the cable tray is designed to carry the weight, is it permitted to secure the rigid metal conduit to the cable tray in this manner? (Yes or No)

Code reference _____

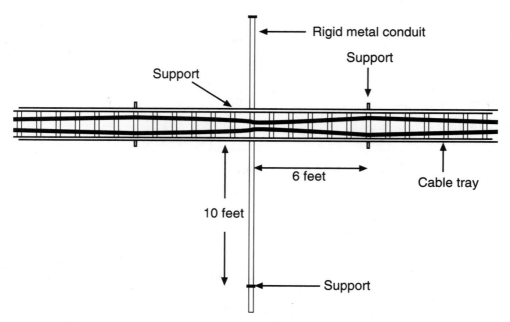

Figure 12.15   A rigid metal conduit runs at a right angle to and is supported by a cable tray where the cable tray is designed to support the weight.

2. Ladder-type cable tray is only permitted to be installed for the support of electrical single conductor cables within industrial-type buildings. (True or False)

   Code reference _____

3. A ladder-type cable tray supports TC multiconductor power cables with type XHHW copper conductors. There are two 3-conductor cables size 500-kcmil with a diameter of 2.26 inches, three 4-conductor cables size 250-kcmil with a diameter of 1.93 inches, and five 4-conductor cables size AWG number 3/0 with a diameter of 1.58 inches and a cross-sectional area of 1.96 square inches. An air space between the conductors as shown in Figure 12.16 is not maintained. The minimum width of cable tray permitted for the installation is:
   A. 9 inches.
   B. 12 inches.
   C. 18 inches.
   D. 24 inches.
   E. 30 inches.

   Code reference _____

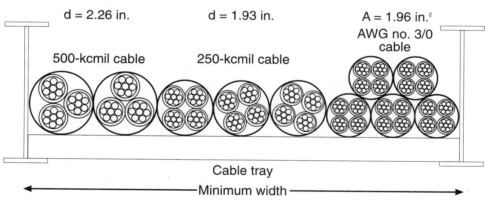

Figure 12.16   A cable tray contains only type TC multiconductor cables, several of which are larger than AWG number 4/0 and several of which are AWG number 3/0.

4. An industrial building does not have a maintenance electrician or a maintenance supervisor continuously employed, nor does the building supervisor have an arrangement with an electrical contractor to inspect and maintain the electrical system in the building. Is a metal cable tray permitted to serve as the sole equipment grounding conductor for electrical equipment in the building? (Yes or No)

   Code reference _____

5. Ventilated-trough cable tray contains only single copper conductor cables installed in a single layer with an air space maintained between the cables not less than the width of the largest adjacent cable, as shown in Figure 12.17. There are three 1,000-kcmil conductors, four 500-kcmil conductors, and four AWG number 3/0 conductors. The minimum permitted width of cable tray is:

A. 9 inches.
B. 12 inches.
C. 18 inches.
D. 24 inches.
E. 30 inches.

Code reference _____

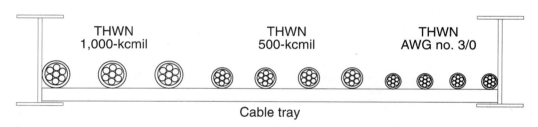

Figure 12.17 A cable tray contains single conductor cables arranged in a single layer with an air space between the cables with a width equal to the diameter of the largest adjacent cable.

6. An aluminum ladder-type cable tray has a marked total cross-sectional area of both side rails of 0.72 square inch. The maximum circuit rating for which this ladder cable tray is permitted to serve as the equipment grounding conductor is _____ amperes.

Code reference _____

7. Type TC power and control cable is permitted to be installed in rigid metal conduit. (True or False)

Code reference _____

8. Type TC multiconductor cables with three copper type XHHW insulated power conductors are run in uncovered ladder-type cable tray in a single layer with no maintained spacing between the conductors, as shown in Figure 12.18. There are two other cables, one cable with copper AWG number 3/0 conductors. The minimum size copper conductor in the other cable required for a 225-ampere feeder if all terminations are rated 75°C is _____.

Code reference _____

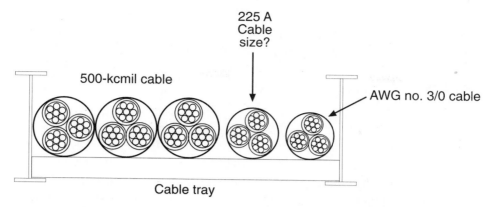

Figure 12.18   A cable tray contains several multiple conductor cables arranged in a single layer, but without a maintained air space between the cables. One of the cables is to be sized as a 225-ampere feeder circuit.

9. A cablebus system is used as feeder conductors from a disconnect at a transformer to the main distribution panelboard. The cablebus will contain one set of 3-phase, 480/277-volt conductors as shown in Figure 12.19. Determine the minimum size type THWN copper conductors permitted to be installed for a 1,200-ampere feeder, with a demand load not more than 960 amperes.

Code reference _____

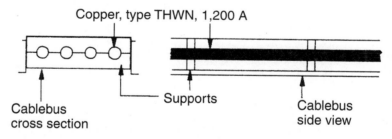

Figure 12.19   A 1,200-ampere cablebus is used as a feeder from a transformer to a distribution panelboard.

10. A cablebus without added support for long spans shall be supported at intervals not to exceed _____ feet.

Code reference _____

11. A 460-volt, 3-phase transformer arc welder with a duty cycle of 50 percent has a primary current listed on the rating plate as 40 amperes. The conductors supplying the welder are AWG number 8 copper with THWN insulation, and conductor terminations are rated 75°C. A single overcurrent device protects both the welder and the circuit. The maximum rating permitted for the overcurrent device for this welder circuit is _____ amperes.

Code reference _____

12. A resistance welder makes repetitive spot welds and has a duty cycle of 4 percent. The actual primary supply current during a weld is 620 amperes at 460 volts. If the supply conductors for the welder are copper, type THWN with 75°C terminations, the minimum permitted size conductors is _____.

    Code reference _____

13. Six transformer arc welders are supplied by one feeder. The welders are 3-phase, 460-volt, with a primary supply current of 46 amperes marked on the rating plate, and the duty cycle is 30 percent, as shown in Figure 12.20. The minimum recommended size type THWN copper feeder conductors to supply these welders, if all terminations are rated 75°C and no other load pattern information is available, is _____.

    Code reference _____

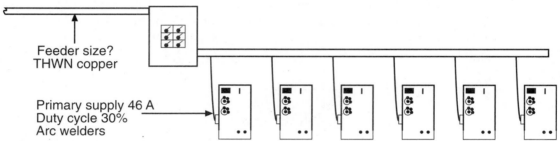

Figure 12.20   Six transformer-type arc welders in an industrial plant are supplied by a common feeder, and the welders have a primary supply current of 46 amperes and a duty cycle of 30 percent.

14. Integrated gas spacer cable trade size 4,500-kcmil is to be run underground in parallel for the service entrance of an industrial building. What is the maximum ampacity of the cable, assuming that the cable is installed such that derating the ampacity is not required?

    Code reference _____

15. A resistance heating tracing cable attached to a metal pipeline, shall be provided with equipment ground-fault protection. (True or False)

    Code reference _____

16. A 120-volt resistance heating element is attached to a pipeline with a maximum current rating of 12 amperes. If this heating element is cord-and-plug connected, the attachment plug and receptacle are permitted to serve as the disconnecting means for the heating element. (True or False)

    Code reference _____

17. A crane runway track is permitted to serve as the grounded phase conductor of a 3-phase circuit with one phase grounded and the other two phases insulated. (True or False)

    Code reference _____

18. Several components of a motor-generator induction heating system output circuit for an industrial annealing process are to be connected with properly sized cable with copper conductors. The cables shall be installed in rigid steel conduit or intermediate metal conduit. (True or False)

    Code reference _____

19. Two direct current power supplies are connected to an electrolytic cell. The required direct current line disconnect to the electrolytic cell from each power supply is permitted to be a removable conductor or circuit conductor link rather than a mechanical switch. (True or False)

    Code reference _____

20. One direct current power supply serves several electroplating tanks. A means shall be provided to disconnect the direct current electrical supply from the electroplating tanks. (True or False)

    Code reference _____

21. The maximum rating of fuse or circuit-breaker permitted to provide short-circuit and ground-fault protection for an industrial machine is required to be provided on the nameplate of the machine. (True or False)

    Code reference _____

22. An industrial machine has a full-load current rating of 42 amperes, 3-phase, 480 volts, and the largest motor of the machine is 15 horsepower with a full-load current of 21 amperes. One additional motor has a full-load current of 11 amperes. The remainder of the load is 10 amperes of resistance heating. A copper wire in conduit with overcurrent protection at the supply end of the supply conductor is properly sized for the conductor and the machine. All conductor terminations are 75°C rated. The minimum size type THWN supply conductor permitted for the industrial machine is AWG number _____.

    Code reference _____

23. The industrial machine in the previous problem is provided with over-current protection rated at 50 amperes at the point where the supply conductors attach to the machine. The ceiling of the industrial area is 40 feet in height, and power for the machine originates at a tap box from a feeder, which is copper type THWN wire size AWG number 3/0 protected with 200-ampere fuses. The manufacturing plant has electrical maintenance personnel on duty at all times. The tap box is 36 feet above the floor, and the tap conductors are copper type THWN in rigid metal conduit. If the total length of the tap is 45 feet, the minimum size tap conductor permitted is AWG number _____.

Code reference _____

24. Conductive optical fiber cable is permitted to be run in the same metal conduit with conductors of light, power, and heating circuits. (True or False)

Code reference _____

25. Nonconductive optical fiber cable for general-purpose use type OFNG installed within a building is not permitted to be run as open cable in plenums or used as a riser cable. (True or False)

Code reference _____

ANSWER SHEET   Name _____

No. 12   INDUSTRIAL ELECTRICAL APPLICATIONS

Answer	Code reference
1. _____	_____
2. _____	_____
3. _____	_____
4. _____	_____
5. _____	_____
6. _____	_____
7. _____	_____
8. _____	_____
9. _____	_____
10. _____	_____
11. _____	_____
12. _____	_____
13. _____	_____
14. _____	_____
15. _____	_____
16. _____	_____
17. _____	_____
18. _____	_____
19. _____	_____
20. _____	_____
21. _____	_____
22. _____	_____
23. _____	_____
24. _____	_____
25. _____	_____

UNIT 13

Commercial Wiring Applications

OBJECTIVES

Upon completion of this unit, the student will be able to:

- state the requirements when multioutlet assembly passes through a partition.
- give a general description of underfloor raceway, cellular metal floor raceway, and cellular concrete floor raceway.
- explain the purpose of an insert for a floor raceway.
- explain what shall be done with wires remaining in use when a receptacle outlet is removed for cellular metal or concrete floor raceway.
- explain how wires are to be connected to outlet devices for underfloor raceway installations.
- describe a flat cable assembly.
- state the maximum branch circuit rating for a circuit of flat cable assembly.
- explain a method used to connect building section wiring when assembling a manufactured building where the connection will be concealed.
- describe the wiring materials from which a manufactured wiring system is constructed.
- explain the requirements for conductor insulation when used for elevators, dumbwaiters, escalators, and moving walks.
- explain the minimum wire size and the number of circuits required for elevator car lighting.
- answer wiring installation questions from *Articles 353, 354, 356, 358, 363, 518, 545, 604,* and *620.*
- state at least three changes that occurred from the 1996 to the 1999 Code for *Articles 353, 354, 356, 358, 363, 518, 545, 604, 620,* or *Appendix D Examples D9* and *D10.*

CODE DISCUSSION

The Code articles studied in this unit deal with installations most frequently encountered in commercial areas and buildings. These applications are not necessarily limited to commercial areas and buildings. Obviously, other articles of the Code also apply to commercial areas and buildings.

Article 353 covers the installation of multioutlet assemblies, which are raceway assemblies containing wiring and outlets. Multioutlet assembly is available factory-assembled, or it may be assembled on location. This wiring method is permitted only in dry locations. *Section 353-2* provides a list of uses permitted and not permitted. Metal multioutlet assembly is permitted to pass through a dry partition provided no outlet is within the partition, and caps or coverings on all exposed portions of the assembly can be removed.

Article 354 deals with the installation of raceways installed under the floor. A raceway listed for the purpose is permitted to be installed in concrete floors. Installation is not limited to concrete floors. Underfloor raceway is permitted to be installed in a floor with a 3/4-inch minimum thickness wood covering, provided the raceway is not more than 4 inches wide. There is a minimum covering for underfloor raceway depending on the width and type of installation. More than one raceway may be run parallel in the floor, with one containing power wires and another containing communications wires. Underfloor raceway listed for the purpose is permitted to be run flush with the floor of an office building and covered with linoleum or an equivalent surface.

When an outlet is to be installed on the floor and have access to the underfloor raceway, an insert is installed. Splices are not permitted to be made at outlets. Connection to a receptacle outlet, for example, is done by means of stripping the insulation from the wire and looping the wire around the terminal of the device, as shown in Figure 13.1. Splices are permitted only in junction boxes, except in the case of flush-mounted underfloor raceway with removable covers.

Article 356 covers the use of the cells of cellular metal floor as electrical raceways. Cellular metal floor decking is sometimes used as the structural floor support between main support beams. Concrete is then poured on this decking to form the finished floor. The Code permits these metal floor cells to be used as electrical raceways. A header is installed before the concrete is poured to provide access to the desired cells. Figure 13.2 is a cross-sectional view of a cellular floor used as an electrical raceway with the header shown connected to two of the cells. The header may connect directly to the supply panelboard, or this connection may be made with a suitable raceway. Splices and taps in wires are only permitted to be made in a header or in a floor-mounted junction box. Wiring requirements for cellular metal floor raceway are similar to those for underfloor raceway. An important requirement of this type of installation is that when an outlet is abandoned, the wire supplying the outlet shall be removed. Reinsulation of the wire is not permitted. If the wire is needed to supply other outlets, it will have to be removed and replaced with a new wire.

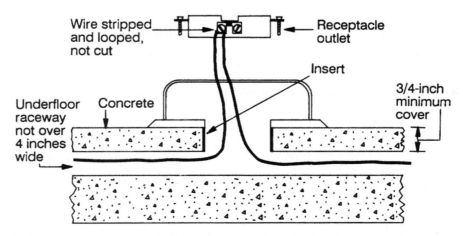

Figure 13.1   A receptacle outlet installed on the floor and supplied with underfloor raceway.

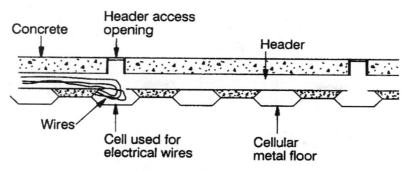

Figure 13.2   Cross-sectional view of cellular metal floor used as an electrical racway.

Article 358 deals with the use of the hollow cells of cellular concrete floor slabs as electrical raceways. Once the cellular concrete slabs are in place, a header is installed to provide access to the desired cells. A separate header and cells are used when communications wires are to be run in the floor. Splices and taps are only permitted to be made in header access units and floor junction boxes. When an outlet is abandoned, the wire supplying the outlet shall be removed. Reinsulation of the wire is not permitted. If the wire is needed to supply other outlets, it will have to be removed and replaced with a new wire.

The raceway in cellular concrete floor slabs is not metal; therefore, an equipment grounding conductor is required to be run from insert receptacles to the header and connected to an equipment grounding conductor. This requirement is found in *Section 358-9*. Figure 13.3 is a cross-sectional view of a cellular concrete floor slab and a header making access to two of the cells.

Article 363 covers the use and installation requirements for type FC, flat cable assembly installed in surface metal raceway identified for use with type FC cable. Figure 13.4 shows type FC cable installed in channel to supply lighting fixtures. This type of cable is permitted only for branch circuits rated at not more than 30 amperes.

Article 518 specifies the type of wiring method that is permitted in buildings or portions of structures where 100 or more persons are likely to assemble. Examples of places of assembly are given in the article. Extra care is taken in these areas to make the wiring system more resistant to fire ignition, fire spread, or the production of toxic vapors. The critical factor for emergency egress is the number of people in an area. It takes time to get a large number of people out of a building during an emergency. If less than 100 persons are in the building or area, egress can be expedited rapidly with a lesser danger to human life. The primary intent here is to buy extra time to get the people out of the building. Electrical nonmetallic tubing

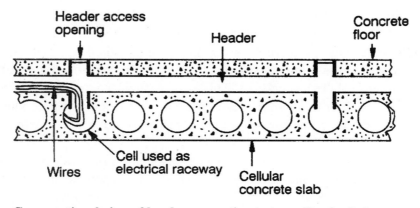

Figure 13.3   Cross-sectional view of header connecting to two cells of cellular concrete floor slab.

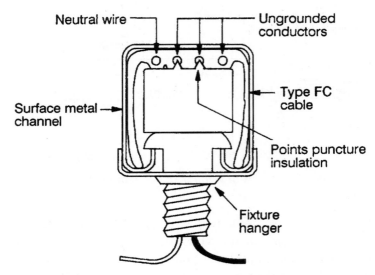

Figure 13.4   Cross-sectional view of type FC cable in channel-type surface metal raceway.

and rigid nonmetallic conduit is permitted to be installed as concealed wiring in walls, ceilings, and floors of some types of areas where separated from the public by a listed 15-minute finish fire-rated surface, *Section 518-4(c)*.

Article 545 covers the installation of wiring components in manufactured buildings. Building components, such as complete wall sections, are constructed (including the wiring) in a manufacturing facility and assembled at a building site. The components are of closed construction, which means the wiring is not accessible for inspection at the building site without disassembling the building components. All raceway and cable wiring methods in the Code are permitted for use in manufactured buildings. Other wiring methods shall be listed for the purpose. Receptacle outlets and switches with integral boxes are permitted if listed for the purpose. At the point where building sections are joined, fittings listed for the purpose and intended to be concealed at the time of on-site assembly shall be permitted to join the conductors of one section to the conductors of another section.

Article 604 covers the use and construction of manufactured wiring systems. These are assemblies of sections of type AC cable, type MC cable, or wires in flexible metal conduit or liquidtight flexible conduit with integral receptacles and connectors for connecting electrical components in exposed locations, as shown in Figure 13.5. These manufactured wiring systems are permitted to extend into hollow walls for direct termination at switches and other outlets. The conductors in the manufactured wiring system are not permitted to be smaller than AWG number 12 copper, except for taps to single lighting fixtures. A common application has been the connection of lay-in fluorescent fixtures in suspended ceilings. Manufactured wiring systems can be listed for installation in outdoor locations as illustrated in Figure 13.6.

Article 620 deals with the wiring of elevators, dumbwaiters, escalators, and moving walks. A major issue is that wiring associated with this equipment shall have flame-retardant insulation, even when installed in conduit. This is of particular importance to minimize the spread of fire from one floor to another. When wireways are installed for conductors supplying elevator circuits, the fill requirements are greater than for other circuit wires in wireways of the building.

Requirements for conductors, wiring, and the installation of conductors are covered in *Parts B, C, D,* and *E. Parts F* and *G* cover requirements for disconnecting means and overcurrent protection. *Section 620-61(b)* states that elevator and dumbwaiter driving motors shall be classed as intermittent duty. In the case of escalator and moving walk driving motors, they

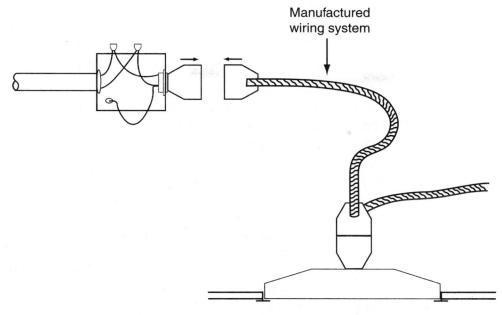

Figure 13.5 A manufactured wiring system is a prewired assembly of conductors and connectors in type AC or type MC cable, flexible metal conduit, or liquidtight flexible conduit.

shall be considered as continuous duty. Another code contains the requirement that a 3-phase drive motor be prevented from starting in the event there is a phase rotation reversal. If there is a phase rotation reversal, the drive motor will reverse direction. This can cause the elevator to run in reverse. In the case of a hydraulic drive, overheating will occur. Interchanging any 2-phase conductors on the building wiring system or on the primary electrical system supplying power to the building will result in a phase rotation reversal.

Requirements for the wiring of wheelchair lifts and stairway chair lifts are also contained in *Article 620.* One complex task involved in the wiring of an elevator system is the determination of the minimum size of feeder required in the system. *Example D9* and *Example D10* show how the feeder conductor is sized in two different situations. Several factors concerning elevator installations are not obvious by reading *Article 620.* Example 13.1 will help illustrate how the feeder conductor selection current is determined.

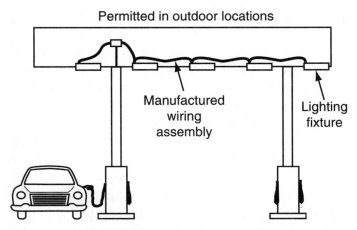

Figure 13.6 Manufactured wiring systems listed for the purpose are permitted to be installed in an outdoor location.

Example 13.1 Four passenger elevators are powered with motor-generator sets with the motor generator set powered by 30-horsepower, 3-phase, 460-volt motors. In addition to this load, 12 amperes continuous load per elevator are required for operation control equipment. The motor-generator sets have a continuous duty rating. Determine the minimum feeder conductor selection current.

Answer: The last sentence of *Section 620-13* states that for a motor-generator set, the feeder conductor shall be determined based on the ampacity of the drive motor. *Section 620-13(b)* can be misleading because passenger elevator motors are actually rated as intermittent duty motors. Start by looking up the 30-horsepower, 3-phase, 460-volt motor full-load current in *Table 430-150,* which is 40 amperes. The load consists of several motors plus other load; therefore, *Section 430-24* is used to determine the feeder current rating. The feeder current is the sum of the full-load current of all the motors, plus other load, plus 25 percent of the full-load current of the largest motor. If all motors are the same size, then take one of them as the largest. But it is not quite so easy. *Exception 1* to *Section 430-24* states that if the motors are used for short-time, intermittent, periodic, or varying duty, the motor current shall be as determined by *Section 430-22(b). Table 430-22(b)* states that a passenger elevator motor is considered to be intermittent duty. If the actual motor is a continuous duty motor, then it can be overloaded without damage if operated intermittently. Therefore, the continuous full-load current rating of the motor is multiplied by 1.4 for use in the feeder calculation because it will be overloaded for short periods of time. In this case, the current to use for the motors is 56 amperes.

$$40 \text{ A} \times 1.4 = 56 \text{ A}$$

The feeder selection current calculation is determined according to *Section 430-24, Exception 1* as follows:

$$4 \times 56 \text{ A} = 224 \text{ A}$$

According to *Section 620-14,* a demand factor can be applied to that portion of the feeder current determined according to *Section 620-13.* The demand factor for four elevators as determined in *Table 620-14* is 0.85. To get the actual feeder selection current, multiply the motor load calculated by this demand factor and then add the operation control load according to *Sections 430-24* and *215-2(a).*

$$
\begin{aligned}
0.85 \times 224 \text{ A} &= 190 \text{ A}\\
4 \times 12 \text{ A} \times 1.25 &= \underline{60 \text{ A}}
\end{aligned}
$$

250 A feeder selection current

Another important safety issue is servicing of elevators and similar equipment. Personnel are frequently in a position to be exposed to shock hazard; therefore, in *Section 620-85,* receptacles are required to be ground-fault circuit-interrupter protected in some locations. All 120-volt, 15- and 20-ampere receptacle outlets in an elevator hoistway pit, on top of an elevator car, and in an escalator or moving walk wellway are required to be of the ground-fault circuit-interrupter type. Receptacles rated 120 volt, 15 or 20 amperes in a machine room or a machine space are also required to be ground-fault circuit-interrupter protected. It is important that lighting in elevator hoistway pits and in machine rooms or machine spaces, if on the same circuit with receptacles, be tapped ahead of the ground-fault circuit-interrupter so the area will not be placed in darkness if the interrupter trips. This is illustrated in Figure 13.7.

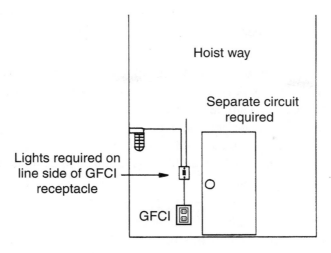

Figure 13.7 **The lighting serving an elevator hoist way is required to be tapped on the line side of any ground-fault circuit-interrupter protecting receptacles.**

MAJOR CHANGES TO THE 1999 CODE

These are the changes to the 1999 *National Electrical Code®* for electrical systems that correspond to the Code sections studied in this unit. The following analysis explains the significance of only the changes from the 1996 to the 1999 Code, and this analysis is not intended to be used in place of the Code. Refer to the actual section of the 1999 Code for the exact wording and meaning of each section discussed. Changes are indicated in the Code with a vertical line in the margin. If material has been deleted or moved to another part of the Code, the location of the deletion is indicated with a dark dot in the margin.

Article 353 **Multioutlet Assembly**

No changes were made to this article.

Article 354 **Underfloor Raceways**

No changes were made to this article.

Article 356 **Cellular Metal Floor Raceways**

No changes were made to this article.

Article 358 **Cellular Concrete Floor Raceways**

No changes were made to this article.

Article 363 **Flat Cable Assemblies**

No changes were made to this article.

Article 518 **Places of Assembly**

518-2: This section provides building and occupancy groups that are considered by the Code to be places of assembly. This section was reorganized into three separate paragraphs. The only significant change that occurred in this section is in subsection *(c)*, for theatrical areas, which was the last paragraph of *Section 518-2* in the previous edition of the Code. The phrase **including associated audience seating areas** was added to make it clear that the requirements of *Article 520* apply to those areas as well as the projection booth or stage area. In the past, there was confusion over which article applied to these areas.

518-3(b): The provisions on temporary wiring installed in exhibition halls for display booths are contained in this section. A new exception was added permitting hard or extra hard usage cords to be placed in a single layer in cable tray for temporary use only, provided qualified personnel will service and maintain the installation, as shown in Figure 13.8. The cable tray is permitted to contain only temporary wiring, and a permanent sign stating **Cable Tray for Temporary Wiring Only** is required to be attached to the cable tray at intervals not exceeding 25 feet.

518-4(a): This section specifies the wiring methods permitted in buildings or portions of buildings classified as places of assembly. Flexible metal raceway is specifically mentioned so it will be clear that material such as flexible metal conduit is permitted to be installed in an area classified as a place of assembly. Type AC cable is now permitted to be installed in an area rated as a place of assembly as shown in Figure 13.9.

Article 545   **Manufactured Buildings**

No significant changes were made to this article.

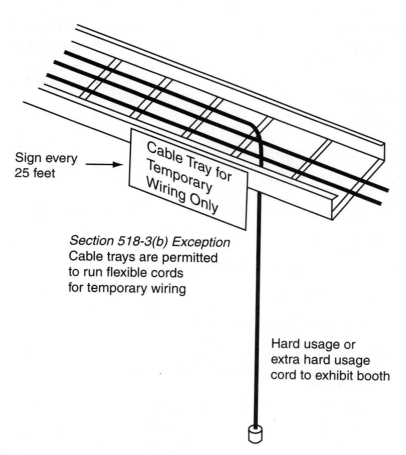

Figure 13.8   Cable tray is permitted to be installed in buildings such as exhibition areas for temporary wiring such as extra hard usage service cords to exhibition booths.

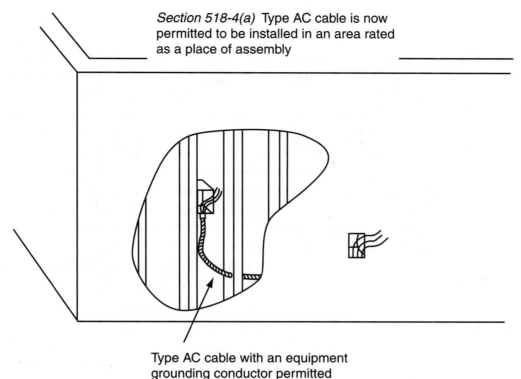

Section 518-4(a) Type AC cable is now permitted to be installed in an area rated as a place of assembly

Type AC cable with an equipment grounding conductor permitted

Figure 13.9   Type AC cable is permitted to be installed in areas classified as places of assembly.

Article 604   **Manufactured Wiring Systems**

604-6(a) Exception 3: When connecting components together, hard usage flexible cords are permitted to be part of a listed factory-made assembly. This new exception allows for such uses provided the conductors are at least AWG number 12. The cable is not permitted to be more than 6 feet in length, concealed, or subjected to damage or strain. The manufactured wiring systems that utilize these cords are not permitted to be permanently secured to any part of the building. This is a practice that can be used to supply power to equipment such as display cases that are freestanding in open areas.

Article 620   **Elevators, Dumbwaiters, Escalators, Moving Walks, Wheelchair Lifts, and Stairway Chair Lifts**

620-21(a)(2)(d): This is a new paragraph that specifies the types of flexible wiring methods permitted on elevator cars to supply power to a driving machine or a driving machine brake located on the car.

620-22: This section requires a separate branch circuit for elevator car lighting, receptacles, auxiliary lighting, and ventilation. Now the overcurrent device for this elevator car circuit is required to be located in the machine room.

620-23(a): This section requires a separate branch circuit to supply lighting and receptacles in the machine room. The receptacles were required to be ground-fault circuit-interrupter protected. The change deals with how the lighting is connected in the circuit. The previous edition of the Code did not permit the lighting to be connected on the load side of the ground-fault circuit-interrupter **receptacle.** If the ground-fault circuit-interrupter is a circuit-breaker, then an alternate means of wiring the circuit will need to be used so the lighting is not on the load side of any type of ground-fault circuit-interrupter device, as illustrated in Figure 13.10.

620-24(a): This section requires a separate circuit to be installed in an elevator hoistway pit with the same change as in *Section 620-23(a)* with regard to the connection of the lighting on the supply side of the circuit ground-fault circuit-interrupter **device.**

620-53: The disconnecting means requirements for car lights, receptacles, and ventilation equipment are given in this section. A new sentence was added that permits the disconnect for this equipment to be placed in the machinery space for the elevator. Some elevators are designed so that a machinery room is not required.

620-54: The disconnecting means requirements for heating and air-conditioning equipment are given in this section. A new sentence was added that permits the disconnect for this equipment to be placed in the machinery space for the elevator. Some elevators are designed so that a machinery room is not required.

620-85: Receptacles installed on 125-volt, 15- and 20-ampere circuits and located in a machinery space are no longer required to be of the ground-fault circuit-interrupter type. These receptacles need only be ground-fault circuit-interrupter protected.

620-91(c): When there is a power source on the load side of the disconnecting means, auxiliary contacts are required to be provided to disconnect power from the auxiliary power source when the disconnecting means is opened. The change in this section is that the means by which this auxiliary contact opens must be mechanical and cannot rely only on springs.

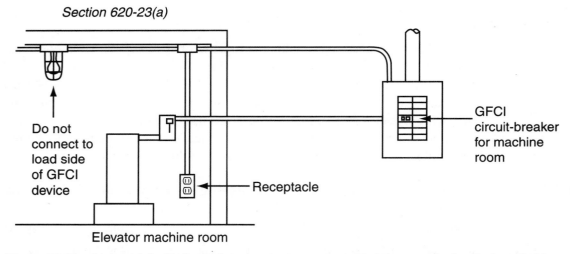

Figure 13.10   A ground-fault circuit-interrupter type receptacle is not required to be installed in an elevator machine room or area, but the receptacle must be ground-fault circuit-interrupter protected. The lighting in the room or area is not permitted to be connected on the load side of a ground-fault circuit-interrupter device.

WORKSHEET NO. 13
COMMERCIAL WIRING
APPLICATIONS

These questions are considered important to understanding the application of the *National Electrical Code®* to electrical wiring, and they are questions frequently asked by electricians and electrical inspectors. People working in the electrical trade must continue to study the Code to improve their understanding and ability to apply the Code properly.

DIRECTIONS: Answer the questions and provide the Code reference or references where the necessary information leading to the answer is found. Space is provided for brief notes and calculations. An electronic calculator will be helpful in making calculations. You will keep this worksheet; therefore, you must put the answer and Code reference on the answer sheet as well.

1. An insert is installed into underfloor raceway of an office building for electrical outlets at workstation locations. The raceway is covered with concrete. The wires to the receptacle outlets are permitted to be cut and spliced with a short pigtail for each outlet as shown in Figure 13.11. (True or False)

    Code reference _____

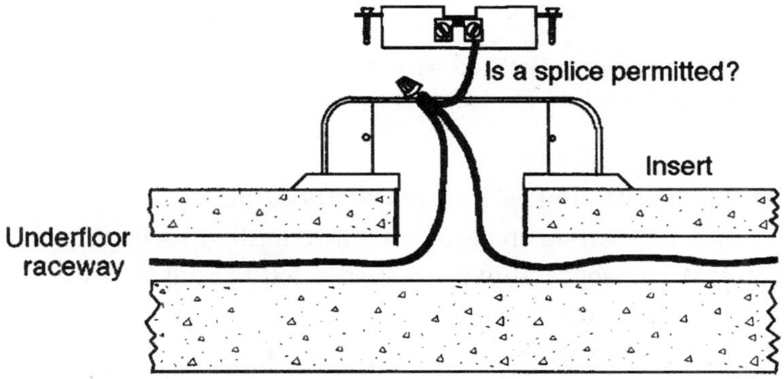

Figure 13.11   Is the circuit conductor permitted to be cut and spliced for a receptacle at an insert for underfloor raceway?

2. Type AC cable is permitted to be installed as permanent wiring in areas classified as places of assembly. (True or False)

    Code reference _____

3. If metallic underfloor raceway does not provide for the termination of an equipment grounding conductor and approved fittings allowing for equipment grounding are not used, the connection to the metallic underfloor raceway shall be by an approved metallic conduit. (True or False)

   Code reference _____

4. A manufactured wiring system is permitted to be concealed in walls to connect a switch to a lighting fixture as illustrated in Figure 13.12. (True or False)

   Code reference _____

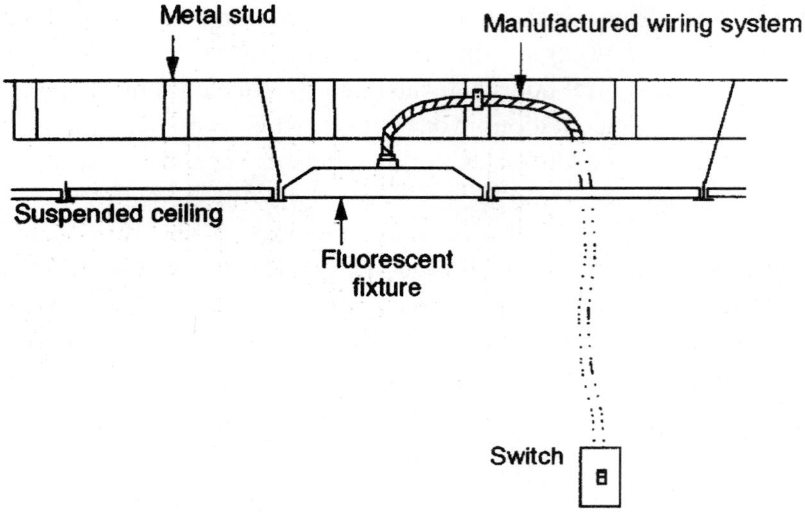

Figure 13.12   Is a manufactured wiring system permitted to extend from a lighting fixture down a concealed wall to a switch?

5. The suspended ceiling of a restaurant does not have a 15-minute finish rating, and the ceiling material is not identified in listings of fire-rated assemblies. The space above the suspended ceiling is not used for environmental air. Is electrical nonmetallic tubing (ENT) permitted to be installed in the space above the suspended ceiling? (Yes or No)

   Code reference _____

6. Eight passenger elevators are powered with motor-generator sets with each motor-generator set powered by a 30-horsepower, 3-phase, 460-volt continuous duty motor. In addition to this load, the motion-operation controller for each elevator has a **continuous** load of 9 amperes. The minimum feeder conductor ampacity for the eight elevators is _____ amperes.

   Code reference _____

7. Metallic-multioutlet assembly is permitted to be run through a partition as long as all exposed covers can be removed, and no outlet is contained within the partition. (True or False)

   Code reference _____

8. A receptacle outlet is removed at an insert in cellular metal floor raceway, and the outlet is abandoned. It is permitted to tape the bare conductors once used to connect to the receptacle outlet, as long as the tape restores the original insulation value of the conductors. (True or False)

   Code reference _____

9. Except by special permission of the authority having jurisdiction, the maximum size conductor permitted to be installed in cellular metal floor raceway is AWG number _____.

   Code reference _____

10. A floor is constructed of precast cellular concrete slabs. Provided approved fittings are used, are these concrete cells in the floor permitted to be used to run electrical wires as shown in Figure 13.13? (Yes or No)

    Code reference _____

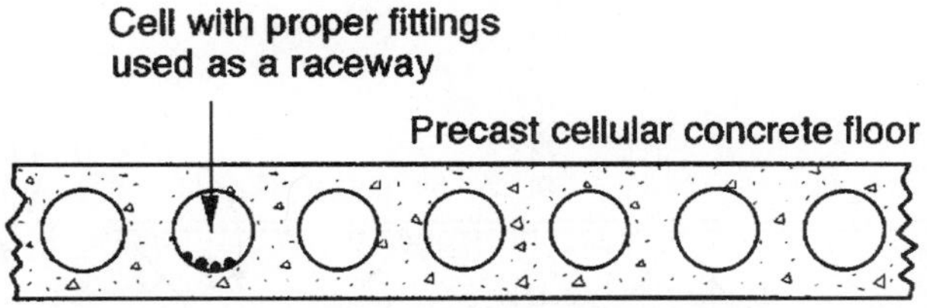

Figure 13.13   Are the cells of precast cellular concrete slabs permitted to serve as raceways for electrical conductors?

11. An insert is installed into cellular concrete floor raceway for a receptacle outlet at a workstation. An equipment grounding wire is required to be run from the receptacle and bonded to the equipment grounding conductor at the header for the cell. (True or False)

    Code reference _____

12. Type FC, flat cable assembly is permitted to be run as exposed cable on nonmetallic surfaces without raceway protection for distances not to exceed 6 feet. (True or False)

    Code reference _____

13. Type FC, flat cable assembly is permitted to be installed in channel-type surface metal raceway before mounting the surface metal raceway to the exposed surface of the building. (True or False)

    Code reference _____

14. Is it permitted to install a tap device in a flat cable assembly for a lighting fixture operating at 240 volts? (Yes or No)

    Code reference _____

15. General-use, 125-volt, 15- or 20-ampere receptacles installed in an elevator machine room are required to be of the ground-fault circuit-interrupter type. (True or False)

    Code reference _____

16. In lengths not exceeding 6 feet, type SO flexible cord with AWG number 12 conductors is permitted to be installed as part of a listed factory-made assembly to connect the manufactured wiring system to the lighting fixture not permanently secured to the building as shown in Figure 13.14. (True or False)

    Code reference _____

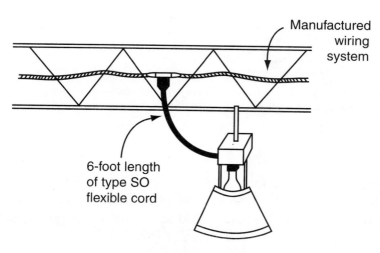

Manufactured
wiring
system

6-foot length
of type SO
flexible cord

Figure 13.14   A high-intensity discharge lighting fixture is connected to a manufactured wiring system using a listed fixture with a factory-installed type SO flexible cord that is not more than 6 feet in length.

17. The wiring methods of *Chapter 3* of the Code are permitted to be installed in manufactured buildings. (True or False)

    Code reference _____

18. Cable wiring is permitted to be joined in a manufactured building without the use of junction boxes at points where wall sections and other building components are joined, using fittings and connectors listed for the purpose and concealed once the building components have been joined. (True or False)

    Code reference _____

19. Type THWN copper wire for a circuit associated with an escalator is permitted to be installed in electrical metallic tubing within the escalator wellway. (True or False)

    Code reference _____

20. A building or a portion of a building is considered a place of assembly if it is intended for the assembly of _____ or more persons.

    Code reference _____

21. A 7-foot run of liquidtight flexible nonmetallic conduit, securely fastened in place, is permitted to be installed on an elevator car. (True or False)

    Code reference _____

22. A single separate branch circuit is required for the supply of lighting auxiliary lighting power source, receptacles, ventilation, and accessories for each elevator car. (True or False)

    Code reference _____

23. A wireway containing only wires for elevator circuits is permitted to have the sum of the cross-sectional area of the individual conductors in the wireway exceed the 20 percent fill limitation of *Article 362*. (True or False)

    Code reference _____

24. The disconnecting means for the motor driving a pumping unit for a hydraulic elevator is permitted to be a circuit-breaker capable of being locked in the open position. (True or False)

Code reference _____

25. The car light and auxiliary equipment circuit disconnecting means for each of four elevators are in a machine room. The disconnecting means are required to be numbered with numbers corresponding to the number on each elevator car. (True or False)

Code reference _____

ANSWER SHEET   Name _____

No. 13   COMMERCIAL WIRING APPLICATIONS

Answer	**Code reference**
1. _____	_____
2. _____	_____
3. _____	_____
4. _____	_____
5. _____	_____
6. _____	_____
7. _____	_____
8. _____	_____
9. _____	_____
10. _____	_____
11. _____	_____
12. _____	_____
13. _____	_____
14. _____	_____
15. _____	_____
16. _____	_____
17. _____	_____
18. _____	_____
19. _____	_____
20. _____	_____
21. _____	_____
22. _____	_____
23. _____	_____
24. _____	_____
25. _____	_____

UNIT 14

Special Applications Wiring

OBJECTIVES

Upon completion of this unit, the student will be able to:

- state the wiring methods permitted for a theater or similar location.
- explain the location of the controls and overcurrent protection for stage lighting outlets and receptacles.
- explain the special requirements for switching lighting and receptacles in a theater dressing room.
- explain the wiring methods permitted to be used in a motion picture or television studio.
- state the minimum size of wire permitted to supply an arc or xenon projector in a motion picture projection room.
- define an equipotential plane, as related to agricultural building wiring.
- explain the grounding requirements for water pump on a farm.
- describe which article of the Code applies to the various types of communications circuits that can be installed within a building.
- describe the separation requirements of various types of communications wires in relation to circuit wires for normal light and power.
- explain the grounding of various types of communications circuits and equipment.
- answer wiring installation questions relating to *Articles 520, 525, 530, 540, 547, 553, 555, 625, 640, 650, 780, 800, 810, 820,* and *830.*
- state at least five significant changes that occurred from the 1996 to the 1999 Code for *Articles 520, 525, 530, 540, 547, 553, 555, 625, 640, 650, 780, 800, 810, 820, or 830.*

CODE DISCUSSION

This unit deals with special applications where conditions exist resulting in special requirements for wiring installations. Generally, the requirements of the Code apply except for specific requirements of these articles.

Article 520 covers wiring installations in buildings or portions of buildings used as theaters for dramatic and musical presentations and motion picture projection. This article also applies to portions of motion picture and television studios used as assembly areas.

Parts A, F, and *G* apply to the fixed wiring in the stage, auditorium, dressing rooms, and main corridors leading to the auditorium. The fixed wiring for lighting and power shall be in metal raceway, nonmetallic raceway encased in at least 2 inches of concrete, or type MI or type MC cable. Many runs of wires are required for control of lighting. Many wires will be subjected to continuous load at times, but not all wires will be subjected to continuous load at the same time. For this reason, the derating factors of *Section 310-15(b)(2)* do not apply when there are more than 30 wires in wireway or auxiliary gutter, as stated in *Section 520-6.*

Dressing rooms of theaters and studios are areas where portable equipment is used, and there are usually high lighting levels. The potential for fire is great; therefore, special precautions are taken to minimize the chances of fire. Permanently attached open-ended guards are required to be installed around all incandescent lamps that are located less than 8 feet above the floor, as indicated in *Section 520-72.* According to *Section 520-73,* all receptacles located adjacent to a mirror or serving a dressing table counter top are required to be controlled by a switch located in the dressing room. In addition, a pilot light must be installed outside of the dressing room adjacent to the door indicating when the dressing room receptacles are energized. These requirements are shown in Figure 14.1. Other receptacle outlets installed in the dressing room not adjacent to a mirror or not serving a dressing table counter top are not required to be controlled by a switch.

Part B applies to fixed stage switchboards and the feeders for switchboards. The requirements for portable switchboards are covered in *Part D. Section 520-23* requires that receptacles intended to supply cord-and-plug stage lighting equipment, whether the receptacles are on the stage or located elsewhere, shall have their circuit overcurrent protection at the stage lighting switchboard location. *Part E* applies to portable stage equipment, but the receptacles for portable equipment are considered to be a fixed part of the building wiring and are covered in *Section 520-45.* Conductors supplying receptacles shall be sized and protected for overcurrent

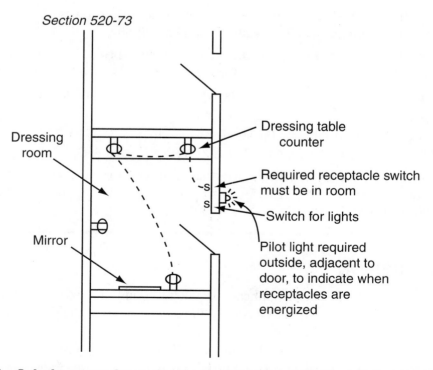

Figure 14.1   Only the receptacles serving the dressing table counter top and the receptacles located adjacent to the mirror are required to be controlled by a switch, and the pilot light to indicate when the receptacles are energized is located outside the dressing room adjacent to the door.

in accordance with the provisions of *Article 310*. The receptacles for connection of equipment shall supply continuous loads not in excess of 80 percent of the ampere rating of the receptacle. A receptacle outlet is permitted to supply a noncontinuous load rated not more than 100 percent of the receptacle ampere rating.

    Article 525 deals with wiring of a temporary nature, installed in adverse conditions, where the public is exposed to the temporary wiring and equipment supplied. This article deals with electrical power for equipment at carnivals, circuses, fairs, and similar events. *Article 305* does not adequately cover the conditions that exist at these events. *Section 525-12(b)* specifies the clearances of overhead wires from amusement rides and attractions. For example, a horizontal clearance of not less than 15 feet must be maintained from amusement rides or attractions to overhead conductors as illustrated in Figure 14.2. *Section 525-30* requires a disconnecting means for the power to all rides to be located within sight and not more than 6 feet from the operator's location. Flexible cords and cables shall be listed for extra hard usage, wet locations, and be sunlight-resistant, *Section 525-13(a)*. Where accessible to the public, flexible cords and cables shall be covered by approved nonmetallic mats. Receptacles shall have overcurrent protection that does not exceed the rating of the receptacle, *Section 525-15(c)*. According to *Section 525-18*, the ground-fault circuit-interrupter protection requirement of *Section 305-6* shall not apply to installations covered by this article. Grounding and bonding are covered in *Part C*, which essentially complies with *Article 250*. It is pointed out that the grounded circuit conductor and the equipment grounding conductor are to be maintained separate on the load side of the service disconnecting means or the separately derived system, *Section 525-22*. In *Article 305*, 90 days is generally defined as the maximum length of time permitted for temporary wiring. There is no time limit for wiring to equipment at carnivals, circuses, fairs, and similar events.

    Article 530 applies to buildings or portions of buildings used as studios, using motion picture film or electronic tape more than 7/8 inch in width. This article also applies to areas where film and tape are handled, or where personnel are working with the film or tape for various purposes. An area for an audience is not present in the facilities covered by this article. The same requirements as for a theater, as covered in *Article 520*, should be applied. This article contains special requirements for areas where quantities of highly flammable materials,

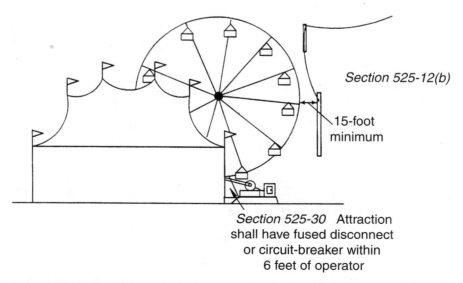

Section 525-12(b)

15-foot minimum

Section 525-30 Attraction shall have fused disconnect or circuit-breaker within 6 feet of operator

Figure 14.2 Overhead conductors are required to be kept at least 15 feet horizontally from rides and attractions, and a disconnect is to be located within sight and not more than 6 feet from the operator's location.

such as motion picture film and electronic recording tape, are present. The permanent wiring in stage and set areas shall be type MC cable, type MI cable, or approved raceway. The authority having jurisdiction will decide what wiring methods are approved. Generally, metal raceway wiring is the approved method for raceway wiring.

There is a type of electrical system referred to as **technical power** covered in *Part G.* The purpose is to supply a 120-volt circuit that does not have a grounded conductor, yet it is a grounded electrical system. Noise in audio and video production can be reduced with this type of system, which actually uses a transformer to create a new 60/120-volt system as illustrated in Figure 14.3. The receptacles are required to be labeled **technical power,** and to reduce the chances of connecting normal 120-volt equipment into the technical power outlet, a receptacle supplied from the normal power system is required to be located not more than 6 feet from the technical power outlets.

Article 540 deals with the wiring for motion picture projectors and the projection rooms. NFPA standard, number 40-1997 provides additional information about the storage and handling of cellulose nitrate motion picture film. This material is highly flammable; therefore, special requirements are placed on the wiring of projection rooms. Local building codes will provide additional information as to which areas are considered part of the projection room. For example, a film rewinding room would be considered part of the projection room because it opens directly into the projection room.

Article 547 covers the wiring of agricultural buildings where excessive dust, dust with water, or corrosive atmospheres are present. Buildings in agricultural areas in which these conditions are not present are permitted to have the wiring installed according to the requirements elsewhere in the Code. The wiring methods permitted are cables and raceways suitable for the conditions. The type of raceways permitted is not stated, but rigid nonmetallic conduit is frequently used where not exposed to physical damage. Several cables are permitted, but the cables most commonly used for interior wiring of agricultural buildings are type UF and copper type SE. Where flexible connections are needed, liquidtight flexible conduit or flexible cord listed for hard usage is permitted. Boxes and enclosures used in agricultural buildings shall be weather-proof and dust-proof in areas where there is excessive dust and dust with water. In areas where the atmosphere is corrosive, boxes and enclosures shall be suitable for

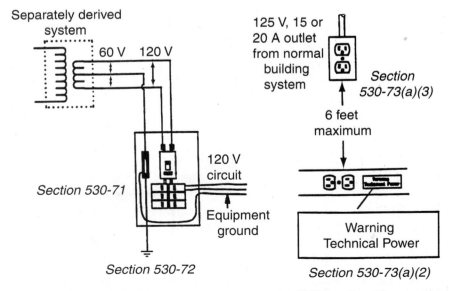

Figure 14.3   A technical power system operates at 60/120 volts with a common conductor grounded to supply power to audio and video production equipment.

the conditions. Motors used in these areas shall be totally enclosed or designed to minimize the entrance of dust, particles, and moisture. Lighting fixtures in agricultural buildings shall be a type that minimizes the entrance of dust, moisture, and other foreign material.

Grounding and bonding for an agricultural building are done according to the requirements of *Article 250,* but there are some exceptions. Physical damage by animals and equipment can interrupt equipment grounding when the equipment grounding is metal raceway. A copper equipment grounding conductor is required by *Section 547-4(f),* even when metal raceway is the wiring method. If a portion of the circuit is underground, the equipment grounding conductor is required to be an insulated or covered copper conductor.

The neutral conductor of a grounded electrical system carrying electrical current will usually have a measurable voltage from the neutral conductor to a ground rod driven into the earth in an isolated location. This is called neutral-to-earth voltage. The term **stray voltage** describes the condition where a neutral-to-earth voltage can be measured between points with which livestock may make contact, such as from a watering device in the barn to the concrete floor. A source of this stray voltage is the neutral wire of the feeder to a farm building. This article of the Code permits the neutral of the feeder to a building to be run separate from, and insulated from, the equipment grounding conductor to the building. This procedure will prevent the neutral-to-earth voltage from the feeder neutral from getting to the areas where an animal may make contact. This is accomplished by running four wires to a farm building in the case of a single-phase, 120/240-volt feeder. An equipment grounding terminal is installed at the building service panel, which is separate from the neutral terminal, as shown in Figure 14.4.

It is extremely important to note that when the neutral conductors and the equipment grounding conductors are separated in a building, an equipment grounding conductor must be run from that building back to the source of power. That source of power may be another building, or it may be a center distribution pole. Simply grounding the equipment grounding bus to the earth and not running the equipment grounding bus back to the main source of power is a

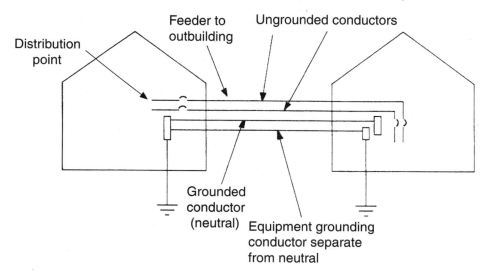

Section 547-8(c) The equipment grounding conductor shall have an ampacity not less than that of the ungrounded circuit conductors when overcurrent protection is not provided at the distribution point.

Figure 14.4 The equipment grounding conductor for a feeder on a farm is permitted to be run separated from the grounded circuit conductor, and is required when there is a metal connection between buildings such as a metal water pipe, *Section 250-32(b)(2).*

violation of *Section 250-2(d)*. The earth is not permitted to be the only equipment grounding conductor.

Electrical wiring in agricultural buildings often is subject to physical damage and corrosive conditions. *Section 547-4(f)* requires that a copper equipment grounding conductor be run for all circuits even when run in metal conduit. This is to ensure that noncurrent-carrying parts of equipment will remain grounded even in the harsh environments found in agricultural areas. *Section 250-112(l)* and *(m)* require that a water pump installed on a farm be grounded. There is an additional requirement that in the case of a submersible water pump, as shown in Figure 14.5, a metal well casing shall be bonded to the pump equipment grounding conductor.

Another requirement of agricultural buildings is that concrete-embedded metal elements shall be installed. This is called an equipotential plane. This equipotential plane shall be bonded to the grounding electrode system of the building. By bonding all metal objects in an animal confinement area together, and then bonding the metal to the service grounding electrode system, an animal is not exposed to stray voltage in that confinement area. This article requires that an equipotential plane shall be bonded to the building grounding electrode system. An equipotential plane in a milking parlor is illustrated in Figure 14.6.

Article 553 covers the wiring of an electrical service to floating buildings. A floating building is a structure moored in a permanent location, provided with electrical wiring, and permanently connected to an electrical system not associated with the floating building. The service equipment is not permitted to be located on or in the floating building. It shall be located adjacent to the building location. The equipment grounding conductor shall be run separate from the insulated neutral conductor. The equipment grounding conductor and the neutral conductors shall be maintained separate throughout the wiring system and equipment of the floating building. This is similar to the separation of neutral and equipment grounding conductor illustrated in Figure 14.4.

Article 555 covers the wiring of marinas and boatyards where electrical power is used on fixed or floating piers, wharfs, and docks, and where power from these structures supplies shore power to boats and floating buildings. A receptacle outlet intended to supply shore power to a boat shall be of the grounding and locking type. A receptacle outlet with a minimum

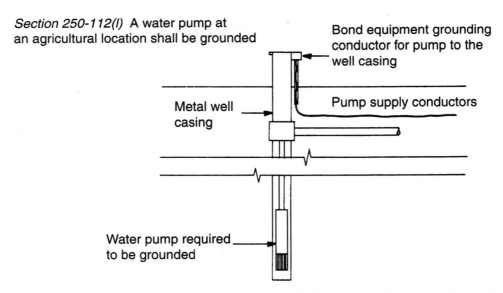

Section 250-112(l) A water pump at an agricultural location shall be grounded

Bond equipment grounding conductor for pump to the well casing

Metal well casing

Pump supply conductors

Water pump required to be grounded

Figure 14.5 A water pump on a farm is required to be grounded, and a metal well casing is required to be bonded to the equipment grounding conductor for a submersible pump.

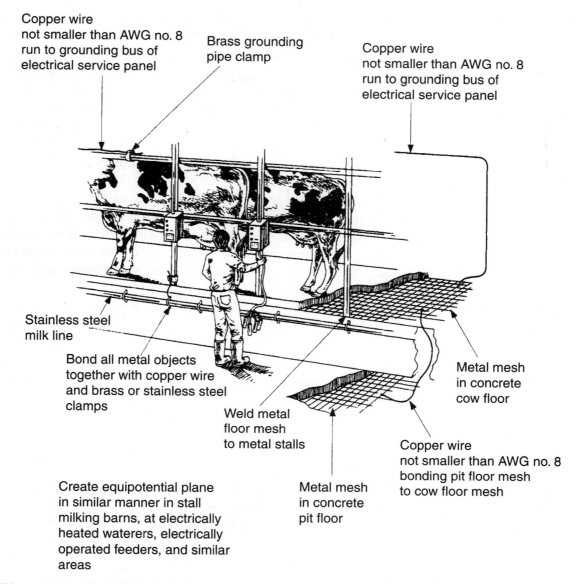

Copper wire
not smaller than AWG no. 8
run to grounding bus of
electrical service panel

Brass grounding
pipe clamp

Copper wire
not smaller than AWG no. 8
run to grounding bus of
electrical service panel

Stainless steel
milk line

Bond all metal objects
together with copper wire
and brass or stainless steel
clamps

Weld metal
floor mesh
to metal stalls

Metal mesh
in concrete
cow floor

Copper wire
not smaller than AWG no. 8
bonding pit floor mesh
to cow floor mesh

Create equipotential plane
in similar manner in stall
milking barns, at electrically
heated waterers, electrically
operated feeders, and similar
areas

Metal mesh
in concrete
pit floor

Figure 14.6 **The installation of metal mesh in the floor connected to all exposed metal forms an equipotential plane that helps prevent animals from being exposed to stray voltage. Courtesy of Consumers Energy, Jackson, MI.**

rating of 20 amperes shall be provided for boats not longer than 20 feet. For boats longer than 20 feet, the receptacle outlet shall have a minimum rating of 30 amperes. If a receptacle rated at 15 or 20 amperes supplies 120-volt power to boats, ground-fault circuit-interrupter protection shall be provided. Wiring methods and grounding are covered. Extra hard usage cord is permitted to be used as a feeder to boats provided it is listed for use in wet locations and is marked as sunlight-resistant. Extra hard usage cord also can be used as a feeder to floating docks and similar structures. But the electrical service is not permitted to be located on the floating dock. The limits of the classified hazardous location near gasoline dispensers are found in *Article 514.* The wiring within these classified areas shall be as required in *Articles 501* and *514.*

Article 625 deals with the wiring and equipment external to a vehicle used for charging of electric vehicles. *Section 625-5* specifies that all materials and equipment shall be listed for the purpose. Specifications for the connection to the electric vehicle are covered in

Part B. Electric vehicle charging loads are to be considered continuous loads, *Section 625-14*. Electric vehicle supply cable shall be a type with the letters EV as stated in *Section 626-17*, or a type listed as being suitable for the purpose. Basically this is a cable suitable for hard usage and wet locations. The electric supply equipment for the vehicle shall have a listed shock protection system for personnel. It shall also have a ground-fault protection system that will disconnect power to the equipment at a level less than required to operate the circuit or feeder overcurrent device, *Section 625-22*. A minimum ground-fault level is not specified. *Section 625-23* requires a disconnecting means capable of being locked in the open position when the equipment is rated more than 60 amperes or more than 150 volts to ground.

Attached and detached residential garages are included in the rules for supply to electric vehicle charging equipment. *Section 625-29(c)* provides specifications for ventilation. The batteries will give off vapors during charging that can create a hazard if not vented. Ventilation is not required where nonvented storage batteries are utilized. Systems for charging electric vehicles shall have provisions to disconnect power in the event of loss of normal power as stated in *Section 625-25*.

Article 640 applies to the wiring of sound recording and sound reproduction systems, such as public address systems, and it even applies to sound reproduction systems of electronic organs. The power wiring to the components of a system follow the rules of the Code that apply to the particular type of wiring or type of conditions. The wiring of the amplifier output and similar wiring shall follow the requirements of *Article 725* for Class 2 or Class 3 wiring, whichever applies. When these wires are installed in wireways or auxiliary gutters, ampacity derating factors do not apply, and the cross-sectional area of the wires is permitted to fill up to 75 percent of the cross-sectional area of the wireway or auxiliary gutter.

Article 650 applies to the circuit wiring for the keyboard of an electrically operated pipe organ and to the controls of the sounding devices. The wiring of electronic organs is to be done following the requirements of *Article 640*. The energy source for electrically operated pipe organs is not permitted to exceed 30 volts and shall originate from a self-excited generator, a rectifier supplied from a two-winding transformer or a battery. When a motor-generator set is used, the bonding is important to prevent the 120- or 240-volt supply voltage from getting onto the generator output circuit wires, which are limited to not more than 30 volts. Either the generator shall be effectively insulated from the motor, or they shall be bonded together. The conductor size, insulation, overcurrent protection, and installation of the wiring operating not over 30 volts are specified in this article.

Article 780 covers closed-loop and programmed power distribution systems. A hybrid cable, which may consist of power conductors, communication conductors, and signaling conductors, is used to wire the convenience outlets in the building. A flat cable is used to connect the service panel and controller to the convenience outlets of each circuit. Figure 14.7 shows how a typical circuit may be installed. The hybrid cable runs from the power control center to the convenience outlets of the circuit. From the convenience outlets, appropriate cables or wiring methods are used to connect to lights, security sensors, telephone, fire detectors, and other outlets. It is intended that individual outlets can be controlled from the control center. For example, outlets in one room can be deenergized to prevent power use. An individual outlet could also be controlled from anywhere in the building. Switches would communicate to the control center, and not directly control power to an outlet. This would permit the outlet and switching arrangement to be reprogrammed as desired. The wiring system can be installed such that each switch, sensor, and power outlet has a unique electronic address. It would therefore be possible to control any outlet in the building, at any time, from any location in the world where there is a communication terminal. After going on vacation, a person can

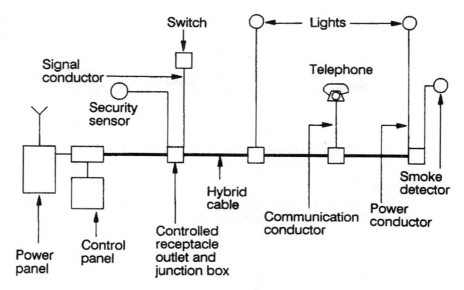

Figure 14.7 **A schematic diagram of how a typical closed-loop and programmed power distribution circuit may be installed.**

determine if something was left energized. But that may not be necessary if an occupancy detector has been programmed to deenergize certain outlets if the building is not occupied.

Article 800 deals with the installation of communication circuits, such as telephone wiring within a building. *Section 800-1* provides a description of the types of communications circuits covered by this article. Some important terms are described in the fine print notes to *Section 800-30(a)*. *Part B* of the article deals with the requirements for the installation of a primary protector on the communications circuit at or near the point of entry to a building or structure. Point of entry is defined in *Section 800-2*. The minimum grounding requirements at the primary protector are covered in *Section 800-40(b)(3)*. In the case of a mobile home, the primary protector is permitted to be grounded at the grounding electrode of the mobile home service or disconnecting means, which may be up to 30 feet from the mobile home. The grounding requirements at the primary protector when a grounding means is not available, are illustrated in Figure 14.8.

Part E provides the requirements for the installation of communications cables on the inside of buildings. It should be noted the requirements of *Section 300-22* are followed when these cables pass through areas where environmental air is contained. The installation of

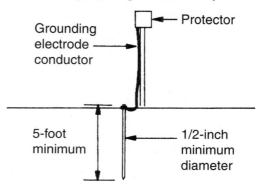

Figure 14.8 **The grounding electrode for communication equipment is permitted to be a pipe or rod not less than 1/2 inch in diameter and a minimum of 5 feet in length when there is no grounding means in the building.**

communication conductors outside and entering buildings is covered in *Part B*. A local jurisdiction may have special rules different from the Code. Communications cables used for wiring in buildings are required to be marked as communications cable or a substitute listed in *Section 800-53(f)* or *Figure 800-53*. The listings and markings of cables for inside installation are covered in *Sections 800-50* amd *800-51*. The grounding of the communication equipment, communication cable sheath, and the protector is covered in *Part D*. There is a provision in the Code that requires that communications systems be installed in a neat and workmanlike manner. Also, raceway is not permitted to be used as means of support for communications cables. This is illustrated in Figure 14.9.

Article 810 deals with radio and television receiving equipment, and receiving and transmitting equipment used in amateur radio. A point of emphasis is the avoidance of contact between the normal electric circuits for light and power and the antenna and wiring of the radio or television lead-in circuits. The normal power circuits for equipment shall follow the provisions of the appropriate portions of the Code. Specific clearances are given in the article for inside and outside installations. These requirements must be studied carefully before making an installation.

Article 820 covers the installation of coaxial cable for community antenna distribution systems for radio and television. Cables used for these purposes shall be listed for use as community antenna television cable, type CATV. *Sections 800-50* and *800-51* list the specific cable markings required for particular installations. The requirements for installation of the various cable types are covered in *Sections 820-52* and *820-53*. The clearances of outside conductors are given in *Part B*. The grounding of the cable, protector, and equipment shall be done following the provisions of *Part D*.

Article 830 covers wiring requirements for network-powered broadband communications systems. The fine print note to *Section 830-1* provides some insight into these systems. Broadband data communication uses a modem to introduce carrier signals onto the transmission system. The carrier signals are modulated by a digital signal. Broadband systems can be subdivided into multiple carrier signals, each carrying a different digitized signal. Broadband operates at the high end of the radio range with frequencies generally between 10 megahertz and 400 megahertz. Multiple carrier signals each transporting a digital signal can operate simultaneously, thus allowing large quantities of digital data to be transported. Teleconferencing including audio, video, and data communication is made possible with a broadband communication system.

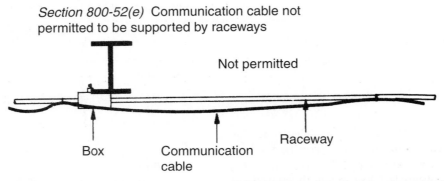

Section 800-52(e) Communication cable not permitted to be supported by raceways

Not permitted

Box   Communication cable   Raceway

Figure 14.9   Communications cables are not permitted to use raceways as a means of support.

MAJOR CHANGES TO THE 1999 CODE

These are the changes to the 1999 *National Electrical Code*® for electrical systems that correspond to the Code sections studied in this unit. The following analysis explains the significance of only the changes from the 1996 to the 1999 Code, and this analysis is not intended to be used in place of the Code. Refer to the actual section of the 1999 Code for the exact wording and meaning of each section discussed. Changes are indicated in the Code with a vertical line in the margin. If material has been deleted or moved to another part of the Code, the location of the deletion is indicated with a dark dot in the margin.

Article 520 **Theaters, Audience Areas of Motion Picture and Television Studios, and Similar Locations**

520-2: There is a new definition of a breakout assembly, which is a type of connector that has multiple poles on one end and **breaks out** or splits into separate connectors such as standard 15- or 20-ampere grounded plug receptacles.

520-25(d): If a solid-state dimmer is to be used where there is more than 150 volts nominal between conductors, it now must be **listed** for the higher voltage.

Table 520-44: Three paragraphs dealing with when the neutral conductor is to be considered a current-carrying conductor were placed as a footnote to this table that determines the ampacity of flexible cords. Actually the table and the notes are the same as found in *Article 400*.

520-53(h)(2): The previous edition of the Code stated that when single conductor cables supplied power to a portable switchboard on stage, all of the conductors were required to be of the same type, size, and length. The change is that this requirement applies only to paralleled conductors.

520-53(h)(5): This is a new provision that permits conductors to pass through holes in walls specifically designed for the purpose to supply power to portable switchboards on stage. If the wall is rated as a fire wall, then appropriate steps must be taken to prevent the spread of fire as provided in *Section 300-21*.

520-53(k): Listed single-pole cable connectors, such as the locking pin and sleeve type, are not required to provide additional strain relief between the cable and connector, as illustrated in Figure 14.10.

520-73: Now only the receptacles in dressing rooms located adjacent to mirrors and serving the dressing table counter tops are required to be controlled by a switch. The pilot light to indicate when the receptacles are energized is now required to be located outside the dressing room adjacent to the door.

Article 525 **Carnivals, Circuses, Fairs, and Similar Events**

525-3(c): This new section requires that wiring for audio signal processing, amplification, and reproduction equipment be installed according to the rules in *Article 640*.

525-12(a): The vertical clearance of 18 feet to overhead conductors applies only to conductors installed outside of tents and concessions.

525-13(a): This section provides specifications on flexible cords. The conditions vary depending upon the location. Cords are not required to be listed for **extra-hard** usage if not subject to physical damage. In this case, they can be listed only for **hard-usage.** Only if cords are used outside are they required to be of a type listed as **sunlight-resistant.**

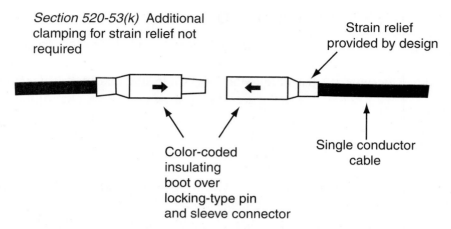

Section 520-53(k) Additional clamping for strain relief not required

Strain relief provided by design

Color-coded insulating boot over locking-type pin and sleeve connector

Single conductor cable

Figure 14.10   Listed single conductor cable connectors, such as the pin and sleeve type, that provide for strain relief by design are not required to provide additional clamping or to utilize other means of providing strain relief to prevent tension from being transmitted to the connection.

525-13(d): Connectors and cable connections are not permitted to be located in audience traffic paths or in areas accessible to the public unless suitably guarded.

525-13(e): Wiring to a ride, tent, or other attraction shall not be supported by another ride, tent, or attraction unless specifically designed to provide such support.

525-13(g): Temporary lighting for general illumination installed inside tents and concessions must have the lamps protected from breakage with a suitable fixture or guard.

525-18: This is a new section that describes specifically the ground-fault circuit-interrupter requirements for receptacles. The previous edition of the Code made reference to *Section 305-6,* which only applied to receptacle outlets that were not a part of the permanent wiring of a building or structure. This new requirement applies to all 125-volt, 15- and 20-ampere receptacles in use by personnel. If receptacles that are a part of permanent wiring are used, they must be ground-fault circuit-interrupter protected. There is a provision for personnel ground-fault protected listed cord sets to be used where receptacles are not protected. Receptacles supplying cooking and refrigeration equipment are not required to be ground-fault circuit-interrupter protected. There is also a reminder that any lighting required for emergency egress for areas is not permitted to be connected to the load side of a ground-fault circuit-interrupter.

525-40: This is a new section that requires equipment to meet the applicable sections of *Article 680* when the attraction has a contained volume of water.

Article 530   Motion Picture and Television Studios and Similar Locations

530-12(a): Stage set wiring for portable lighting and equipment shall be listed for **hard usage,** and if subject to physical abuse, it shall be listed as extra hard usage.

530-12(b): Stage effects and similar equipment is permitted to be wired with listed single conductor or multiconductor flexible cords and must be fastened to scenery by approved cable ties or insulated staples.

530-18(c): This is a new paragraph that requires protection of cables when they pass through enclosures. They are to be protected from abrasion, and secured so that tension will not be applied to terminations. Portable feeder cables are permitted to pass through a fire-rated wall, floor, or ceiling for temporary power provided a seal is used to prevent the spread of fire.

530-22(b): Single-pole separable connectors, such as the pin and sleeve type, are permitted to be used for either ac or dc portable power circuits provided the connector is listed for use with ac or dc circuits.

Article 540   Motion Picture Projectors

540-11(a): Equipment associated with a professional projector that may produce arcs or sparks must be protected. The change is that the acceptable methods of protection are now specifically described in paragraphs (1) through (6).

Article 547   Agricultural Buildings

547-4(a): This section states the wiring methods that are approved for agricultural buildings. The previous edition of the Code stated **other raceways suitable for the location.** That same statement is still in the Code, but now the Code specifically mentions rigid non-metallic conduit and liquidtight flexible nonmetallic conduit as two types of raceway suitable for installation in agricultural buildings.

547-4(e): All wiring and electrical equipment subject to physical damage is now required to be protected. In the past, this requirement applied only to lighting fixtures.

547-8(a): A disconnecting means is required to be installed at all electrical distribution points on farms where two or more buildings or structures are served power from a distribution point, as illustrated in Figure 14.11. This is in direct conflict with *Section 230-21,* which makes reference to a central distribution point on a farm where only a meter is provided. This is a significant change in the practice of supplying electrical power to farms where frequently only a meter at the distribution point is provided.

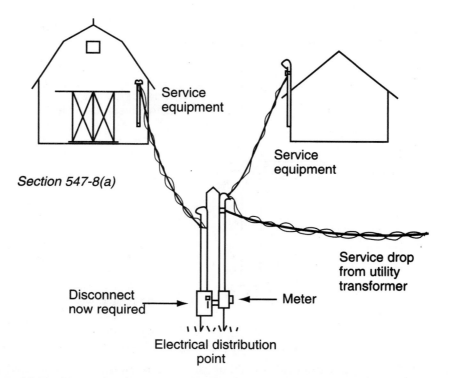

Section 547-8(a)

Service equipment

Service equipment

Service drop from utility transformer

Disconnect now required

Meter

Electrical distribution point

Figure 14.11   Now it is required to provide a disconnect at the electrical distribution point on a farm.

547-9(b): This was *Section 547-8(b)* in the previous edition of the Code, and it is the requirement that an equipotential plane be installed in agricultural buildings covered by this article. It no longer requires that a voltage gradient be established at every entrance and exit for the building. A voltage gradient transition is needed only at entrances that are used on a daily basis by the same livestock. The fine print note further implies that a voltage gradient ramp is only needed when an actual voltage gradient may exist that is greater than 1 volt per linear foot across the earth. A voltage gradient ramp is illustrated in Figure 14.12.

547-9(b) Exception 1: If there is no electrical power in an agricultural building, then an equipotential plane is not required to be installed. In the past, all agricultural buildings were required to have equipotential planes.

547-9(b) Exception 2: In the case where livestock buildings have slatted floors of metal or concrete, an equipotential plane is established by installing a metal equipotential plane element in the supporting walls for the slatted floor. If the supporting floor is an equipotential plane, then any concrete or metal element placed on that supporting structure will also be an equipotential plane.

547-9(c): All 125-volt, 15- and 20-ampere general-purpose receptacles in areas where an equipotential plane is installed must be ground-fault circuit-interrupter protected. The key words for this requirement are **general-purpose** receptacle. If the receptacle is on a separate circuit for a specific purpose, such as supplying feed handling equipment, it is not a general-purpose branch circuit, and ground-fault circuit-interrupter protection for personnel is not required.

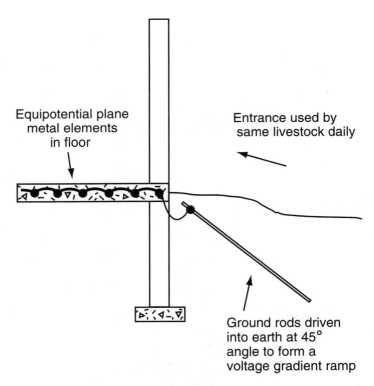

Equipotential plane
metal elements
in floor

Entrance used by
same livestock daily

Ground rods driven
into earth at 45°
angle to form a
voltage gradient ramp

Figure 14.12 A voltage gradient transition ramp can be constructed of ground rods driven into the earth at an angle at entrances to livestock buildings only at the entrances where the same livestock pass in and out of the building on a daily basis.

Article 553 **Floating Buildings**

553-7(b): This section specified flexible cord or liquidtight flexible conduit as the method of connecting shore power to a floating building. Now other wiring methods suitable for the conditions are permitted where flexibility is not required.

Article 555 **Marinas and Boatyards**

555-4: A readily accessible disconnecting means is required for each boat supplied with shore power. This is in addition to the shore power connection. The disconnecting means is required to be located in sight from the shore power connection, and it shall consist of a circuit-breaker or disconnect switch.

Article 625 **Electric Vehicle Charging System Equipment**

625-2: There is a new definition of an electric vehicle inlet and of a personnel protection system.

625-9(b): There is a new requirement that the electric vehicle coupler is required to have a configuration so that it cannot be interchanged with other electrical devices.

625-15(c): Electrical charging equipment must be marked by the manufacturer where ventilation is required for the area in which the equipment is installed.

Article 640 **Audio Signal Processing, Amplification, and Reproduction Equipment**

640-2: This is a new section that provides definitions of terms that are necessary for understanding the requirements in this article.

640-5: Access to portions of the building cannot be prevented due to the installation of sound equipment and cables.

640-6: Installation of cables and equipment is required to be done in a neat and workmanlike manner.

640-10: Audio systems near bodies of water have new installation requirements given in this section.

Part B: There are new wiring methods for permanently installed audio systems.

Part C: These are new wiring requirements for portable and temporary audio system installations.

Article 650 **Pipe Organs**

No changes were made to this article.

Article 780 **Closed-Loop and Programmed Power Distribution**

No changes were made to this article.

Article 800 **Communications Circuits**

800-2: This section contains definitions used in *Article 800* for communication circuits. Now definitions for **block, exposed,** and **premises** are found in this section. These three definitions are not new to this article, they were found in the first two fine print notes to *Section 800-30(a)* in the previous edition of the Code.

800-11(a): Communication cables entering a building from underground are required to be separated from conductors for electric light and power, Class 1 conductors, and nonpower-limited fire-alarm circuit conductors. The change in this section permits the use of a **suitable barrier** as means used to provide separation.

800-48: There were no references in *Article 800* for installation of communications circuits in raceways in the previous edition of the Code. This new section requires that raceways be installed in accordance with the requirements of *Chapter 3* when they contain a communication circuit. Because *Section 90-3* makes *Chapter 8* independent for the remainder of the Code, a specific reference to *Chapter 3* was needed if raceway installation requirements were to be enforced.

The exception to this section permits the installation of a listed nonmetallic communications raceway. This type of raceway can be listed for installation in a plenum, as a riser, or for general-purpose installations. Installation requirements for this type of raceway are the same as those for electrical nonmetallic tubing.

800-50: This section contains requirements on the listing, marking, and installation of communication cables and wires. To clarify that the marking requirements of under-carpet communications wires are the same as other types of communication cables, the phrase **under-carpet communications wires** was added to this section. These wires and other communication cables are to be marked in accordance with *Table 800-50*.

800-50 Exception 3: Provided that the length of communications cable installed in a building is not greater than 50 feet, no listing or marking requirements apply to the cable. The change in this section was to clarify how the 50 feet are to be measured. The measurement is made from the point of entry to the termination point at an enclosure or primary protector.

800-51: In the previous edition of the Code, this section provided the listing requirements for communication wires and cables. This section now recognizes and provides listing requirements for communication raceways. Communication raceways are listed as either general-purpose, riser, or plenum.

800-52: In the previous edition of the Code, this section provided the separation requirements for communication circuits and electric light, power, remote control, signaling, and nonpower-limited fire-alarm circuits. Now communication conductors must also be separated from medium-power broadband communication circuits. Low-power broadband communication circuits, on the other hand, are permitted to occupy the same raceway or enclosures as communication conductors.

800-53: The permitted uses of listed communication cables are found in this section. Now this section contains the permitted uses of communication raceways, and more importantly, it contains the types of communication cables that are permitted to be used in these raceways. Raceways and types of cables permitted to be run in them are summarized in Table 14.1.

Article 810 Radio and Televison Equipment

810-1: The scope of this article still covers antenna systems for radio, television, and amateur radio equipment. The scope of this article was modified to include a multitude of

Table 14.1 Communications raceway types and the types of communications cables they are permitted to contain.

Raceway Type	Permitted Cable Type
Plenum	CMP
Riser	CMR or CMP
General-purpose	CMG, CM, CMR, or CMP

antenna systems such as multielement, vertical rod, and dish. Satellite antenna systems are now clearly covered by this article.

810-2: This section lists other articles of the Code whose requirements pertain to certain aspects of radio or television equipment. In the previous edition of the Code, the provisions of *Article 820* were used for installation of coaxial cables that are installed for televison equipment. Now the requirements in *Article 820* are used for coaxial cables that extend from any antenna and its associated equipment.

Article 820 Community Antenna Television and Radio Distribution Systems

820-2: This section contains definitions used in *Article 820*. Now definitions for **exposed** and **premises** are found in this section. If an exposed cable could possibly contact other wiring systems by means such as insulation damage or due to failure of supports, it is considered to be exposed according to this article.

820-3: This new section lists requirements found in other articles of the Code that cover wiring methods that must be used for CATV and radio distribution systems. Wiring methods in *Article 830* are permitted to be used instead of those in *Article 820*.

820-50 Exception 3: Provided that the length of coaxial cable installed in a building is not greater than 50 feet, no listing or marking requirements apply to the cable. The change in this section was to clarify how the 50 feet are to be measured. The measurement is made from the point of entry to the termination point at the grounding block.

820-52: In the previous edition of the Code, this section provided the separation requirements for coaxial cables and electric light, power, Class 1, and nonpower-limited fire-alarm circuits. Now coaxial cables must also be separated from medium-power broadband communication circuits. Low-power broadband communication circuits, on the other hand, are permitted to occupy the same raceway or enclosures as communication conductors.

820-53: Common cable substitutions for antenna television cables are given this section. Some cable types were deleted because they are generally not used for these applications. A new figure was also added to this section that illustrates the permitted cable substitutions.

Article 830 Network-Powered Broadband Communications Systems

This is a new article covering a type of network-powered communications system when data, audio signals, and video signals can be simultaneously conveyed over a single conductor. The wiring within a building is connected to the network through a network interface unit. Community antenna television distribution systems in buildings may be installed using materials and wiring methods meeting the requirements of *Article 830* so the system will be ready for broadband hookup when it becomes available in the future. This is covered in *Section 820-3(g)*.

830-4: Network-powered broadband communications systems are classified as having low-power sources or medium-power sources.

830-5: Cable installed as a part of a network-powered broadband communications system is required to be listed for the purpose.

830-5(a)(1): Network-powered broadband communications system cable permitted to be installed where the source is medium-powered is marked as type BM, and will have an insulation rated not less than 300 volts. If the cable is suitable for underground installation, it will have the type letters BMU. If it is permitted to be installed between floors as a riser cable, it will have the type letters BMR.

830-5(a)(2): Network-powered broadband communications system cable permitted to be installed where the source is low-powered is marked with the prefix letters BL, and will have an insulation rated not less than 300 volts. Type BLX is limited-use cable and is permitted to be installed outside, or inside dwellings as a cable without raceway protection. If installed in raceway, it is permitted to be installed inside any type of building including as a riser between floors. Type BLP is permitted to be installed as an exposed cable in ducts and plenums.

830-9(a): For a system with a medium-powered source, cables installed outside and used as an entrance cable are required to be of types BM, BMU, or BMP.

830-9(b): For a system with a low-powered source, cables installed outdoors and entering buildings are required to be labeled type BLX of BLU.

830-10: This section covers the installation of aerial installations of network-powered broadband communications cables. Minimum overhead conductor clearances are given in *Section 830-10(d)* and following paragraphs.

Table 830-11: The minimum cover requirements are given for underground installation of network-powered broadband communications cables. Direct burial cables without protection are required to have a minimum cover of not less than 18 inches below grade unless installed at a one- or two-family dwelling, in which case the minimum depth of burial is 12 inches.

830-40: Grounding is required for network interface units and primary protectors. The grounding conductor is required to be insulated copper not smaller than AWG number 14. Acceptable grounding electrodes are described in *Section 830-40(b)*. The electrical service grounding electrode is used, but if not available, a rod or pipe not less than 1/2 inch in diameter and not less than 5 feet long is permitted to serve as the grounding electrode.

Part E: This portion of the article covered the installation of wiring within buildings. Cables that are permitted to be substituted for network-powered broadband communications cables are listed in *Table 830-58*. Only coaxial cables of the types listed in the table are permitted to be used as substitutions.

WORKSHEET NO. 14
SPECIAL APPLICATIONS
WIRING

These questions are considered important to understanding the application of the *National Electrical Code*® to electrical wiring, and they are questions frequently asked by electricians and electrical inspectors. People working in the electrical trade must continue to study the Code to improve their understanding and ability to apply the Code properly.

DIRECTIONS: Answer the questions and provide the Code reference or references where the necessary information leading to the answer is found. Space is provided for brief notes and calculations. An electronic calculator will be helpful in making calculations. You will keep this worksheet; therefore, you must put the answer and Code reference on the answer sheet as well.

1. Fixed wiring for lighting and power circuits in an area of a theater required to be of fire-rated construction is permitted to be exposed rigid nonmetallic conduit. (True or False)

    Code reference _____

2. Wireway is used for wiring in the stage area of a theater used for musical and dramatic productions. Only a few wires in the wireway are energized at any one time. Derating factors for wire ampacity shall apply if there are more than 30 current-carrying wires contained at any cross section in the wireway. (True or False)

    Code reference _____

3. Receptacle outlets in a dressing room of a theater adjacent to the mirror and above dressing table counters shall be wall-switch-controlled, and are required to have a pilot light to indicate when the receptacles are energized. (True or False)

    Code reference _____

4. The permanent power wiring for a set in a commercial television studio where there is no provision for an audience is permitted to be run as type MC metal-clad cable. (True or False)

    Code reference _____

5. The switch located outside of the vault and controlling the 120-volt lighting in a cellulose nitrate film storage vault shall simultaneously open the neutral conductor to the lighting fixtures as well as the ungrounded conductor. (True or False)

    Code reference _____

6. The projection room of a motion picture theater is classified as a Class III hazardous location. (True or False)

    Code reference _____

7. The minimum copper conductor size permitted to supply an outlet for an arc or xenon projector in the projection room of a theater is AWG number _____.

    Code reference _____

8. The service entrance for a theater is not permitted to be installed in the projection room of the theater. (True or False)

    Code reference _____

9. An electric motor installed in a shop of a farm where the shop is not considered to be an area of excessive dust, dust with water, or a corrosive atmosphere shall have a totally enclosed type motor without ventilation openings to the internal moving parts of the motor. (True or False)

    Code reference _____

10. Type UF cable is permitted as a wiring method in a building housing animals where there is excessive dust with water. (True or False)

    Code reference _____

11. Rigid steel conduit is used to supply a motor in an agricultural building where strength is needed and chances of corrosion are minimal. The steel conduit is permitted to serve as the equipment grounding conductor for the motor. (True or False)

    Code reference _____

12. A metal grid of welded steel rods is installed in the concrete floor of a barn where dairy cows are to be milked, and the grid is welded to the metal support poles for the equipment in the barn. This metal grid in the concrete floor is required to be bonded to the building equipment grounding electrode system with copper wire not smaller than AWG number _____. .

    Code reference _____

13. The feeder conductor of a single-phase, 120/240-volt electrical supply for a floating building shall have an insulated neutral conductor run separate from and insulated from the equipment grounding conductor of the feeder and in the floating building panelboard. (True or False)

    Code reference _____

14. The service equipment for a floating building shall be located on the floating building. (True or False)

    Code reference _____

15. A 30-ampere locking and grounding type receptacle is permitted to supply shore power to a boat not more than 20 feet in length. (True or False)

    Code reference _____

16. General use receptacles installed on 15- and 20-ampere, 120-volt circuits in agricultural buildings are required to be ground-fault circuit-interrupter protected. (True or False)

    Code reference _____

17. A duplex receptacle outlet used on a normal wiring system is permitted to be used as a receptacle outlet for general loads on a branch circuit of a closed-loop and programmed power distribution system in a dwelling. (True or False)

    Code reference _____

18. A group of control wires to electric pipe organ sounding devices mounted on a common frame is permitted to have a common copper return conductor not smaller than AWG number _____.

    Code reference _____

19. An auxiliary gutter contains only Class 3 wires of a public address system for a building, and it is permitted to be grounded to a common ground for the public address system with a copper wire not required to be larger than AWG number _____ or the equivalent.

    Code reference _____

20. The amplifier output wires to the speakers in a building are to be installed in accordance with requirements of *Article 725*. (True or False)

    Code reference _____

21. A ground rod is driven for a communications cable primary protector that is separate from the electrical service entrance ground rod, and they are located 6 feet apart. It is permitted to bond these two ground rods together with copper wire not smaller than AWG number 6. (True or False)

    Code reference _____

22. Telephone cable is run from outlet to outlet in a one-story commercial building and not run in plenums or air handling spaces. Type CM telephone communication cable is permitted for this application, but it is not permitted to be used as a riser or in air handling spaces without being run in raceway. (True or False)

    Code reference _____

23. Communication cables and wires from light and power circuits are permitted to be run in the same conduit. (True or False)

    Code reference _____

24. It is not permitted to install an outdoor antenna lead-in cable such that it passes above outside open conductors used for electric light and power. (True or False)

    Code reference _____

25. Coaxial cable supplied from a community television system is installed within a commercial building beyond the point where the lead-in cable terminates. Type CATV cable is permitted to be used as an exposed riser cable not run in metal conduit. (True or False)

    Code reference _____

ANSWER SHEET   Name _____

No. 14   SPECIAL APPLICATIONS WIRING

Answer   **Code reference**

1. _____   _____

2. _____   _____

3. _____   _____

4. _____   _____

5. _____   _____

6. _____   _____

7. _____   _____

8. _____   _____

9. _____   _____

10. _____   _____

11. _____   _____

12. _____   _____

13. _____   _____

14. _____   _____

15. _____   _____

16. _____   _____

17. _____   _____

18. _____   _____

19. _____   _____

20. _____   _____

21. _____   _____

22. _____   _____

23. _____   _____

24. _____   _____

25. _____   _____

UNIT 15

Review

OBJECTIVES

Upon completion of this unit, the student will be able to:

- evaluate ability to solve basic electrical fundamentals calculations.
- evaluate ability to find answers to questions from the Code.
- evaluate ability to make electrical wiring calculations.
- determine which articles of the Code require further work to increase the student's level of understanding.

EVALUATION PROCESS

This review is designed to serve as a self-evaluation. Mark only the answer on the answer sheet. It is best to do this evaluation in a specified period of time. A suggested time interval is two hours. If the review test is taken in a two-hour time limit, it will not be possible to look up the answer to every question in the Code. Usually, the answer can be narrowed to two possible answers by elimination. The following example of an electrical fundamentals problem will illustrate how the possible answers can be reduced to only two by using basic understanding of the concept.

Example   Three resistors with the values 4 ohms, 6 ohms, and 12 ohms are connected in parallel. The total resistance of the circuit is:

A.   1 ohm.
B.   2 ohms.
C.   6 ohms.
D.   22 ohms.

Answer: The circuit resistance is less than the smallest resistor in the group; therefore, the answer must be either response A or B. Working out the answer shows that the answer is response B.

$$\frac{1}{R_T} = \frac{1}{4} + \frac{1}{6} + \frac{1}{12} = \frac{3}{12} + \frac{2}{12} + \frac{1}{12} = \frac{6}{12}$$

$$R_T = \frac{12}{6} = 2 \text{ ohms}$$

REVIEW TEST

1. An incandescent lamp operating at 120 volts and drawing 0.8 amperes has a resistance of:
   A. 95 ohms.
   B. 121 ohms.
   C. 150 ohms.
   D. 240 ohms.

2. A resistance-type baseboard electric heater rated at 1,200 watts and operating at 240 volts draws:
   A. 0.2 ampere.
   B. 4.1 amperes.
   C. 5.0 amperes.
   D. 8.4 amperes.

3. An electrical device has a resistance of 5.2 ohms, and it is located 200 feet from the circuit-breaker panelboard. If the resistance of 400 feet of the wire is 0.2 ohm, the total resistance of the circuit is:
   A. 4.8 ohms.
   B. 5.0 ohms.
   C. 5.4 ohms.
   D. 5.6 ohms.

4. Wire type THWN copper is run in raceway from a wall outlet in a building under the concrete floor where the conduit is in contact with earth. Then the conduit runs up another wall to an outlet. Terminations are rated at 75°C. If the calculated load is 58 amperes, the minimum wire size permitted is AWG number:
   A. 2.
   B. 4.
   C. 6.
   D. 8.

5. A 15-ampere, 120-volt branch circuit consisting of type UF cable is supplied from the service panel in a dwelling, and the entire circuit is protected with a ground-fault circuit-interrupter. If the direct burial type UF cable supplies lighting outlets in the yard, the minimum permitted depth of burial is:
   A. 6 inches.
   B. 12 inches.
   C. 18 inches.
   D. 24 inches.

6. A straight horizontal run of 2-inch intermediate metal conduit (IMC) with threaded couplings is required to be supported within 3 feet of a box or termination and supported at intervals not exceeding:
   A. 10 feet.
   B. 12 feet.
   C. 14 feet.
   D. 16 feet.

7. Nonmetallic single gang 2-inch by 3-inch device boxes are used in the new construction of a dwelling. The nonmetallic-sheathed cable is not required to be secured to the box provided at least 1/4 inch of cable sheath extends into the box, and measured along the cable sheath, the cable is stapled a distance from the box not exceeding:
   A. 8 inches.
   B. 10 inches.
   C. 12 inches.
   D. 36 inches.

8. A rigid metal conduit nipple connects two enclosures. Six new wires, type THWN, AWG number 2 with a total cross-sectional area of 0.69 square inch are being run for a rewire job. The minimum permitted trade diameter nipple is:
   A. 1-inch.
   B. 1 1/4-inch.
   C. 1 1/2-inch.
   D. 2-inch.

9. A pull box is installed with a 3-inch conduit entering one end and leaving the opposite end. Wires in the conduit are larger than AWG number 6. The minimum permitted length of the pull box is:
   A. 18 inches.
   B. 20 inches.
   C. 24 inches.
   D. 30 inches.

10. The minimum branch circuit rating permitted for a 10-kilowatt dwelling electric range is:
    A. 20 amperes.
    B. 30 amperes.
    C. 40 amperes.
    D. 50 amperes.

11. A single-family dwelling has a living area of 1,340 square feet. The minimum number of 15-ampere general-purpose lighting circuits required for the dwelling not including small appliance, bathroom, and laundry is:
    A. 2.
    B. 3.
    C. 4.
    D. 5.

12. The controls of an 80-gallon storage-type electric water heater permits a maximum load of 4,500 watts at 240 volts. The minimum copper supply conductor size, type NM cable based on 60°C terminations permitted is AWG number:
    A. 4.
    B. 6.
    C. 8.
    D. 10.

13. Receptacle outlets shall be installed along the wall of a room of a dwelling such that no point along the wall shall be located from a receptacle more than:
    A. 6 feet.
    B. 8 feet.
    C. 10 feet.
    D. 12 feet.

14. The only electrical loads in a 540-square-foot addition to a dwelling are lights and receptacles. If 15-ampere branch circuits are installed, the minimum number required is:
    A. 1.
    B. 2.
    C. 3.
    D. 4.

15. Type SE cable is used for a 120/240-volt, 3-wire, 150-ampere dwelling service. The minimum size aluminum ungrounded conductors permitted is AWG number:
    A. 2.
    B. 1.
    C. 0.
    D. 2/0.

16. A commercial building has a demand load of 130 amperes; however, the electrician will install a 200-ampere service. The minimum size type THWN aluminum service entrance conductors permitted, if the wires are in conduit that is in free air, is:
    A. AWG number 2/0.
    B. AWG number 3/0.
    C. AWG number 4/0.
    D. 250-kcmil.

17. Service entrance cable installed for a dwelling service entrance shall be supported by straps or similar means within 12 inches of the service head or gooseneck and at intervals not to exceed:
    A. 30 inches.
    B. 3 feet.
    C. 4.5 feet.
    D. 6 feet.

18. A single-family dwelling unit in an apartment building has a total living area of 720 square feet. Laundry facilities are provided elsewhere in the buildings available to all residents. The living unit contains a 12-kilowatt electric range, 3.5-kilowatt electric water heater at 240 volts, two air conditioners each of which draws 8 amperes at 240 volts, a space heater that draws 6 amperes at 120 volts, and a 1.2-kilowatt dishwasher at 120 volts. The minimum calculated demand load in amperes of the 120/240-volt subfeeder to each dwelling unit using Code *Article 220, Part B* is:
    A. 60 amperes.
    B. 86 amperes.
    C. 110 amperes.
    D. 147 amperes.

19. Rigid nonmetallic conduit, schedule 40, is used to connect the forming shell of a wet-niche low-voltage light of a permanent swimming pool to the transformer enclosure. The minimum size copper bonding conductor permitted to be run inside the conduit is AWG number:
    A. 6.
    B. 8.
    C. 10.
    D. 12.

20. A commercial building is served by a 400-ampere, 120/208-volt, 3-phase wye electrical system, and the service entrance conductors are 500-kcmil copper with type THWN insulation. The 3/4-inch steel reinforcing bars in the foundation are welded together and used as a grounding electrode for the service. The minimum size copper grounding electrode conductor permitted to the reinforcing bars is AWG number:
    A. 6.
    B. 4.
    C. 2.
    D. 0.

21. The maximum number of single-pole circuit-breakers permitted in a panelboard is:
    A. 42.
    B. 30.
    C. 24.
    D. not limited.

22. In a recreational vehicle park, a site with electrical power shall have a 125-volt receptacle outlet rated not less than:
    A. 15 amperes.
    B. 20 amperes.
    C. 30 amperes.
    D. 15 or 20 amperes.

23. It is permitted to plug a listed type SP extension cord with copper wire without integral overcurrent protection into a receptacle outlet of a 20-ampere, 120-volt circuit, of an unspecified length in sizes not smaller than AWG number:
    A. 20.
    B. 18.
    C. 16.
    D. 14.

24. A 300-ampere service entrance consists of a 200-ampere panelboard and a 100-ampere fused disconnect switch tapped from the 500-kcmil, type THWN aluminum wires. The minimum size type THWN aluminum tap wire permitted for the 100-ampere disconnect is:
    A. AWG number 1.
    B. AWG number 3/0.
    C. 250-kcmil.
    D. 500-kcmil.

25. An attachment plug and receptacle are permitted to serve as the controller for a portable motor with a rating of not more than:
    A. 1/4 horsepower.
    B. 1/3 horsepower.
    C. 1/2 horsepower.
    D. 2 horsepower.

26. A remote start-stop station wired with AWG number 14 copper wire is permitted to be protected from overcurrent by the motor branch circuit fuses provided the fuse rating is not greater than:
    A. 15 amperes.
    B. 20 amperes.
    C. 30 amperes.
    D. 45 amperes.

27. The maximum standard rating inverse-time circuit-breaker permitted to serve as branch circuit short-circuit and ground-fault protection when the motor starts the load without difficulty for a 10-horsepower, 240-volt, 3-phase design B motor is:
   A. 80 amperes.
   B. 70 amperes.
   C. 60 amperes.
   D. 50 amperes.

28. A single-phase, 230-volt, 3-horsepower design B electric motor powering an easy starting load has a nameplate full-load current rating of 16 amperes and a service factor of 1.15. If time-delay fuses are to be used to serve both as short-circuit and ground-fault protection, and as running overload protection for the motor, the maximum standard rating of fuse permitted is:
   A. 20 amperes.
   B. 25 amperes.
   C. 30 amperes.
   D. 35 amperes.

29. The full-load current rating of a 240-volt winding of a 50-kVA single-phase transformer is:
   A. 104 amperes.
   B. 186 amperes.
   C. 200 amperes.
   D. 208 amperes.

30. A single-phase 25-kVA transformer rated 480 volts primary and 240 volts secondary is provided with secondary overcurrent protection of 100 amperes, and type THWN, copper, size AWG number 3 secondary wires. The primary wires are copper, type THWN, size AWG number 8, and the distance from the primary overcurrent device to the transformer is 120 feet. All terminations are 75°C rated. The maximum permitted standard rating primary overcurrent device is:
   A. 50 amperes.
   B. 60 amperes.
   C. 100 amperes.
   D. 125 amperes.

31. An overcurrent device protecting the primary winding of a transformer at not more than 125 percent of the primary full-load current, and operating at less than 600 volts, is permitted to protect the secondary winding of the transformer:
   A. if the transformer is 3-phase.
   B. if the transformer is single-phase, 3-wire, 120/240 volts.
   C. only for copper wire AWG number 12 and smaller.
   D. if the secondary is 2-wire, single voltage.

32. Areas within a grain elevator containing electrically operated grain handling equipment where dust is frequently in the air are considered to be locations classified as Class II Division:
   A. 1.
   B. 2.
   C. 3.
   D. G.

33. The Class I Division 2 hazardous area extends outward from the edge of the gasoline dispenser in all directions a minimum distance of:
    A. 4 feet.
    B. 10 feet.
    C. 18 feet.
    D. 20 feet.

34. The circuits within the anesthetizing location, classified as a hazardous location, shall be isolated from any supply system. The single-phase circuit wires supplying receptacles and equipment shall be identified by the colors orange and:
    A. green.
    B. brown.
    C. red.
    D. blue.

35. An intensive care unit of a hospital is considered:
    A. a critical care area.
    B. a general care area.
    C. an emergency unit.
    D. a hazardous location.

36. The nameplate continuous current rating of a 120/240-volt single-phase alternating current generator permanently installed in a building is 60 amperes. The minimum type THWN copper wire size permitted to be installed between the generator and the first overcurrent device based on 75°C terminations is AWG number:
    A. 8.
    B. 6.
    C. 4.
    D. 2.

37. A cable run in raceway for a power-limited fire-alarm system of a commercial building from one floor to another is permitted to be marked type:
    A. FPL.
    B. FAC.
    C. FLX.
    D. ALF.

38. Twelve 3-cell lead-acid batteries are connected in series. The Code-rated output of this battery set is:
    A. 24 volts.
    B. 60 volts.
    C. 72 volts.
    D. 120 volts.

39. A receptacle outlet is removed at an insert in cellular metal floor raceway, but the circuit wire is needed to feed other outlets. It is:
    A. required to replace the wire with new wire.
    B. permitted to tape the wire at the point where the receptacle was removed.
    C. permitted to cut, solder splice, and tape the wire where the receptacle was removed.
    D. not permitted to remove a receptacle after it has been installed.

40. The conductive optical fiber cable from the following list permitted to be installed as wiring within a building is type:
    A. OPTCA.
    B. FCOX.
    C. OFC.
    D. EOPTCA.

41. Type MC cable with three AWG number 12 copper conductors installed exposed in a commercial building shall be supported at intervals not to exceed:
    A. 6 feet.
    B. 3 feet.
    C. 12 inches.
    D. 8 inches.

42. An aluminum ladder-type cable tray has a marked total cross-sectional area of both side rails of 0.62 square inch. The maximum circuit rating for which this ladder-type cable tray is permitted to serve as the equipment grounding conductor is:
    A. 100 amperes.
    B. 200 amperes.
    C. 600 amperes.
    D. 1,000 amperes.

43. A legally required standby power system is permitted to supply power:
    A. only for emergency systems.
    B. other than for emergency systems.
    C. for exit signs.
    D. with manual start-up.

44. A cablebus without added support for long spans shall be supported at intervals not to exceed:
    A. 6 feet.
    B. 10 feet.
    C. 12 feet.
    D. 20 feet.

45. Except by special permission of the authority having jurisdiction, the maximum size conductor permitted to be installed in cellular metal floor raceway is AWG number:
    A. 4.
    B. 2.
    C. 1/0.
    D. 4/0.

46. A building or area within a building is considered a place of assembly if the area is occupied by:
    A. 50 people or more.
    B. 100 people or more.
    C. 200 people or more.
    D. 500 people or more.

47. The maximum size circuit wire permitted to be installed in conduit as solid wire except as a bonding wire is AWG number:
    A. 12.
    B. 10.
    C. 8.
    D. 6.

48. The color of insulation of a grounded circuit conductor AWG number 6 and smaller shall be:
    A. white or gray.
    B. green.
    C. yellow or green.
    D. black or red.

49. A fusible switch used as the disconnect for a 40-horsepower, 3-phase design B motor shall be rated in:
    A. amperes.
    B. foot-pounds.
    C. horsepower.
    D. degrees temperature rise.

50. The design letter marked on a motor nameplate is used to size the:
    A. branch circuit wires.
    B. branch circuit short-circuit and ground-fault protection.
    C. running overcurrent protection.
    D. branch circuit disconnect.

ANSWER SHEET

Name _____

No. 15   REVIEW

DIRECTIONS: Please indicate the answer by making a dark dot over the correct letter. (Example: A   B   C   D)

1.	A B C D		26.	A B C D
2.	A B C D		27.	A B C D
3.	A B C D		28.	A B C D
4.	A B C D		29.	A B C D
5.	A B C D		30.	A B C D
6.	A B C D		31.	A B C D
7.	A B C D		32.	A B C D
8.	A B C D		33.	A B C D
9.	A B C D		34.	A B C D
10.	A B C D		35.	A B C D
11.	A B C D		36.	A B C D
12.	A B C D		37.	A B C D
13.	A B C D		38.	A B C D
14.	A B C D		39.	A B C D
15.	A B C D		40.	A B C D
16.	A B C D		41.	A B C D
17.	A B C D		42.	A B C D
18.	A B C D		43.	A B C D
19.	A B C D		44.	A B C D
20.	A B C D		45.	A B C D
21.	A B C D		46.	A B C D
22.	A B C D		47.	A B C D
23.	A B C D		48.	A B C D
24.	A B C D		49.	A B C D
25.	A B C D		50.	A B C D

INDEX